Artificial Intelligence Techniques for a Scalable Energy Transition

Moamar Sayed-Mouchaweh
Editor

Artificial Intelligence Techniques for a Scalable Energy Transition

Advanced Methods, Digital Technologies, Decision Support Tools, and Applications

Springer

Editor
Moamar Sayed-Mouchaweh
Institute Mines-Telecom Lille Douai
Douai, France

ISBN 978-3-030-42728-3 ISBN 978-3-030-42726-9 (eBook)
https://doi.org/10.1007/978-3-030-42726-9

This Springer imprint is published by the registered company Springer Nature Switzerland AG.
The registered company address is: Gewerbestrasse 11, 6330 Cham, Switzerland

Preface

Energy transition aims also at integrating more efficient technologies, practices, and services in order to reduce energy losses and wastes. However, the increasing penetration rate of renewable energy into the grid is a challenging task because of their intermittency due to their strong dependence on weather conditions. To this end, the traditional electricity grid requires undergoing a transition to be more resilient, reliable, and efficient. This can be achieved by a transition towards a smart grid (SG) in which a two-way flow of power and data between suppliers and consumers is provided. SG includes an intelligent layer that analyzes these data volumes produced by users and production side in order to optimize the consumption and production according to weather conditions and the consumer profile and habits. The goal of this analysis and treatment is to maximize the grid flexibility, stability, efficiency, and safety.

The use of SG within the context of energy transiting is facing several challenges related to the grid stability, reliability, security as well as the energy production/consumption optimization in the presence of renewable energy resources. Addressing these challenges requires the development of scalable advanced methods and tools able to manage and process efficiently and online the huge data streams produced by heterogeneous technologies and systems in order to extract useful knowledge, recommendations, or rules. The latter are then used to optimize the energy consumption, to facilitate the penetration (integration) of distributed/centralized renewable energy systems into electric grids, to reduce the peak load, to balance and optimize generation and consumption, to reinforce the grid protection as well as cybersecurity and privacy issues.

This book focuses on the use of Artificial Intelligence (AI) techniques in order to develop decision support tools allowing to tackle and address these challenges in different applications and use-cases. The gathered methods and tools are structured into three main parts: Artificial Intelligence for Smart Energy Management, Artificial Intelligence for Reliable Smart Power Systems, and Artificial Intelligence for Control of Smart Appliances and Power Systems. In each part, different AI methods (artificial neural networks, multi-agent systems, hidden Markov models, fuzzy rules, support vector machines, first-order logic, etc.), energy transition challenges (availability of data, processing time, kind of learning, sampling frequency, time-horizon, and physical-scale granularity), operational conditions (centralized, distributed), and application objectives (prediction, control, optimization) as well as domains (demand side management, energy management, flexibility maximization, load monitoring, battery configuration, conversion system or power system monitoring, etc.) are discussed.

Finally, the editor is very grateful to all the authors and reviewers for their valuable contributions. He would like also to acknowledge Mrs. Mary E. James for establishing the contract with Springer and supporting the editor in any organizational aspects. The editor hopes that this book will be a useful basis for further fruitful investigations for researchers and engineers as well as a motivation and inspiration for newcomers in order to address the challenges related to energy transition.

Douai, France Moamar Sayed-Mouchaweh

Contents

About the Editor

Moamar Sayed-Mouchaweh received his Master's degree from the University of Technology of Compiegne, France in 1999, PhD degree from the University of Reims, France in December 2002, and the Habilitation to Direct Researches (HDR) in Computer Science, Control and Signal Processing in December 2008. Since September 2011, he is working as a Full Professor in the High National Engineering School of Mines-Telecom Lille-Douai in France. He edited and wrote several Springer books, served as member of Editorial Board, IPC, conference, workshop, and tutorial chair for different international conferences, an invited speaker, a guest editor of several special issues of international journals targeting the use of advanced artificial intelligence techniques and tools for digital transformation (energy transition and Industry 4.0). He served and is serving as an expert for the evaluation of industrial and research projects in the domain of digital transformation. He is leading an interdisciplinary and industry-based research theme around the use of advanced Artificial Intelligence techniques in order to address the challenges of energy transition and Industry 4.0.

Chapter 1
Prologue: Artificial Intelligence for Energy Transition

Moamar Sayed-Mouchaweh

1.1 Energy Transition: Definition, Motivation, and Challenges

Facing the problem of global climate change and the scarcity of fossil sources requires the transition to a sustainable energy generation, distribution, and consumption system. Energy transition [1] aims at pushing towards replacing large, fossil-fuel plants with clean and renewable resources, such as wind and solar energy, with distributed generation. The latter allows generating electricity from sources, often renewable energy sources, near the point of use in contrary to centralized generation from power plants in traditional power grids.

Energy transition aims also at integrating more efficient technologies, practices, and services in order to reduce the energy losses and wastes. However, the increasing penetration rate of renewable energy into the grid entails to increase the uncertainty and complexity in both the business transactions and in the physical flows of electricity into the grid because of their intermittency due to their strong dependence on weather conditions. As an example, the electricity production of solar photovoltaic panels is high in the morning when demand is low, while it is low in the evening when the demand is high. This can impact the grid stability, i.e., the balance between supply and demand.

The energy consumption is continuing to increase as electrification rate grows. Therefore, the use of distributed energy resources (DERs) [2] through shifting users from being only consumers to be also producers, called prosumers, leads to increase significantly the gridcapacity. A DER is a small-size energy generator used locally

M. Sayed-Mouchaweh (✉)
Institute Mines-Telecom Lille Douai, Douai, France
e-mail: moamar.sayed-mouchaweh@mines-douai.fr

M. Sayed-Mouchaweh (ed.), *Artificial Intelligence Techniques for a Scalable Energy Transition*, https://doi.org/10.1007/978-3-030-42726-9_1

and is connected to a larger energy grid at the distribution level. DERs include solar photovoltaic panels, small natural gas-fuel generators, electric vehicles and controllable loads, such as electric water heaters. The major interest of a DER is that the energy it produces is consumed locally, i.e., close to the power source. This allows reducing the transmission wastes. A set of DERs that operates connected to the main grid or disconnect (island mode) is called a microgrid [3].

However, the energy transition faces multiple challenges such as ensuring:

• grid stability with a large penetration of renewable energy resources into the grid,

• active participation of users in order to optimize their energy consumption and to improve the balance between supply and demand,

• maximal use of renewable energy produced locally in particular during peak demand or load periods.

To cope with these challenges, the traditional electricity grid requires undergoing a transition to be more resilient, reliable, and efficient. This can be achieved by a transition towards a smart grid (SG) in which a two-way flow of power and data between suppliers and consumers is provided. SG [4] includes an intelligent layer that analyzes these data volumes produced by users and production side in order to optimize the consumption and the production according to weather conditions and the consumer profile and habits. The goal of this analysis and treatment is to maximize the grid flexibility, stability, efficiency, and safety.

Flexibility [5] can be defined as the ability of the electricity system to respond to fluctuations of supply and demand while, at the same time, maintaining system reliability. As an example, grid operators can use a set of photovoltaic panels, batteries, electrical vehicles, chargers, etc., in order to modify generation or consumption to stabilize grid frequency and voltage. Power retailers can reduce costs during peak demand periods by using stored energy or deleting (shifting) deferrable loads in order to reduce consumption based on price or incentive signals. Energy storage through distributed batteries can increase the resilience and reliability of grid thanks to their aggregated stored energy that can be used during outages or peak demands hours knowing that the majority of outages are caused by disturbances in the distribution system.

In addition, it is important to detect both internal and external faults during operations and react quickly in order to find a safe state to reach it. Internal faults are related to the system's internal components (generators, converters, actuators, etc.), while external faults are related to environmental interactions not expected or not modeled during system development. Moreover, due to the extensive use of advanced Information and Communication Technologies (ICT), such as Internet of Things, in the SG, the latter becomes vulnerable to hacker attacks. Indeed the use of ICT allows hackers to have multiple entry points to the grid in order to infiltrate the control centers of several power plants. These attacks can impact significantly the reliability of the grid by turning off the power of entire cities (airports, road networks, hospitals, etc.).

Therefore, it is essential to develop advanced management and control tools in order to ensure the safety, reliability, efficiency, and stability of the SG. To this end, the intelligent layer of the SG uses artificial intelligence techniques and tools in order to achieve prediction [6, 7] and/or optimization [8], as it is discussed in the next section.

1.2 Artificial Intelligence for Energy Transition

The AI methods are used for prediction and/or optimization. The goal of prediction is to predict the electrical energy consumption or demand [6, 7], the produced energy by wind turbines or photovoltaic panels, the health state [9] of a component or machine. The optimization aims mainly to perform cost minimization, peak reduction, and flexibility maximization. The cost minimization [8] aims at optimizing the energy consumption or/and the energy bill or price for a customer as well as the potential risk related to a cyber-attack or to a fault. The peak reduction problem [10] aims at ensuring the balance between the energy demand and energy production during the periods where the demand is very high. Flexibility maximization [11] aims at finding the maximal energy that can be deleted at a certain point of time based on energy-consuming and energy-producing devices in residential buildings. The flexibility is then used to ensure the balance between demand and supply.

The use of AI techniques to perform prediction and/or optimization within the energy transition faces several challenges. Prediction of energy consumption as a function of time plays an essential role for the efficiency of the decision strategies for energy optimization and saving. The variability introduced by the growing penetration of wind and solar generations hinders significantly the prediction accuracy. In addition, this energy prediction is performed at different time horizons and levels of data aggregation. The learnt models must be enough flexible in order to be easily extendable to these different time scales and aggregation levels.

In addition, the learnt model requires historical data about the building/user/renewable energy resources consumption behavior (users' energy consumption behavior, building's energy performance, weather conditions, etc.) such as energy prices, physical parameters of the building, meteorological conditions or information about the user behavior. However, sometimes there is no historical data available due, as example, to the appearance of new buildings. Therefore, the energy prediction must be performed without the use of historical data about the energy behavior of the building under consideration.

Finally, the built model in order to perform prediction and optimization requires adapting in response to building renovation and/or introduction of new technologies as well as user's consumption behavior. This adaptation in the model's parameters and/or structure is necessary in order to maintain the prediction accuracy and optimization efficiency.

1.3 Beyond State-of-the-Art: Contents of the Book

According to the aforementioned challenges discussed in the previous section, the book is structured into three main parts, where in each of them different AI methods (Artificial Neural Networks, Multi-Agent Systems, Hidden Markov Models, Fuzzy rules, Support Vector Machines, first order logic, etc.), energy transition challenges (availability of data, processing time, kind of learning, sampling frequency, time horizon and physical-scale granularity, etc.) operational conditions (centralized, distributed), and application objectives (prediction, control, optimization) as well as domains (demand side management, energy management, flexibility maximization, load monitoring, battery configuration, conversion system or power system monitoring, etc.) are discussed:

- Artificial intelligence for Smart Energy Management (Chaps. 2, 3, 4, and 5),
- Artificial intelligence for Reliable Smart Power Systems (Chaps. 6, 7, 8, and 9),
- Artificial intelligence for Control of Smart Appliances and Power Systems (Chaps. 10, 11, 12, and 13).

1.3.1 Chapter 2: Large-Scale Building Thermal Modeling Based on Artificial Neural Networks: Application to Smart Energy Management

This chapter proposes a smart building energy management system (SBEMS) in order to help users to reduce their consumption, in particular by optimizing their use of heating, ventilation, and air condition system. The proposed SBEMS is based on the prediction of thermal dynamics in different instrumented and non-instrumented zones in a large scale building. This prediction is based on the use of neural networks. The latter have as inputs the electric power consumption for heating, in instrumented (equipped with sensors) and non-instrumented zones as well as the weather conditions (outdoor temperature, outdoor humidity, and solar radiation). The output of the neural networks is the estimated indoor temperature for both instrumented and non-instrumented zones. The inputs of the neural networks are discretized into segments, with minimal and maximal values, indicating different meaningful behaviors (states). Moving from one segment to another generates an event. A recommender is built as a finite state automaton composed by these states and events. At each state, a recommendation is provided to users in order to invite them to adopt a "green behavior and/or activity." As an example, if the indoor temperature is in the upper level, then the recommendation could be to lower the thermostat for one graduation to save energy. The proposed approach is applied for smart energy management of student residential building. Different thermal behaviors are recorded in order to obtain a rich learning and testing data set. The proposed approach (neural networks and the recommender) is implemented and tested using a smart interactive interface composed of different levels (webpages) allowing users to obtain information about their thermal zone (energy consumed, its cost, average temperature, trend of electrical consumption, etc.) and to display the recommendations linked to the user activity and the quantity of energy saved thanks to the application of these recommendations.

1.3.2 Chapter 3: Automated Demand Side Management in Buildings: Lessons from Practical Trials

This chapter discusses the problem of demand side management (DSM) in particular within the context of energy transition (smart grids, distributed energy resources, etc.). It focuses on the use of artificial intelligence techniques to answer the challenges (response time, data available, privacy issues, etc.) related to this problem. It presents the DSM's motivations and objectives around demand reduction (energy efficiency), demand response (local or self-consumption of energy generated by distributed energy resources), user engagement (interaction between energy companies and building occupants), engagement on investments, engagement on operations, price optimization, and providing ancillary services (load and production modulation, frequency regulation). The chapter divides the DSM problem into problem of forecasting and problem of automated control. Then, it divides the methods of the state of the art used for forecasting and control into artificial intelligence (data-driven, model-free) and model-based (model predictive control, model-based reinforcement learning, etc.). The goal is to discuss the advantages and drawbacks of these methods according to the challenges related to the problem of DSM within the context of energy transition. The chapter highlights the use of transfer learning in order to avoid the problem of lack of availability of data and to improve the model accuracy as well as its training or learning time. It provides also guidance to select the adapted methods for forecasting and control.

1.3.3 Chapter 4: A Multi-Agent Approach to Energy Optimization for Demand-Response Ready Buildings

This chapter proposes a distributed energy optimization approach as a multi-agent system in order to perform energy management in buildings that are equipped with a wide range of energy-consuming and energy-producing devices such as household appliances in residential buildings, photovoltaic, and local generators. The consuming energy-devices consist of fixed and flexible (sheddable and shiftable) load while energy-producing generators are curtailable local energy sources. Each type of devices is represented by an agent an objective function that incorporates user constraints and demand-response incentives. The optimization of the objective function is performed using the alternating direction method of multipliers. The goal of the optimization is to obtain the optimal energy flow (i.e., consumption and generation) that takes into account the incentives for demand-responses schemes, the electricity prices (real-time pricing or time-of-use pricing) while respecting user constraints (inconveniences). The advantage of the proposed approach is twofold: it takes into account both price-based demand response as well as incentive-based demand response together with consumers' inconvenience when applicable and preserves user privacy since each agent performs its local optimization based on its local model. The proposed approach is applied to a prosumer building with a connection to an energy supplier (i.e., external tie) and equipped with a photovoltaic (PV) for local uses. Several scenarios simulating different consumption, production, and demand-response requests with time horizon of 24 h divided into 96 time periods (TP) of 15-min interval are conducted. The goal is to measure the reduction in energy bill and imported energy from grid for different fixed amounts of consumption (i.e., fixed load) that must be satisfied, and amounts of flexible consumption (i.e., shiftable load and sheddable load) that can be shifted or shed to some extent over a given time frame.

1.3.4 Chapter 5: A Review on Non-Intrusive Load Monitoring Approaches Based on Machine Learning

This chapter presents a survey about the problem of residential non-intrusive load monitoring (NILM) and discusses its challenges and requirements. Residential NILM aims at recognizing the individual household appliances that are active (consuming) from the total load in the house (the total consumption). This recognition is performed without the need for any additional sensor but only the total consumption provided by the smart meter. The goal of residential NILM is twofold. First, it allows inviting consumers to adapt a conservative "green" consumption behavior by optimizing their consumption according to their profile or activity (e.g., his presence and behavior). Second, it can improve their involvement in the demand-response program by scheduling their activities (consumption) while respecting their comfort. The chapter presents the three steps of a NILM framework: data acquisition, feature extraction, and inference and learning. Data acquired about the user consumption can be samples either at low or at high frequency. The sampling frequency determines the features that can be extracted in the second step of NILM framework. Indeed, the extracted features can be related either to the appliance stable consumption states or conditions such as the active power. In this case, low frequency sampling is adapted. However, features related to the transition dynamics between different stable states require a high frequency sampling rate because of the very short time laps of the transition. Contextual features, such as time of use or its duration, can also be extracted in order to better distinguish appliances of close consumption behaviors. Then, the chapter studies the machine learning approaches used to perform the appliance active state recognition. It divides them into event-based and probabilistic model-based. The chapter focuses on probabilistic model-based approaches, in particular hidden Markov models. For the latter, the chapter discusses their performances around their kind of learning (supervised, unsupervised, semi-supervised) and the used features (stable, transient, and contextual).

1.3.5 Chapter 6: Neural Networks and Statistical Decision Making for Fault Diagnosis in Energy Conversion Systems

This chapter presents a model-free approach based on the use of feed-forward neural networks (FNNs) in order to perform the fault diagnosis of DC-DC conversion systems. Indeed, DC-DC conversion systems, widely used in many applications, such as photovoltaic power pumps or in desalination units, undergo different faults impacting their efficiency, reliability, and lifetime. These faults can impact either their mechanical part (DC motor) or electrical part (DC-DC conversion). This chapter proposes the development of a model based on the use of FNNs with Gauss-Hermite activation functions in order to model the power conversion systems' dynamics in normal operation conditions. In FNNs with Gauss-Hermite, the activation function satisfies the property of orthogonality as the case of Fourier series expansions. A fault is detected when the difference between the FNNs output and the real output is greater than a certain threshold. The latter is determined using the $\chi 2$ statistical change detection test with 98% confidence interval. For the fault isolation, the $\chi 2$ statistical change detection test is applied to the individual components of the DC-DC converter and DC motor energy conversion system. The fault is isolated by finding out the individual component that exhibits the highest score. The proposed approach has been tested using several simulation experiments in normal operation conditions and in presence of faults generated using an energy conversion system turning solar power into mechanical power.

1.3.6 Chapter 7: Support Vector Machine Classification of Current Data for Fault Diagnosis and Similarity-Based Approach for Failure Prognosis in Wind Turbine Systems

This chapter proposes a data-driven approach based on the combination of physical and reasoning models in order to perform the fault diagnosis and prognosis of wind turbine systems. The goal is to decrease the maintenance costs of wind turbines.

The physical model is built using the Bond Graph (BG) methodology. This allows to exploit the already available knowledge about the wind turbine dynamics (the phenomena of transformation of wind power into mechanical power and then into electrical power, the phenomena of power conservation and dissipation, etc.). This model is then used to generate data sets about faults in critical components for which it is hard to obtain enough of data. The reasoning model is based on the use of a Multi-Class Support Vector Machine (MC-SVM) classifier. The goal of the latter is to detect online the occurrence of degradation. When degradations are detected, the fault prognosis is activated. The goal is to estimate the remaining useful life before the wind turbine reaches the failure threshold (end of life). This estimation is based on the geometrical degradation speed in the feature space. The chapter evaluates the proposed approach using two evaluation metrics: the prognosis horizon and α—λ performance. The obtained results show that the proposed approach is able to perform the fault diagnosis and prognosis of four tested faults: unbalance caused by a deformation of the blade, unbalance in high speed shaft, stator eccentricity in the generator, and electrical faults in the stator resistance.

1.3.7 Chapter 8: Review on Health Indices Extraction and Trend Modeling for Remaining Useful Life Estimation

In this chapter, an overview of approaches used for fault prognosis is presented. These approaches aim at estimating the remaining useful life before the failure (end of life) occurs. The interest of fault prognosis is twofold: alerting supervision operators of the future occurrence of a failure, and giving them a sufficient time to plan the maintenance actions. The chapter focuses on fault prognosis as a horizontal approach allowing to link fault diagnosis and fault prognosis. It classifies these approaches into three major categories: expert, physical model-based, and data-driven approaches. Then, it focuses on data-driven approaches by showing how the health indices are built and evaluated. Indeed, health indices are used to follow the evolution (decrease) of the system health (ability) to perform a task. Therefore, it is used to estimate the remaining useful life. The chapter presents some meaningful criteria (monotonicity, trendability, prognosability, prognosis horizon, relative accuracy) in order to evaluate the built health indices and the estimated remaining useful life. The chapter compares the performances of several major approaches of fault prognosis and discusses their limits as well as their future challenges.

1.3.8 Chapter 9: How Machine Learning Can Support Cyber-Attack Detection in Smart Grids

This chapter provides an overview of the major components of smart grids, kinds of attacks against them, and the machine learning techniques used for the attack detection. Through the presentation of the different components (generation, transmission, distribution, communication, consumption) of a smart grid, the chapter highlights its vulnerability to cyber-attacks against its cyber and physical layers. The chapter classifies these attacks around three categories: attacks on confidentiality (gaining access to data belonging to others), attacks on integrity (someone other than the legitimate device fraudulently claims to be that component), and attacks on availability (generating lots of traffic to overwhelm the capacity of target devices to render the services). Then, the chapter presents the detection methods that are used for the attack detection. They are classified into signature-based, anomaly-based, and specification-based detection. It highlights the advantages and drawbacks of three decision (attack detection) structures: centralized, partially distributed (hierarchical), and fully distributed. Then, the chapter focuses on Machine Learning approaches (Support Vector Machine, Neural Networks, K means, Hoeffding tree, etc.) used for the attack detection. It discusses the use of these methods (classification and regression) by dividing them into supervised, unsupervised, and semi-supervised learning approaches. The chapter shows the advantages and drawbacks of these different categories of approaches for the detection of the different attacks that can occur in smart grids. The chapter ends by discussing the open problems and the challenges to be addressed related to the problem of cyber security in smart grids.

1.3.9 Chapter 10: Neurofuzzy Approach for Control of Smart Appliances for Implementing Demand Response in Price Directed Electricity Utilization

This chapter proposes an approach in order to conduct the demand-response program at the appliance level by considering the evolution of electricity prices. Indeed, the amount of consumption of a set of aggregated loads is determined by a set of used appliances by the corresponding consumers. The proposed approach is based on two steps. In the first step, an extreme learning machine (ELM) is used in order to predict the future price of electricity during the time use of an appliance. To this end, a rolling time window of the ten previous prices of electricity is used as well as the current price. After the reception of a new current price, it replaces the oldest one in order to keep tracking the prices evolution with a fixed size of training set. The second step is a set of fuzzy rules. The goal of these rules is to determine the period of use (full, reduced) of an appliance according to the predicted and current electricity prices as well as the appliance's operational variables. The advantage of this approach is its short time of training and processing in order to provide the output (the decision on the period of use of an appliance). Therefore, it is adapted for the implementation of demand-response program since the latter requires quick decision making. The proposed approach is applied to the demand-response program for heat, ventilation, and air condition appliance. The approach considers as input the actual temperature, the minimum desired temperature, the maximum desired temperature, and time for reaching the minimum desired temperature from the current temperature. Its output is the operational time of the appliance.

1.3.10 Chapter 11: Using Model-Based Reasoning for Self-Adaptive Control of Smart Battery Systems

This chapter discusses the use of model-based reasoning, in particular the first order logic, for the fault diagnosis and configuration of smart battery systems. The latter become increasingly important within the context of energy transition through their use in distributed energy resources, electric and autonomous cars. The chapter highlights the interest of performing the fault diagnosis and reconfiguration for smart battery systems in order to guarantee their safety during operation and to extend their lifetime. The chapter describes in detail a smart battery system comprising n batteries and k-1 wire cells. The different valid and invalid configurations are presented. The invalid configurations correspond to the ones that cause harm on side of batteries or the electronic or do not deliver the specified properties. Then, the chapter details the use of a model-based reasoning, in particular the first order logic, in order to avoid these invalid configurations in presence of faults (a battery run out of power when being used causing the given voltage to drop, or the required current cannot be delivered due to a faulty battery). The chapter shows how the developed approach sets up the right reconfiguration (connect, disconnect, or recharge batteries) for fulfilling given electrical requirements (required voltage and current) and a diagnosis problem during operation of such battery systems.

1.3.11 Chapter 12: Data-Driven Predictive Flexibility Modeling of Distributed Energy Resources

This chapter treats the problem of the use of distributed energy resources (DERs) in order to assist conventional generators in providing ancillary grid services such as instantly overcoming local supply shortages, reducing costs during peak price hours, maintaining grid stability, etc. To this end, the chapter proposes an approach allowing quantifying the available load flexibility of an ensemble of DERs, in particular air conditioners and electric water heaters, to provide grid services. The proposed approach uses the notion of virtual battery (VB) in which the aggregated load flexibility of the ensemble of DERs is represented (stored) in the form of thermal energy. The modeling of the VB (first order) is based on the following components: Stacked Auto Encoder (SAE), Long-Short-Term-Memory (LSTM) network, Convolution Network (ConvNet), and Probabilistic Encoder and Decoder. They are used to find the VB model's parameters (self-dissipation rate and lower and upper power limits) allowing quantifying the load flexibility that best tracks the regulation signal by respecting the consumer comfort. The aim of this model is to estimate the state of charge (soc) of the VB at time t with the initial soc and the regulation signal as input. In addition, the proposed approach combines two transfer learning Net2Net strategies, namely Net2WiderNet and Net2DeeperNet in order to update the VB's parameters in the case of adding or removing DERs. Both of these two strategies are based on initializing the "target" network to represent the same function as the "source" network. The proposed approach is applied in order to quantify the soc of a set of 100 air conditioner devices and 150 electric water heaters by considering uncertainties in the water draw profile.

1.3.12 Chapter 13: Applications of Artificial Neural Networks in the Context of Power Systems

This chapter treats the use of machine learning techniques, in particular artificial neural networks (ANNs), to predict power flows (the bus voltage and the line current magnitudes), in power grids. The chapter focuses on the interest of using ANNs in order to provide an accurate estimation of line loadings and bus voltages magnitudes in distribution grids with a high percentage of distributed energy resources (DER). The estimation accuracy of power flows is crucial in order to identify fast and in reliable manner the critical loading situations and the energy losses in particular in low voltage (LV) grids. The goal is to improve the real-time monitoring of power systems, in particular at low and medium voltage level, for grid planning and operation. The chapter uses the open-source simulation tool Pandapower in order to generate suitable training and test sets for the built ANNs. Two ANNs are trained, one to estimate line loading and the other to estimate voltage magnitude. The estimation accuracy and the computation time of these ANNs are considered as performance criteria. The chapter presents the obtained results around the use of the trained ANNs for two case-studies: the estimation of grid losses and the grid equivalents. The grid equivalent aims at approximating the interaction at the interconnection of two interconnected areas operated by two different grid operators. The chapter highlights clearly the interest of using ANNs in order to address the challenges of power systems analysis related to the intermittent nature of DER and their increasing rate in power grids within the context of energy transition, the changing load behavior in particular with the increasing role of users as local producers, and the huge number of grid assets and their incomplete measurements.

References

1. M. Sayed-Mouchaweh. *Artificial Intelligence Techniques for Smart Energy Management within the Context of Energy Transition*, Editions Techniques de l'Ingénieur (2019). https://www.techniques-ingenieur.fr/base-documentaire/energies-th4/energie-economie-et-environnement-42593210/intelligence-artificielle-et-gestion-intelligente-de-l-energie-be6000/
2. J. Li, G. Poulton, G. James, Coordination of distributed energy resource agents. Appl. Artif. Intell. **24**(5), 351–380 (2010)
3. N. Hatziargyriou, H. Asano, R. Iravani, C. Marnay, Microgrids. IEEE Power and Energy Mag. **5**(4), 78–94 (2007)
4. J.A. Momoh, Smart grid design for efficient and flexible power networks operation and control, in *2009 IEEE/PES Power Systems Conference and Exposition* (IEEE, Piscataway, 2009, March), pp. 1–8
5. Digitalization of the Electricity System and Customer Participation, in *European Technology & Innovation Platforms (ETIP) Smart Networks for Energy Transition (SNET), Workgroup-4 "Digitalization of the Energy System and Customer Participation* (2018). https://www.etip-snet.eu/wp-content/uploads/2018/10/ETIP-SNET-Position-Paper-on-Digitalisation-FINAL-1.pdf)
6. S. Mohamad, M. Sayed-Mouchaweh, A. Bouchachia, Online active learning for human activity recognition from sensory data streams. Neurocomputing (2019 in press). https://doi.org/10.1016/j.neucom.2019.08.092
7. H. Salem, M. Sayed-Mouchaweh, A.B. Hassine, A review on machine learning and data mining techniques for residential energy smart management, in *2016 15th IEEE International Conference on Machine Learning and Applications (ICMLA)*, (IEEE, Piscataway, 2016, December), pp. 1073–1076
8. E. Mocanu, D.C. Mocanu, P.H. Nguyen, A. Liotta, M.E. Webber, M. Gibescu, J.G. Slootweg, On-line building energy optimization using deep reinforcement learning. IEEE Trans. Smart Grid **10**(4), 3698–3708 (2018)
9. E. Lughofer, M. Sayed-Mouchaweh (eds.), *Predictive Maintenance in Dynamic Systems: Advanced Methods, Decision Support Tools and Real-World Applications* (Springer, Cham, 2019)
10. Z. Wen, D. O'Neill, H. Maei, Optimal demand response using device-based reinforcement learning. IEEE Trans. Smart Grid **6**(5), 2312–2324 (2015)
11. C. Joe-Wong, S. Sen, S. Ha, M. Chiang, Optimized day-ahead pricing for smart grids with device-specific scheduling flexibility. IEEE J. Selected Areas Commun. **30**(6), 1075–1085 (2012)

Part I
Artificial Intelligence for Smart Energy Management

Chapter 2
Large-Scale Building Thermal Modeling Based on Artificial Neural Networks: Application to Smart Energy Management

Lala Rajaoarisoa

2.1 Introduction

Today in the world and more particularly in Europe, the building sector consumes more than a third of global energy. Indeed, the residential, academic, and commercial building sector is considered with transportation, as the largest savings energy potential. Thus, the improvement of energy efficiency is a priority characterized by the European Directive [1], which aims in reducing the Greenhouse Gases (GHG) emissions by 2020 while simultaneously increasing the building energy efficiency by 20% [2, 3]. In this way, various national/international level initiatives have been set out by proposing a home energy management systems to minimize the building energy consumption. Typically in smart building, for instance, tools and methods developed in [4–7]. Globally for recent work in smart energy management system applied onto different applications, including the residential building, reader can refer to the literature review [8] and check also all the references herein.

So, most of the time, action to reduce energy consumption is associated with the control and/or the improvement of the thermal performance of the building by finding an optimal controller in combining operational parameters (e.g., climate, building use, owner requirements, control, etc.) with passive (e.g., building envelop materials, insulation, roofing materials, finishing materials, window types, etc.) and particularly the active system components (e.g., heating, air conditioning, cooling, ventilation, alternative energy sources, building management systems, etc.). Therefore, the optimization of the heating, ventilation, and air conditioning (HVAC) systems is particularly important [9], and inefficient operation and maintenance

L. Rajaoarisoa (✉)
IMT Lille Douai, University of Lille, Unité de Recherche en Informatique et Automatique, Lille, France
e-mail: lala.rajaoarisoa@imt-lille-douai.fr

M. Sayed-Mouchaweh (ed.), *Artificial Intelligence Techniques for a Scalable Energy Transition*, https://doi.org/10.1007/978-3-030-42726-9_2

of the HVAC systems can cause energy wastage, customer complaints [10], poor indoor air quality, and even environmental damage [11].

The building services and energy system control had been discussed by several researchers over the last decade. In this framework, two main research directions are pursued in advanced control of building energy management of HVAC systems. Firstly learning-based methods of artificial intelligence [12–14] and secondly model-based predictive control (MBPC) [15–18], which has its own advantages and inconveniences. A major challenge with MBPC is in accurately modeling the dynamics of the underlying physical system. The task is much more complicated and time consuming in case of a large-scale building and often times, the use and exploitation can be more complex than the controller design itself [19]. On the other hand, the energy performance of HVAC systems is impacted by operating conditions as well as the response time (inertia) to a building's heating and cooling but also occupants' energy needs. Indeed, studies in literature have extensively evaluated the sensitivity of models to the building technical design parameters, where the areas of organizational energy management policies/regulations and human factors (i.e., energy users' behavior), are very important elements influencing building energy consumption [20–22].

Moreover, works realized into [23–26] show that more than half of the total building energy is typically consumed during the non-working hours mainly due to occupancy related actions (e.g., equipment and lighting after-hours usage) and can be reduced through behavioral changes. As also argued into [27, 28], occupancy presence and behavior in buildings has shown to have large impacts on space heating, cooling and ventilation demand, energy consumption of lighting and space appliances, and building controls where careless behavior can add one-third to a building's designed energy performance, while conservation behavior can save a third.

As a result, this chapter presents the development of a smart building energy management system with which we would like to reach two main goals. Firstly, by using an artificial neural network (from artificial intelligence technique), we want to estimate and predict the thermal behavior of a large-scale building while including instrumented and non-instrumented thermal zone. This information will be used in the next to appreciate the energy users' behavior. In this way via a human graphical interface, we provide different advice to users to educate and attract them about energy reduction challenges. Thus, the originality of this chapter is close to the nature of the intervention of the system itself. Indeed, it does not act directly on the HVAC building systems, as an automation system could realize, but on the USER. In other terms, we propose an interface which provides for users access to assets, controls, data among any others, for enhancing their life's quality through comfort, convenience, reduced costs, and increased connectivity. This approach will allow us to validate the thermal behavior model developed and also realize different factors analysis which may affect the energy consumption for optimization purposes. This leads in setting well the human interface to be sure that each user sticks to each advice in order to guarantee an efficient smart building energy management (SBEM) system design.

The chapter is organized as follows. Section 2.2 states on the formulation of the problem to predict the thermal dynamics in different zones of the building and also design an efficient SBEM solution. Section 2.3 presents the principle and methodology that allows to identify and estimate the large-scale building model parameters and finally provide a new approach of inviting user to adopt a "green behavior and/or activity." The results in terms of thermal modeling and energy consumption will be presented in Sect. 2.4. It will present also a smart and interactive system which was implemented on the Lavoisier student residential building, located in Douai, in the north of the France, to illustrate the effectiveness of the methodology.

2.2 Problem Formulation

This chapter is specifically aimed at developing a smart building energy management system. In order to reach this goal, the first problem we face and we have to solve is the prediction of indoor temperature (y) in different zone of the large-scale building. After that, comes the formulation of different advice for the user (Ad). So, let us consider that the thermal dynamic of a large-scale building (LSB) is driven by the following expressions:

$$LSB = \begin{cases} y(t) = f(u, w, \beta, t) \\ m_i \quad = \langle Ad, I, O, \delta_{int}, \delta_{ext}, \lambda, t_a \rangle, \end{cases} \tag{2.1}$$

where f is a nonlinear function defining the relationship between the input (disturbances and controlled inputs) u and the indoor temperature y mapping and both observed from the system. This is simply the model used to describe the thermal behavior of the LSB considered. β is an adjustment parameter, w is the parameter of the nonlinear model, and t is the discrete time. The discrete system m_i defines the sequential evolution of the situation in the building. In other terms, with the state m_i we will know if the ambient climate, for instance, is suitable, or not according to the dual comfort-energy efficiency point of view. Ad is the set of the personalized advice issued for user. I is the set of input event values (i.e., all the values that input event can take). The events may be considered as defined by output $y(t)$ of the system and any internal y_k-dependent variable, where k is the output number of the system. λ is the output function which warrants the activity execution. δ_{int} is the internal transition function. It ensures recommended advice evolves even if no exogenous events come out before elapsed time e and the maximum dwell-time into the situation, called also time advance t_a, i.e., $Ad_j = \delta_{int}(Ad_i, e + t_a)$. δ_{ext} is the external transit function that will be used when exogenous events come out, i.e. $Ad_j = \delta_{ext}(Ad_i, e, I_i)$.

Then, the design of SBEM system problem consists in the proposal of methodology that allows to predict and estimate all parameters of the LSB model, and provides a new approach of inviting user to adopt a "green behavior and/or activity."

However, these actions must not be to the detriment of well-being in the building. Hereafter is given more details about the approach.

2.3 Principle and Methodology

Aim of this work is to propose a smart energy management system to reduce energy consumption by focusing on users and their behavior (i.e., thermal behavior). By proceeding in this manner, we could appreciate the buildings' thermal behavior and predict energy consumption according to their usage. In other terms, by estimating the ambient temperature in different building's area, our approach consists in comparing it with a comfort temperature and providing advice and recommendations to users by inviting each of them to have the best behavior in terms of reducing housing energy consumption. However, this implies that buildings must be equipped with sensors to provide usage data. And yet, installing sensors in all thermal zones of a large-scale building quickly becomes economically unfeasible. That is why we propose this methodology as an alternative energy management solution that requires neither investment nor capital expenditures or a huge operation.

So, the system will be built according to the following steps (Fig. 2.1):

- Setting the data project and construction in order to create the relationship between building energy performance and sensors data,
- Connecting the digital tools with the physical system,
- Identifying the thermal model parameters by using artificial intelligence techniques, both in instrumented and non-instrumented thermal zone,
- Defining the dashboard and various advises to the users according to the usage of the building,
- Evaluating feedback from users in order to reduce efficiently energy consumption.

Each part of them is detailed in the following paragraphs.

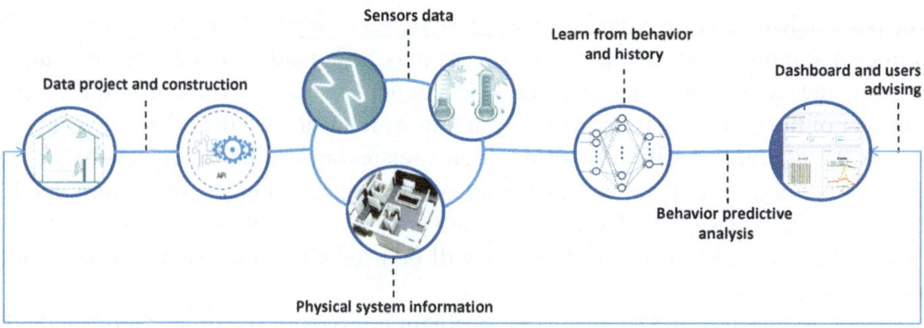

Fig. 2.1 Procedure to implement the SBEM system

2.3.1 Data Project and Construction

Data will be collected onto the buildings according to the nature and objectives of the action. For instance, to evaluate the thermal environment, Fanger in [29] proposes to use six quantities: four physical parameters (air humidity, air temperature, mean radiant temperature, and air velocity) and two subjectives (clothing thermal insulation and metabolic rate). These parameters have to be measured or estimated to evaluate and monitor efficiently thermal comfort [30]. Besides, Fugate et al. [31], for their part, proposed three categories for building sensors and meters for measuring and sensing different building performance parameters: occupants comfort perception and facility characteristics. Temperature, occupancy, humidity, CO_2, and air quality sensors are used to sense occupant comfort and activity. On the other hand; building energy meter, sub-metering, plug-load measurements, natural gas meters, and other sources of energy meters are used to measure energy consumption.

In other terms, the sensor kits should be capable to measure environmental features of an enclosed space, such as an office or hospital room or basement, buildings, etc. and to report any number of factors, including: temperature, humidity, movement, light, and noise. This data can be used to more efficiently deliver needed services on demand or to track utilization of any area. But also should be deployed to enhance the data gathered from external environment. Indeed, by correlating the external temperature and other information with the internal sensor data, a building's complete environment can be monitored.

Basically, the data construction should deliver all kinds of crucial information according to the performance we want to evaluate. As studied previously in [32], the input-output mapping is defined between the performance and the goals as given in Table 2.1. However, for large-scale building it is difficult or impossible to mount all the necessary sensors to all the building parts. That is why in this study we

Table 2.1 Data collection

Data acquisition for performance evaluation		
Performance	Goals	Input/output
Thermal comfort	Temperature range	BIM
		Indoor temperature
		Heating system
	Experimental model	Flux meter
		Indoor humidity
	Adaptive model	Windows and door systems
		Outside temperature
	Controller design	Solar radiation
		Outside temperature
Indoor environment quality	Renovation guide	Wind speed
		Outside humidity

Table 2.2 Wireless sensor characteristics

Measured parameter	Measuring range	Accuracy	Type of sensor	Units
Temperature	−40 to 125 °C	±0.4 °C (max), at −10 to 85 °C	Thermistor	°C
Heat flux	0–2000 W/m^2	40 μV/(W/m^2)	HFP	W/m^2
Relative humidity	0–100%	±3% (max), at 0–80%	Capacitive polymer	%
Solar radiation	1 – 65536 Lux	+/− 100 Lux	Light sensor	Lux to W/m^2
Electrical power	Up to 3500 W	Up to 16 A on 220 V	Power meter	W

limit ourselves to using weather sensors and electrical power consumption for few instrumented zones and we estimate by means of a learning technique the thermal behavior of all non-instrumented ones.

Table 2.2 below defines each sensor characteristic used to collect data. These components are developed by CLEODE™ company (French company), and based on wireless system network (WSN) technology. This is today the most popular technology used in smart system. Moreover, many WSN platforms have been developed [33], ZigBee/IEEE 802.15.4 protocols are often part of them. These protocols are a global hardware and software standard designed for WSN requiring high reliability, low cost, low power, scalability, and low data rate [34].

2.3.2 *Physical System and Data Connection*

According to the results obtained in [35] and also argued in [32], the thermal behavior of the building could be mainly influencing by the (1) building services and energy systems, (2) occupant activities and behavior, (3) and location or the orientation of the building. Therefore, we propose in this chapter to subdivide the building into two main zones, the east (green-back part) and the west (red-front part) as shown in Fig. 2.2. Also, we choose the first floor as the reference zone to estimate the thermal behavior of the whole non-instrumented zones. Depending on the size of the building, we deploy a sufficient number of sensors to cover all the monitored zones. In other terms, a thermal zone (marked by red or green polygon) is defined as the mix of several parts of the building, which means the temperature will be considered identical in each of them. Moreover, each zone is delimited by a certain number of walls, themselves divided into meshes, with the assumption that the temperature is homogeneous in each thermal zone. Thus, we can use data collected from this reference instrumented zone (east and west zones in the first floor here) to estimate and predict the thermal behavior in different non-instrumented zones (east and west zone for the second to nth floor). This is particularly interesting because

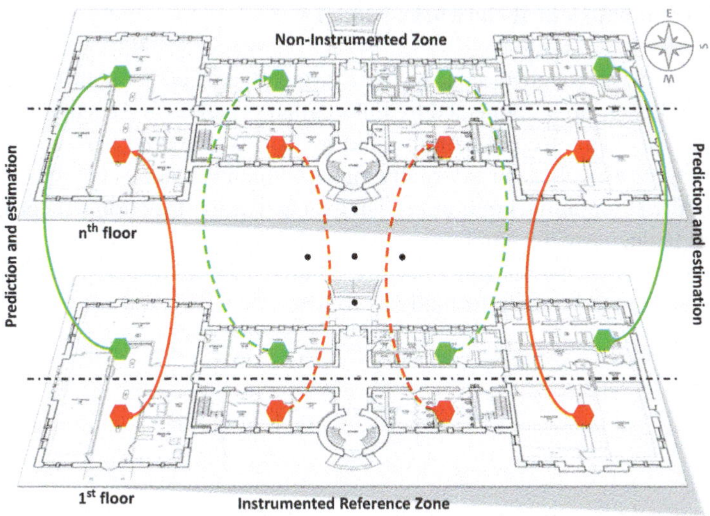

Fig. 2.2 Building and sensors connection

real data collected from buildings is used to reset the model and predict accurately the temperature in all building's areas.

So, data required for following the thermal behavior is complete here. The next step consists in modeling the thermal behavior of the building.

Remark 1 Let us notice to perform the estimation and prediction of the temperature in different non-instrumented zone, we assume here each connected building's area has the same topology whether for instrumented or non-instrumented zone.

2.3.3 Building Thermal Modeling

Artificial intelligence approaches have been used for this last decades to estimate thermal behavior and predict the energy consumption of buildings. As examples artificial neural networks (ANN) and support vector machine (SVM) [36–38]. These methods are well recognized for their accuracy in explaining the complex nonlinear energy consumption behavior in buildings with limited parameters of buildings and has shown better performance than physical and statistical regression methods. In addition due to the ability of ANN model to solve many problems by learning directly from data, they have been widely applied to various fields [39]. Indeed, they are today considered to be a powerful technique and smart tool for modeling, prediction, and optimization of the performance of different engineering systems. This technique will be used hereafter to model the thermal behavior of a LSB defined by Eq. (2.1).

2.3.3.1 Artificial Neural Networks Model

An ANN model is basically composed by a group of neuron model [40] as illustrated in Fig. 2.3. This group is connected by links called synapses and each of them has its own weight w_{kj}. This weight is multiplied by its own input u_j before summing all weighted inputs as well as an externally bias b_k which is responsible for lowering or increasing the summation's output v_k. Then an activation function $\varphi(.)$ is applied to that output to decrease the amplitude range of the output signal y_k into a finite value; different types of activation functions are tabulated in Table 2.3. The computation of the output y_k can be write formally as follows:

$$y_k = \varphi\left(\sum_{j=1}^{n} w_{kj}u_j + b_k\right). \tag{2.2}$$

In this chapter, ANN model is used to lead onto the creation of a global model of a large-scale building. To reach this goal, we use another class of a feedforward neural network distinguishes itself by the presence of one or more hidden layers,

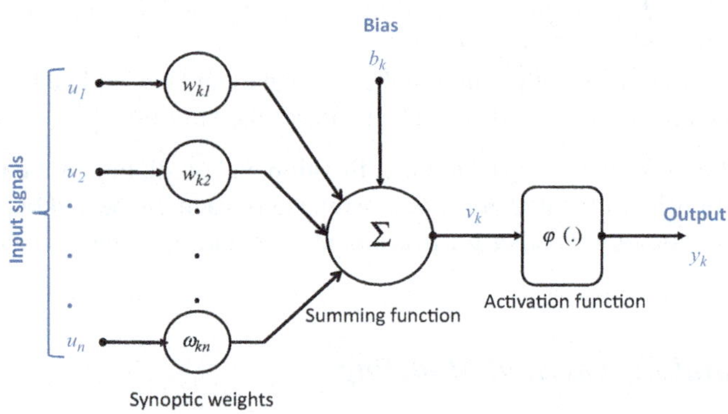

Fig. 2.3 Nonlinear model of neuron

Table 2.3 Activation function examples

Activation function	Formulation	Remark
Threshold function	$\varphi(v) = \begin{cases} 1 & \text{if } v > 0 \\ 0 & \text{si } v = 0 \\ -1 & \text{si } v < 0 \end{cases}$	Equivalent to *Heaviside function*
Identity function	$\varphi(v) = v$	Linear function
Logistic function	$\varphi(v) = \log f(v) = \frac{1}{1+\exp(-av)}$	Sigmoidal nonlinearity and a is the slope parameter
Hyperbolic tangent function	$\varphi(v) = \tan h(v) = \frac{2}{(1+\exp(-av))-1}$	Sigmoidal nonlinearity

known by multilayer perceptron (MLP) neural network model. A MLP are special configurations of neural networks where the neurons are arranged in successive layers. It can then be viewed as a complex mathematical function made of linear combinations and compositions of the function $\varphi(.)$ that links a number of input variables to a number of output variables. The structure of the model used for our study-application is detailed in the following paragraph.

2.3.3.2 Input-Output Mapping

According to the notion introduced in Sect. 2.3.1, the main inputs and outputs of the MLP-thermal model are illustrated in Fig. 2.4. Basically, we have the following elements:

- weather inputs collected by sensors such as outdoor temperature (T_o), outdoor humidity (H_o), and solar radiation (R_a),
- energy measurement given by an electrical consumption meter placed, respectively, in the instrumented (Pzi_m) and non-instrumented zone (Pzn_p),
- the outputs which are the estimation of the indoor temperature, both for the reference zone ($\hat{T}zi_m$) and all non-instrumented zone ($\hat{T}zn_p$). Here the subscripts

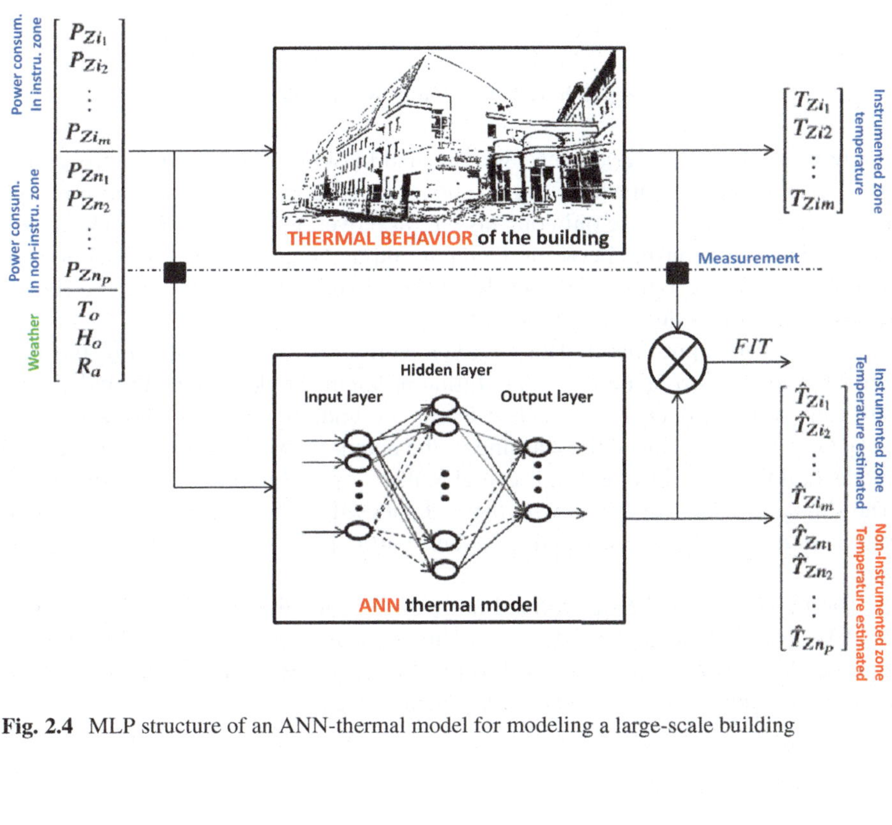

Fig. 2.4 MLP structure of an ANN-thermal model for modeling a large-scale building

m and p are, respectively, the number of instrumented and non-instrumented zones. However, for the sake of ease, we assume $m = p$ for this study.

Briefly, the input-output mapping can be represented by the following expressions:

$$u = \left[Pzi_1 \ldots Pzi_m \; Pzn_1 \ldots Pzn_p \; T_o \; H_o \; R_a \right]^T . \tag{2.3}$$

The temperature measured into the different instrumented zone is given by

$$y = \left[Tzi_1 \; Tzi_2 \ldots Tzi_{m-1} \; Tzi_m\right]^T , \tag{2.4}$$

and the output estimated by the ANN-thermal model is

$$\hat{y} = \left[\hat{T}zi_1 \ldots \hat{T}zi_m \; \hat{T}zn_1 \ldots \hat{T}zn_p\right]^T . \tag{2.5}$$

So, the output of the multilayer perceptron network (MLP) becomes

$$\hat{y}_k(u) = h\left(\sum_{k=0}^{M} w_{ki}^{(2)} g\left(\sum_{j=0}^{D} w_{ij}^{(1)} u_j + b_j^{(1)}\right) + b_k^{(2)}\right), \tag{2.6}$$

and in its compact form, we have

$$\hat{\mathbf{y}}(\mathbf{u}) = \mathbf{h}(\mathbf{w}^{(2)}\mathbf{g}(\mathbf{w}^{(1)}\mathbf{u} + \mathbf{b}^{(1)}) + \mathbf{b}^{(2)}), \tag{2.7}$$

where **w** and **b** are the parameters (weights and biases) for the neuron's processing node. g and h can be threshold (sign) units or continuous ones. However, h can be linear but not g (otherwise only one layer). Moreover, to take into account directly the nonlinear relationship between the input-output data, the activation functions h, and/or g could be a logistic sigmoid ($\log f$) or a hyperbolic tangent sigmoid ($\tan h$) transfer function as defined in Table 2.3.

Thus, given input-output data (u_k, y_k) with $k = 1, \ldots, N$ finding the best MLP network is formulated as a data fitting problem. Model will be provided with sufficient training data from which it learns the underlying input/output mapping [41]. Of course, to achieve the learning of the neural networks model, a 3D thermal model building was designed and several realistic operating ranges were simulated. The simulation data combined with the real data will thus make it possible to build the learning base of the predictive model, particularly for the non-instrumented zone.

The parameters to be determined are (w_{kj}^i, b_j^i) for the ith layer of the neural model. Several algorithms exist to solve this problem. Here we use the Levenberg–Marquardt (LM) algorithm one [42] to identify each parameter. Basically, the LM algorithm optimizes the following performance index:

$$F(\mathbf{w}) = \sum_{k=1}^{K} \left[\sum_{p=1}^{P} (y_{kp} - \hat{y}_{kp}) \right], \tag{2.8}$$

where $\mathbf{w} = [w_1\ w_2\ \ldots\ w_\Phi]$ consists of all weights of the network, y_{kp} is the desired value of the kth output and the pth pattern, $\hat{y}_{kp}$ is the actual value of the kth output and the pth pattern. Φ is the number of the weights, P is the number of patterns, and K is the number of the network outputs.

Readers can refer to [40, 42] for further information about this algorithm. Therefore, it is still important to notify here that the improvement of weighting parameters $\mathbf{w}$ is realized according to the desired level of performance index by minimizing the mean square error:

$$MSE = \frac{1}{N} \sum_{k=1}^{N} (y_k - \hat{y}_k)^2 = \frac{1}{N} \sum_{q=1}^{N} (e_k)^2, \tag{2.9}$$

where $\hat{y}_k$ is the estimated zone temperature.

On the other hand, the multilayer perceptron network model could have different structure according to the number of hidden layer considerate or the number of the perceptron node. To compare one structure to another one the FIT criterion value is adopted. The FIT represents the similarity between the measured output y and the output reconstructed from the model $\hat{y}$. Formally, it is given by the following relation:

$$FIT = \left(1 - \frac{\|y_k - \hat{y}_k\|}{\|y_k - \bar{y}\|}\right) \times 100\%, \tag{2.10}$$

where $\bar{y}$ is the mean of the observation y_k, and $\hat{y}_k$ is the estimate of the measures y_k at each sample time.

Once the MLP-thermal model is validated, we can begin designing the interactive interface on which the estimated temperature will be displayed, as well as recommendations for users to revisit their behavior to save more energy.

2.3.4 *Smart Interactive System Design*

Energy efficiency is one of the primary cost-saving opportunities for building managers. But it can be challenging for owners and managers of multistory offices, schools, etc., who need energy management solutions that do not require huge investments or operating expenses. We suggest for them this SBEM system as an alternative. The implementation of that allows them to collect data gathering and understanding of a wide range of their building. This will help building-stakeholders create policies to control consumption and waste. In other words this system leads to make stronger and more confident all collaborators with regard to their energy footprint, all of this from a smart functional graphical interface.

So, to expect a greater operational performance, the graphical user interface (GUI) should provide important insights on the health of energy-using systems and facilities, alerting users to overheating, for instance. Of course, this is not possible without the knowledge of the thermal behavior in each zone of the building.

Remark 2 For the last point above, the solution is already given in the last section.

2.3.4.1 SBEM System Architecture

The architecture of the smart building energy system is illustrated in Fig. 2.5. Herein, the left part is associated with the graphical user interface used mainly to control energy performance. The middle part is the local area networks (resp. world-wide communication) interface which provides for users access to overall information from the building and its environment anyway. The right part is the control acquisition data system. This last can be based on different technologies. We use especially wireless sensors gathering by CLEODE™ company to capture a number of insights. Mostly, to measure the indoor and outdoor temperature, indoor and outdoor humidity, solar radiation and incident solar radiation on the floor for different periods. A control (ON/OFF) system is also used in this study to control the heating system operation but also to measure and control the electrical power consumption into each thermal zone.

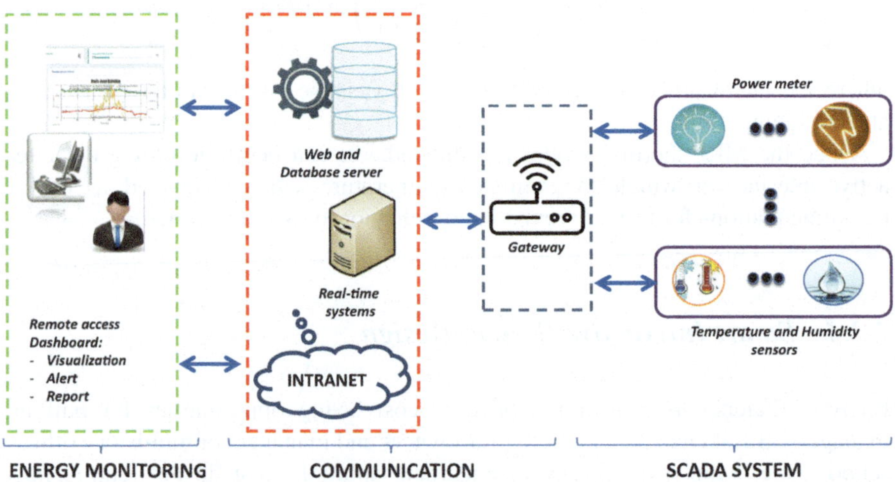

Fig. 2.5 Smart building energy management system architecture

Moreover, aside from sensors and actuators, the application requires two main components. The first one is the real-time system necessary to launch the neural network model. It can be seen as just a laptop where Microsoft Windows operating system runs. The second one is a master coordinator node that controls all connected sensors or elements to perform the data acquisition. For this purpose, WiBee product also made by CLEODE™ company is used in this work.

WiBee is a WIFI-ZigBee™ gateway allowing to generate a wireless mesh home networking based on ZigBee™ protocol. In fact, WiBee is not only a gateway, it is a stand-alone controller capable of managing a ZigBee™ network from the PC using an Ethernet or a WIFI interface, a Web Interface or CleoBee™ and a remote access Cleoweb from a PAD, laptop, iphone, and Smartphone. Besides, WiBee runs natively with an embedded Linux, a database and a Web Server, and integrates embedded Linux. It is also available natively via Ethernet and/or optional WiFi and/or optional GSM/GPRS, 3G. Integrated to the WiBee system, a scenario manager allows for simple and easy configuration, management, and program behavior such as (examples among any others):

- http command, Email on alert (intrusion, drop, . . .),
- Automatic switching off lights,
- Management opening depending on climatic parameters.

Remark 3 Reader can refer to Table 2.2 for sensor and actuator characteristics.

2.3.4.2 Graphical User Interface Design

The interactive graphical user interface (GUI) is the most widely used technique among all others to allow users to access assets, system controls and data. With this GUI, users can improve their life's quality by paying attention to their comfort, their convenience and their energy consumption. As data is critical to understanding how energy is consumed and where it wasted. The GUI (Fig. 2.6) offers complete information on the thermal behavior of the building and its energy performance. We can thus ensure and supply personalized advice and a matter for all stakeholders to measure their energy-saving.

Subsequently, the dashboard will give a persistent information for making best decision for energy consumed reduction, provides different insights, such that: daily and monthly thermal behavior and precise monthly and annual energy consumed.

Moreover, the dashboard could compare users in the workgroup. They can note then the similarity of their energy consumed to others, check standings in energy competitions and assess if their "green behavior" is making an impact globally on the energy efficiency of the large-scale building.

2.3.4.3 Users Recommendation Formulation

There are many different methods and approaches to model a recommender system. Reader can look at [44, 45] and the references herein for further information about

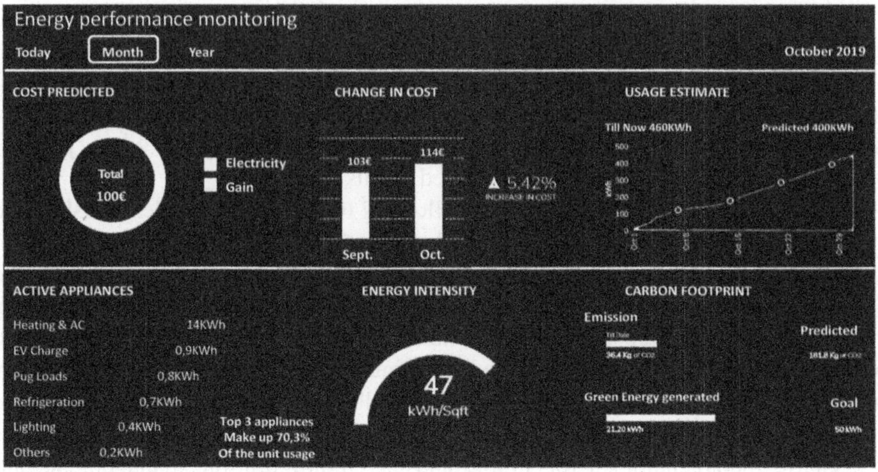

Fig. 2.6 Main page example of the graphical user interface [43]

it. In this chapter, we use a discrete event atomic model m_i to describe the sequential evolution of the situation in the building [46]:

$$m_i = \langle Ad, I, O, \delta_{int}, \delta_{ext}, \lambda, t_a \rangle. \tag{2.11}$$

Let us recall this kind of model represents sequences of events with the condition that the state has a finite number of changes in any finite interval of time.

So, an event is the representation of an instantaneous change in the building that can be associated with one recommendation. But, an event can be characterized by a value and an occurrence time also. To formulate each recommendation, let us consider the following definition and procedure:

Definition 1 (Advice Variables) The discrete state variables are

$$Ad = \{phase, I, t_a\}, \tag{2.12}$$

where $phase$ variable is introduced in order to have a clear and precise interpretation of the model behavior, I and t_a are, respectively, the set of the input variables and the time advance for each recommendation. To obtain the different values $(I_1, I_2, \ldots, I_s)$ of the input variable I, we choose to take up the following definition:

Definition 2 (Input Variables)

$$I = \{ev_1, ev_2, \ldots, I_1, I_2, \ldots, I_s\}, \tag{2.13}$$

where $(ev_1, ev_2, \ldots, ev_s)$ is the set of event variables, $(I_1, I_2, \ldots, I_s)$ are the set of different input values, and s is the maximum number of values that the input can

Table 2.4 Example of threshold definition to discretize a quantitative decision variable

Input \ Values	I_1	I_2	I_3
I	$\leq R_1$	$\leq R_2$ $> R_1$ and	$> R_2$

take. These values are obtained from a discretization of continuous input variables to piecewise continuous segment following the next procedure:

- define a threshold $R_i, i = 1, \ldots, l$, where l is a performance index for quantifying the input descriptive variable (see, for instance, Table 2.4), in order to transform the continuous segment in a piecewise constant segment (note that the discretization procedure is in the variable values regardless of the time, i.e., this is not the output of classical sample and hold),
- a discrete event is then associated with each change of value in the piecewise constant segment.

Definition 3 (Output Variables) An atomic model processes an input event trajectory and, according to that trajectory and its own initial conditions, provokes an output event trajectory. Notice that, when an input event arrives, this implies the building behavior changes instantaneously. Thus, the output variables O values can be associated with one advice defined by the input-output mapping (y_k, u_k). So, we can write

$$O = \{O_1, O_2, \ldots, O_s\} \rightarrow \{y_k(t), u_k(t)\}/Ad. \tag{2.14}$$

Basically, the recommender model works as follows. If we have the recommendation Ad_i at time t_i, after $t_a(Ad_i)$ units of time (i.e., at time $t_a(Ad_i) + t_1$) the recommender system performs an internal transition, going to a new recommendation Ad_j. The new recommendation is estimated as $Ad_j = \delta_{int}(Ad_i)$, where $\delta_{int} : Ad \rightarrow Ad$ is called the internal transition function. When the system goes from Ad_i to Ad_j an output event is produced with value $I_i = \lambda(Ad_i)$, where $\lambda : Ad \rightarrow O \cup \{\varnothing\}$ is called output function. Functions t_a, δ_{int}, and λ define the autonomous behavior of a discrete event model. When an input event arrives, the situation changes instantaneously. The new recommendation value depends not only on the input event value but also on the previous recommendation value and the elapsed time since the last transition. If the system suggests the recommendation Ad_i at time t_i and then an input event arrives at time $t_i + e$ with value O_i, the new recommendation is calculated as $Ad_{i+1} = \delta_{ext}(Ad_i, e, O_i)$ (note that $t_a(Ad_i) > e$). In this case, we say that the recommender system performs an external transition δ_{ext} defined as $(\delta_{ext} : Ad_i \times I \rightarrow Ad_j)$. No output event is produced during an external transition. This allows us to define the transition relation T which is a subset of $(Ad \times O \times Ad)$ as follows: $T = \{\langle Ad_i, e, O_i, Ad_j\rangle : \exists$ transitionfrom Ad_i to Ad_j by $O\}$ computed over arbitrary time intervals (e). Let us notice that this time e takes on values ranging from 0 to $t_a(Ad_i)$.

2.4 Experimental Results

2.4.1 Case Studies

To show the effectiveness of our methodology, we are experimenting it on the Lavoisier student residential building, located in Douai, in the north of the France (Fig. 2.7). The total area of the building is approximately $3500\,\text{m}^2$ and it is subdivided into two sub-building parts. The first part is composed by four floors and the second one by five floors. Moreover, in the first sub-building part each floor consists of ten rooms of $11\,\text{m}^2$, while the second one has 32 rooms with the same surfaces. We focus on the second part of the building during this study. Especially for this part, half of the rooms is in the east side and the other half is in the west side.

To recover a persistent database for identifying the underlying causal thermal model of the building, several experimental scenarios have been taken into consideration. Which are: two different orientations "East-West," heating sequence and different floor levels (first and third floors). The different rooms are subjected to different thermal excitation driven by a heating system supply and the influence of

Fig. 2.7 Top: Lavoisier student residence in Douai. Bottom: Lavoisier student residence in 3D

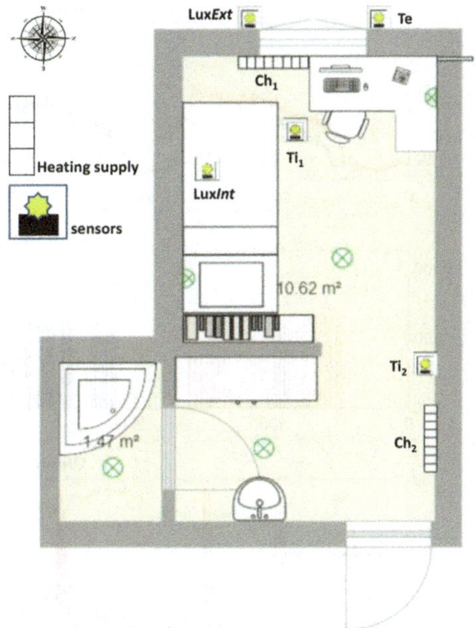

Fig. 2.8 Student room configuration within sensor and actuator location

the solar radiation. Precisely, as the heating power has a strong link with the control system, the heating supply will be controlled on two different sequences such as a controlled sequence and random heating sequence. So, for each scenario we can appreciate and evaluate the thermal environment, thanks to different sensors and actuators placed in each room. Figure 2.8 above gives an example of sensors and actuators placement for the evaluation of thermal environment. Once again, reader can refer to Table 2.2 for sensor and actuator characteristics.

Knowing that during the experiments, the scenarios considered do not worry about thermal comfort. The main purpose of the thermal excitation being to have a different thermal behavior and to enrich the database for obtaining a good estimate of the model parameters. Therefore, for each experiment, each measure is saved in a database in CSV format, thanks to an internally-LAB developed application using LABVIEW™ and NetBeans™ software. So, this historical data is used after that to train the MLP-thermal model from instrumented zone side according to the experimental planning detailed before. Figure 2.9 shows different trends for the indoor temperature (on the top), the electrical power consumption (in the middle) and the weather conditions (on the bottom), respectively, for the east (blue solid line) and the west (red solid line) zone. The weather data is common for the two sides of the building.

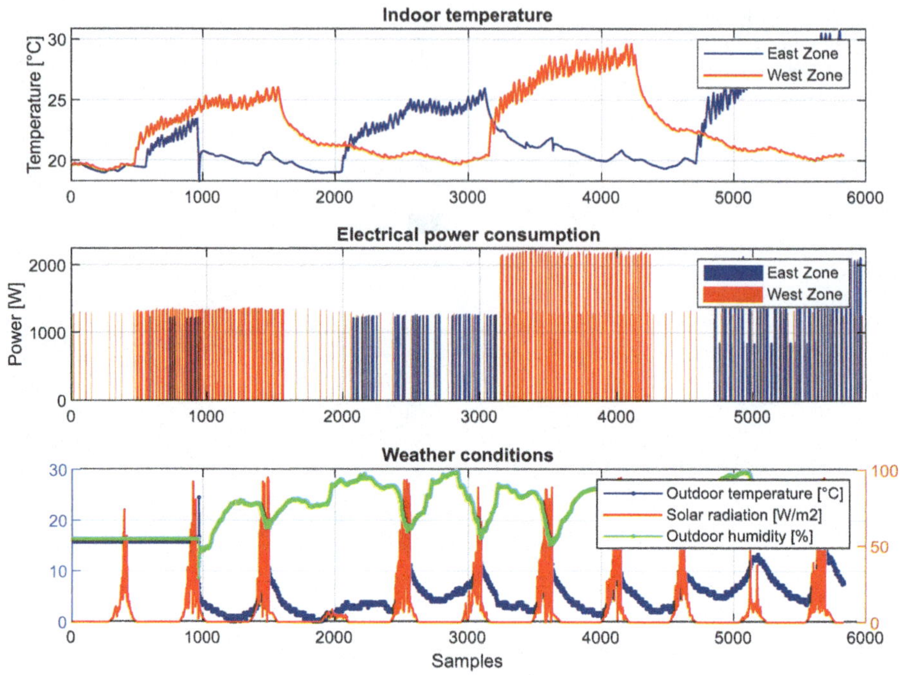

Fig. 2.9 Historical data for training the MLP model from instrumented zone side. Indoor temperature (top), electrical power consumption (middle), and weather conditions: outside temperature, solar radiation, and outside humidity (bottom)

2.4.2 MLP-Thermal Model Identification and Validation

As given and shown in Sect. 2.3.3, the structure of the MLP-thermal model is defined with $(m + p + 3)$-inputs which represent the number of the

- weather inputs collected by sensors such as outdoor temperature (T_o), outdoor humidity (H_o), and solar radiation (R_a),
- energy measurement given by an electrical consumption meter placed, respectively, in the instrumented and non-instrumented zone (Pzi_m, Pzn_p),

and with $(m + p)$-outputs which correspond to

- the estimation of the indoor temperature, both for the reference zone ($\hat{T}zi_m$) and all non-instrumented zone ($\hat{T}zn_p$).

Remark 4 Recall, the subscripts m and p are, respectively, the number of instrumented and non-instrumented zone. However, for the sake of clarity, we assume $m = p = 4$ for this experimental setting, two thermal zones, respectively, for the east and west side. For each of them the first floor is considered as the instrumented reference zone and the third floor is considered as the non-instrumented one.

So, the main parameters of the MLP-thermal model have been defined as follows:

- the architecture is totally connected between each layer;
- the convergence of the learning is guaranteed by gradient descent approach;
- the model has one hidden layer; we use sigmoid type activation function for the hidden layer and identity function for the output layer;
- the learning-rate is fixed at 0.01 and the momemtum at 0.9.

Regarding the learning base, it is composed of 70% of the samples corresponding to the different scenarios of temperature and energy consumed collected during the experiment. Whereas the test base is composed of 30% of remaining samples (Fig. 2.9).

On the other hand, in order to determine the most efficient architecture, several learning procedures have been performed by varying the number of hidden neurons. For a model with one hidden layer, the output error and a FIT criteria are computed to evaluate different numbers of hidden neurons (see Table 2.5). The best parsimony-performance and computational time is established for a number of 56 neurons for the east side and 80 neurons for the west one (resp. performance colored in blue and in cyan in Table 2.5).

For each approved structure of the model (resp. East and West side), in the top of Figs. 2.10 and 2.11, we can see, respectively, the validation of the MLP-thermal model according to each reference instrumented zone (temperature measured in blue and temperature estimated by the MLP-thermal model in red). Whereas the bottom parts of these same figures illustrate, respectively, the evolution of the predicted temperature of the non-instrumented zone (curve in red) according to the reference instrumented one (curve in blue). Moreover, we can see that the temperature is rather homogeneous and correlated from the East side of the building in comparison with the West one. This is just due to the impact of the solar radiation on the thermal behavior of the building. It explains also the results obtained in Table 2.5 why

Table 2.5 Best fitting for the MLP vs hidden neurons number

	East zone		West zone	
Hidden neurons number	MSE (%)	FIT (%)	MSE (%)	FIT (%)
8	92.4311	55.4290	97.2306	68.5097
20	94.2601	67.1542	97.9027	76.1309
32	94.8171	70.7114	98.1812	79.3920
44	66.9926	61.5786	98.6911	84.8270
56	96.6320	80.5964	98.5833	83.8154
68	96.5228	79.9239	98.7373	85.5983
80	95.8277	76.1634	98.9294	87.5942
92	95.9771	76.9680	97.5796	72.3654
104	96.9267	82.0202	98.0041	77.2123
116	94.8809	70.2009	97.9899	77.1547

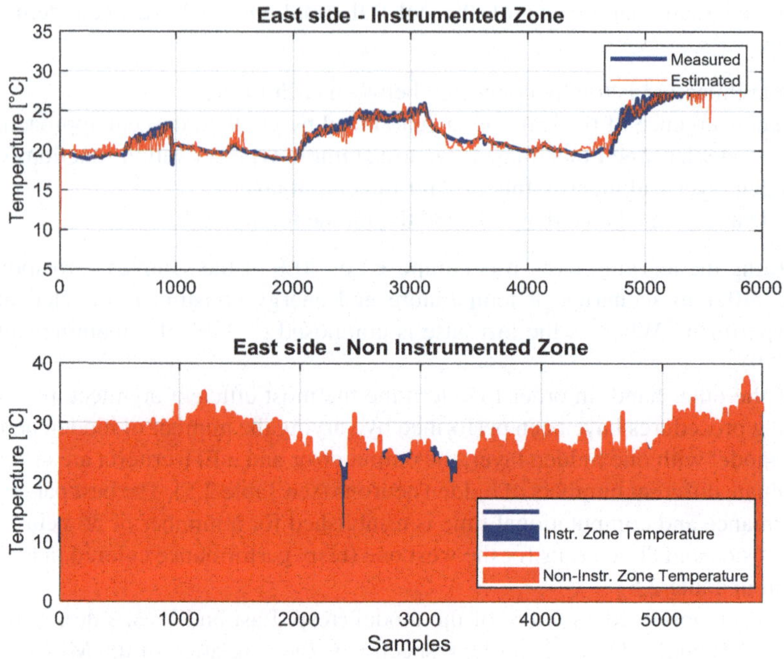

Fig. 2.10 East zone: estimation of the temperature in the instrumented zone (top) and non-instrumented zone (bottom)

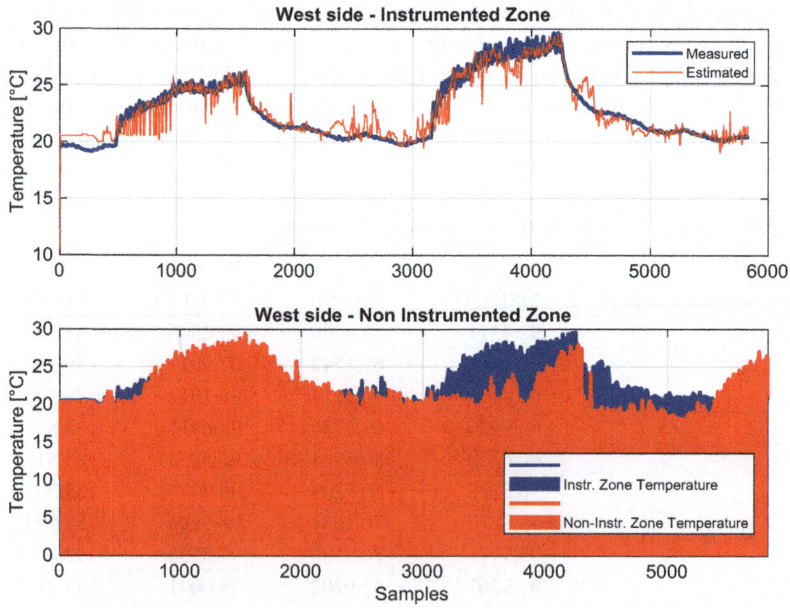

Fig. 2.11 West zone: estimation of the temperature in the instrumented zone (top) and non-instrumented zone (bottom)

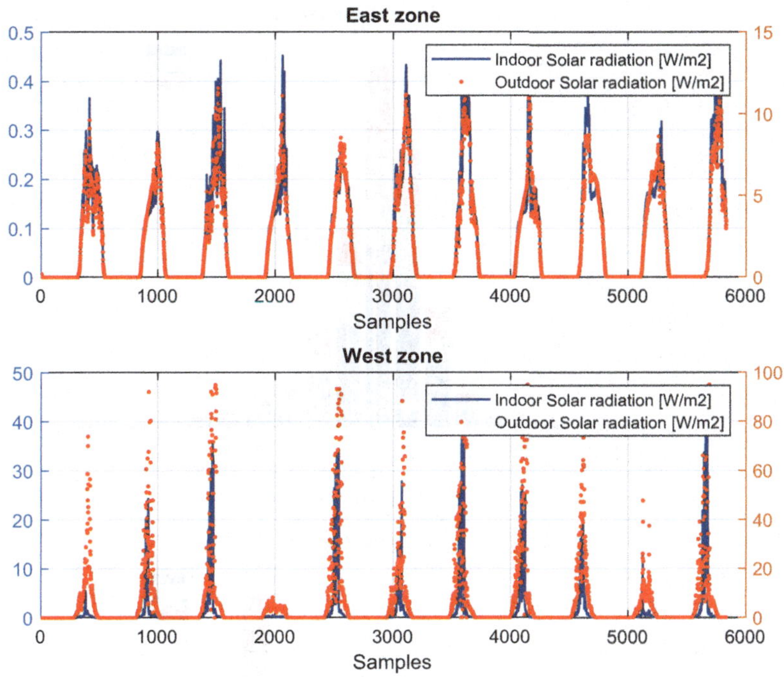

Fig. 2.12 Outdoor solar radiation and its incidence in the floor

we need to considerate different number of hidden layer for each zone to have satisfactory results. Indeed, if we check the evolution of the solar radiation received, respectively, on the east and west wall during this period (Fig. 2.12), it reaches a maximum value of 15 W/m^2 and 95 W/m^2 (red dot line), while the incident solar radiation on the floor is in the order of 0.45 W/m^2 and 40 W/m^2 (blue solid line).

Otherwise, Fig. 2.13 also shows, respectively, the error distribution for the East (left) and West (right) zone of the trained neural network for the training, validation, and testing steps. This one shows that the data fitting errors are normally distributed within a reasonably good range around zero. However, the differences can be more clearly seen from the corresponding error histograms which indicate that the MLP-thermal model was able to accurately generalize the experimental data with only a very small number of predictions exceeding a 3% error margin. Moreover, Fig. 2.14 shows the best performance for the training, test and validation can be reached with about 400 and 800 epochs, respectively, for the East and West data sets.

Remark 5 Let us notice that the performance of the neural network model is sensitive to the amount of representative training data. Thus, the above results can change significantly (negatively or positively) if the learning database presents or not more situations or scenarios in order to appreciate the thermal performance of the overall building.

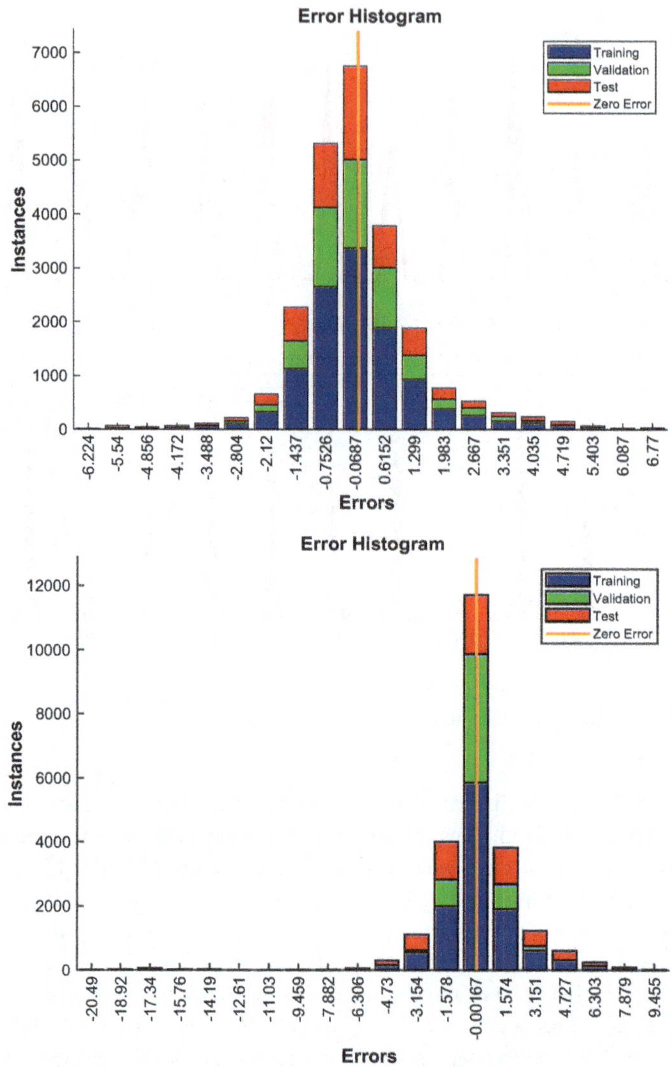

Fig. 2.13 Error distribution for the training, validation and test for the East zone (top) and West zone (bottom)

2.4.3 Smart Interactive Interface

As claimed in the previous paragraph, the human-machine interface should be as efficient as possible. That is, the information disseminated should be complete and precise but simple and easily understood by the user. To continue along the same directive, in this experiment the SBEM system is composed wholly by seven web

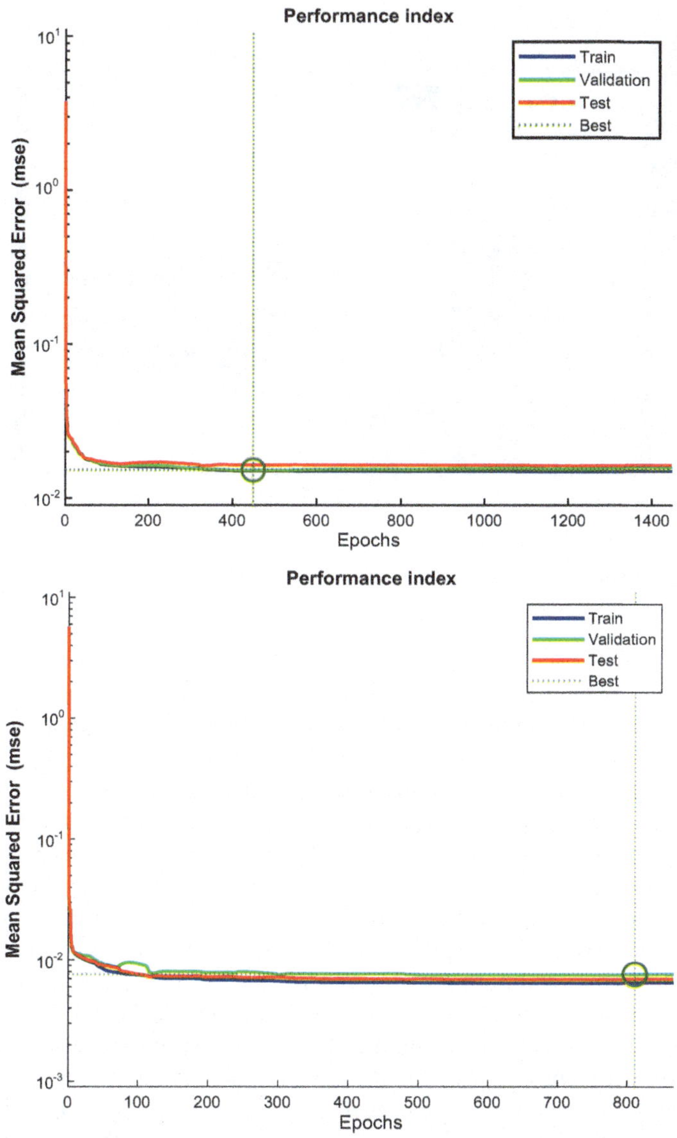

Fig. 2.14 Performance index for the training, validation and test for the East zone (top) and West zone (bottom)

pages. The first one is the login page. Indeed, to access relevant information from anywhere, at any time, users must identify themselves at the login page (Fig. 2.15). Once logged, users can access to the main homepage which presents particularly the main layout of the building.

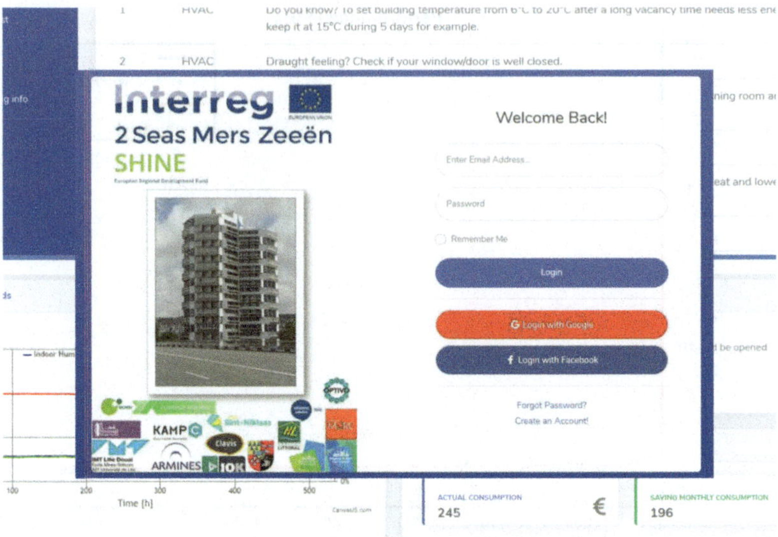

Fig. 2.15 Login page of the graphical user interface

Herein, they can choose their thermal zone, or room or office number to check their energy consumed evolution via the energy performance monitoring page. This last is illustrated in Fig. 2.16. Moreover, part 1 of the energy monitoring page relates the cost of energy consumed (daily and monthly) and the average temperature in the zone. However, part 2 illustrates temperature, humidity, and the electrical consumption trends of the zone considered. The last part is associated to display the recommendation linked to the users activity and information about energy-saving that they can realize by adopting the recommendation displayed. Let us notice that the advice are displayed according to the following parameters: the outside temperature and humidity (T_o and H_o), the indoor temperature estimated (T_z), and the heating power (P_z) for the thermal zone considered. These variables are discretized to compute the recommender model (see specification in Table 2.6).

Here, we just focus onto two situations where each variable can take a lower or upper value. In fact, a lower value is as if $I_L \equiv \{T_{zL} \mid P_{zL} \mid T_{oL} \mid H_{oL}\}$. In this case the situation is not really suitable for comfort purposes. On the other hand, if the variable takes an upper value (i.e., $I_U \equiv \{T_{zU} \mid P_{zU} \mid T_{oU} \mid H_{oU}\}$), that means the situation is maybe critic. By this definition, we can write as follows, for example, for the temperature of the zone T_z (each variable will be discretized in the same manner):

$$T_z = \begin{cases} T_{zL} \text{ if} T_z \leq T_{rz} - \varepsilon_z \\ T_{zU} \text{ if} T_z > T_{rz} + \varepsilon_z, \end{cases}$$

Fig. 2.16 Energy performance dashboard

Table 2.6 Input specification and discretization

Input \ Values	Lower	Upper
Indoor temperature zone (°C)	$T_z \leq T_{zr} - \varepsilon_z$	$T_z > T_{zr} + \varepsilon_z$
Power consumption zone (W)	$P_z \leq P_r - \varrho$	$P_z > P_r + \varrho$
Outdoor humidity (%)	$H_o \leq H_{or} - \eta$	$H_o > H_{or} + \eta$
Outdoor temperature (°C)	$T_o \leq T_{or} - \varepsilon_o$	$T_o > T_{or} + \varepsilon_o$

where ε_z is the hysteresis threshold and T_{rz} is the temperature setpoint in the zone considered.

So, the combination of the descriptive variable values allows us to have 16 tips defining the thermal zone behavior, and for each of them we can associate a personalized advice for the user. Reader can see, for instance, in Table 2.7 an extract of this personalized tips for three particular conditions. The relationship between all tips is described by the recommender model as illustrated in Fig. 2.17.

Besides these main pages, the SBEM system offers other functionalities. For instance, users can set their comfort setpoint in the parameter page. Also, they

Table 2.7 Extract from the table of personalized advice [47]

Advice	State	Inputs value	Personalized advice
Ad_0	S_0	$I_1 = \{T_{oL}, T_{zL}, H_{oU}, P_{zU}\}$	It is T_z (°C) in your zone. Your zone is fresh. If you want to have warmer, ensure that the window is closed and put the heating thermostat by two graduations between the current one
⋮	⋮	⋮	⋮
Ad_6	S_6	$I_6 = \{T_{oL}, T_{zU}, H_{oL}, P_{zL}\}$	It is T_z (°C) in your zone. Your zone is warm. This does not result in additional consumption. If you have too hot, slightly open your window, but think of closing it before leaving
⋮	⋮	⋮	⋮
Ad_{15}	S_{15}	$I_{15} = \{T_{oL}, T_{zU}, H_{oL}, P_{zU}\}$	It is T_z (°C) in your zone. Your zone is warm, you could lower your thermostat for one graduation to save energy

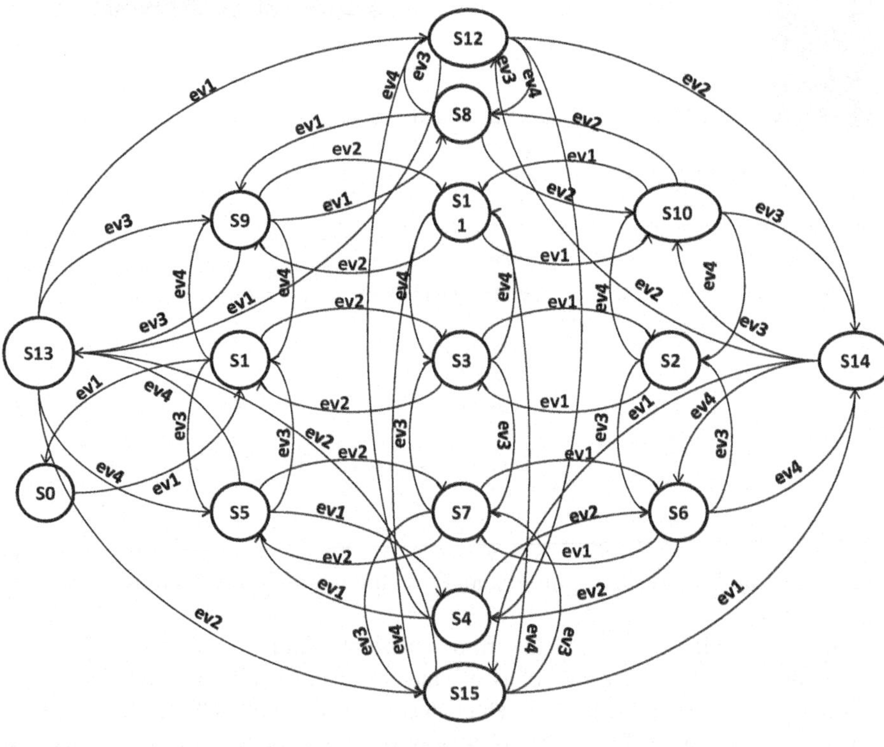

Fig. 2.17 Recommender discrete event model

Table 2.8 Extract from the table generic advice displayed into energy information page

Number	Main field	Advice
1	HVAC	Turn off the radiators when windows are open to ventilate
⋮	⋮	⋮
5	HVAC	Avoid overheating. The recommended temperature is 19 °C in zone with no more activity or people and 16 °C otherwise
⋮	⋮	⋮
10	Electronic devices	Turn off the devices instead of leaving them in standby. Their standby consumption can represent up to 10% of the non-heating electricity bill
⋮	⋮	⋮
25	Electronic devices	Unplug electronic devices systematically: laptop charger, small appliances, PC, etc.

can have an overview of the thermal performance in different zones via the global view page as well as numerical data associated with their zone via data tables page. Finally, to make this system more generic we have implemented a general energy information table. Reader can see an extract of that in Table 2.8 below. Let us notice that this table is used particularly to familiarize the user to have a green-activity and be more confident with regard to their energy footprint. In addition, to make the system scalable, a specific page for the web administrator has been designed to easily add/or remove a generic recommendation provided on the homepage.

Finally, the results in terms of energy consumption are estimated by comparing the overall consumption of the building from its energy balance and the sum of the estimated consumption, thanks to the MLP-thermal model. The difference, which represents the realized energy gain, is estimated by integrating a weighting between the weather indices of the current season and those of the reference one. In terms of energy efficiency, this system should help all building-stakeholders to reduce their energy cost by 10%.

2.5 Conclusion and Perspectives

This chapter presents the development of a smart building energy management (SBEM) system for a large-scale building. A class of building where energy-saving is clearly important and real source of economy. However, designing such a solution does not require huge investments or operating expenses. That is why we present an alternative solution in this work. A solution based especially on the USERS. Indeed, the system does not act directly on the HVAC building systems, as an automation system could realize, but on the USER. However, in order to provide important and persistent insights to control consumption and energy-waste to users, an artificial neural networks model is computed. This plays an important role in estimating and predicting the thermal behavior of the studied building including instrumented and non-instrumented thermal zone.

All of this information is displayed later on smart interactive interface such as assets, controls, data among any others, for enhancing user life's quality through comfort, convenience, and of course reduced energy costs. Experimental test was done on the academic building located in Douai in the North of the France to show the effectiveness of the methodology. We wish to save up to 10% per year if all occupants adopt the system.

In the short term, we plan to extend this system to other academic building in Douai in order to reduce globally the heating energy costs by 10–20%. In the long term, a survey of information on occupancy models may be of interest in redefining control strategies to optimize energy performance. Indeed, it can provide quantitative information, mainly occupants' profile and electric equipment scheduling and operation.

Acknowledgements This work was supported by the European project "SHINE: Sustainable Houses in Inclusive Neighbourhoods." A project granted by Interreg 2 Seas and the European Regional Development Fund.

The author would also like to thank Pr. Luc Fabresse for all the help, advising and support for the development of this smart building energy management system.

References

1. Directive 2012/27/UE, Directive relative à l'efficacité énergétique, modifiant les directives 2009/125/CE et 2010/30/UE, J.O. de l'Union européenne, 14 November 2012
2. L. Pérez-Lombard, J. Ortiz, C. Pout, A review on buildings energy consumption information. Energy Build. **40**(3), 394–398 (2008)
3. R. Missaoui, H. Joumaa, S. Ploix, S. Bacha, Managing energy smart homes according to energy prices: analysis of a building energy management system. Energy Build. **71**, 155–167 (2014)
4. D.M. Han, J.H. Lim, Smart home energy management system using IEEE 802.15.4 and Zigbee. IEEE Trans. Consum. Electron. **56**, 1403–1410 (2010)
5. S. Veleva, D. Davcev, M. Kacarska, Wireless smart platform for home energy management system, in *The 2nd IEEE PES International Conference and Exhibition on Innovative Smart Grid Technologies (ISGT Europe)*, Manchester, 5–7 December 2011, pp. 1–8
6. B. Hamed, Design and implementation of smart house control using LabVIEW. Int. J. Soft Comput. Eng. **1**(6), 98–106 (2012)
7. S. Zhou, Z. Wu, J. Li, X.P. Zhang, Real-time energy control approach for smart home energy management system. Electr. Power Compon. Syst. **42**(3–4), 315–326 (2014)
8. N. Loganathan, P.S. Mayurappriyan, K. Lakshmi, Smart energy management systems: a literature review. MATEC Web Conf. **225**, 01016 (2018)
9. J. Cho, S. Shin, J. Kim, H. Homg, Development of an energy evaluation methodology to make multiple predictions of the HVAC and R system energy demand for office buildings. Energy Build. **80**, 169–183 (2014)
10. C.P. Au-Yong, A.S. Ali, F. Ahmad, Improving occupants' satisfaction with effective maintenance management of HVAC system in office buildings. Autom. Constr. **43**, 31–37 (2014)

11. S. Wu, K. Neale, M. Williamson, M. Hornby, Research opportunities in maintenance of office building services systems. J. Qual. Maint. Eng. **16**(1), 23–33 (2010)
12. P.H. Shaikh, N.M. Nor, P. Nallagownden, I. Elamvazuthi, Intelligent optimized control system for energy and comfort management in efficient and sustainable buildings. Procedia Technol. **11**, 99–106 (2013)
13. R. Yang, L. Wang, Multi-zone building energy management using intelligent control and optimization. Sustain. Cities Soc. **6**, 16–21 (2013)
14. J. Abdi, A. FamilKhalili, Brain emotional learning based intelligent controller via temporal difference learning, in *Proceedings of the 2014 International Conference on Mathematical Methods, Mathematical Models and Simulation in Science and Engineering* (2014)
15. G. Serale, M. Fiorentini, A. Capozzoli, D. Bernardini, A. Bemporad, Model predictive control (MPC) for enhancing building and HVAC system energy efficiency: problem formulation, applications and opportunities. Energies **11**, 631 (2018)
16. A. Parisio, C. Wiezorek, T. Kyntaja, J. Elo, K.H. Johansson, An MPC-based energy management system for multiple residential microgrids, in *Proceedings of the 2015 IEEE International Conference on Automation Science and Engineering (CASE)*, Gothenburg, 24–28 August 2015, pp. 7–14
17. S. Salakij, N. Yu, S. Paolu, P. Antsaklis, Model-based predictive control for building energy management. I: Energy modeling and optimal control. Energy Build. **133**, 345–358 (2016)
18. G. Bianchini, M. Casini, D. Pepe, A. Vicino, G.G. Zanvettor, An integrated model predictive control approach for optimal HVAC and energy storage operation in large-scale buildings. Appl. Energy **240**, 327–340 (2019)
19. J. Achin, B. Madhur, M. Rahul, Data predictive control for building energy management, in *American Control Conference (ACC)*, Seattle, WA (2017)
20. E. Azar, C.C. Menassa, A comprehensive analysis of the impact of occupancy parameters in energy simulation of office buildings. Energy Build. **55**, 841–853 (2012)
21. T. Zhang, P. Siebers, U. Aickelin, Modelling electricity consumption in office buildings: an agent based approach. Energy Build. **43**, 2882–2892 (2011)
22. M. Bonte, F. Thellier, B. Lartigue, Impact of occupant's actions on energy building performance and thermal sensation. Energy Build. **76**, 219–227 (2014)
23. C.A. Webber, J.A. Roberson, M.C. McWhinney, R.E. Brown, M.J. Pinckard, After-hours power status of office equipment in the USA. Energy **31**, 2823–2838 (2006)
24. O.T. Masoso, L.J. Grobler, The dark side of occupants' behaviour on building energy use. Energy Build. **42**, 173–177 (2010)
25. A. Meier, Operating buildings during temporary electricity shortages. Energy Build. **38**, 1296–1301 (2006)
26. H. Staats, E. van Leeuwen, A. Wit, A longitudinal study of informational interventions to save energy in an office building. J. Appl. Behav. Anal. **33**, 101–104 (2000)
27. T.A. Nguyen, M. Aiello, Energy intelligent buildings based on user activity: a survey. Energy Build. **56**, 244–257 (2013)
28. T. de Meestera, A-F. Marique, A. De Herdea, S. Reiter, Impacts of occupant behaviours on residential heating consumption for detached houses in a temperate climate in the northern part of Europe. Energy Build. **57**, 313–323 (2013)
29. P.O. Fanger, *Thermal Comfort* (Danish Technical Press, Copenhagen, 1970)
30. K. Parsons, *Human Thermal Environments: The Effects of Hot, Moderate, and Cold Environments on Human Health, Comfort, and Performance* (Taylor and Francis, Boca Raton, 2014). ISBN 9781466595996
31. D. Fugate, P. Fuhr, T. Kuruganti, Instrumentation systems for commercial building energy efficiency, in *Future of Instrumentation International Workshop (FIIW)* (2011), pp. 21–24
32. M.H. Benzaama, L. Rajaoarisoa, S. Lecoeuche, Data-driven approach for modeling the thermal dynamics of residential buildings using a PieceWise ARX model, in *Proceedings of the 16th International Building Performance Simulation Association Conference*, Rome (2019)
33. R.M. Badreddine, Gestion Energétique optimisée pour un bâtiment intelligent multisources multi-charges: différents principes de validations. Thèse de doctorat, Université de Grenoble (2012)

34. T. Markus, E.N. Morris, *Buildings, Climate and Energy* (Pitman, London, 1980)
35. H. Yoshino, T. Hong, N. Nord, IEA EBC annex 53: total energy use in buildings - analysis and evaluation methods. Energy Build. **152**, 124–136 (2017)
36. H. Zhao, F. Magoules, A review on the prediction of building energy consumption. Renew. Sustain. Energy Rev. **16**, 3586–3592 (2012)
37. S. Paudel, P. Nguyen, W.L. Kling, M. Elmitri, B. Lacarrière, Support vector machine in prediction of building energy demand using pseudo dynamic approach, in *Proceedings of ECOS2015-The 28th International Conference on Efficiency, Cost, Optimization, Simulation and Environmental Impact of Energy Systems*, Pau (2015)
38. R. Yokoyama, T. Wakui, R. Satake, Prediction of energy demands using neural network with model identification by global optimization. Energy Convers. Manag. **50**, 319–327 (2009)
39. Q.J. Zhang, K.C. Gupta, *Neural Networks for RF and Microwave Design* (Artech House, Norwood, 2000)
40. S.S. Haykin, *Neural Networks and Learning Machines* (Pearson, Upper Saddle River, NJ, 2009)
41. S. Haykin, *Neural Networks - A Comprehensive Foundation*, 2nd edn. (Prentice Hall, Upper Saddle River, 1999)
42. B.M. Wilamowski, Y. Chen, A. Malinowski, Efficient algorithm for training neural networks with one hidden layer, in *Proceedings (Cat. No.99CH36339) of the International Joint Conference on Neural Networks, IJCNN'99*, Washington, DC, vol. 3 (1999), pp. 1725–1728
43. K. Collins, M. Mallick, G. Volpe, W.G. Morsi. Smart energy monitoring and management system for industrial applications. 2012 *IEEE Electrical Power and Energy Conference, EPEC 2012*, 92–97 (2012). https://doi.org/10.1109/EPEC.2012.6474987
44. J.S. Breese, D. Heckerman, C. Kadie, Empirical analysis of predictive algorithms for collaborative filtering, in *Proceedings of the Fourteenth Conference on Uncertainty in Artificial Intelligence (UAI'98)* (1998)
45. G. Adomavicius, A. Tuzhilin, Toward the next generation of recommender systems: a survey of the state-of-the-art and possible extensions. IEEE Trans. Knowl. Data Eng. **17**(6), 734–749 (2005)
46. B.P. Zeigler, DEVS theory of quantized systems, in *Advance Simulation Technology Thrust (ASTT)*, DARPA Contract N6133997K-0007 (1998)
47. D. Juge-Hubert, L.H. Rajaoarisoa, S. Lecoeuche, Modélisation thermique des bâtiments et responsabilisation des usagers, in *Conférence Francophone de l'International Building Performance Simulation Association*, Arras, 20–21 mai 2014

Chapter 3
Automated Demand Side Management in Buildings

Hussain Kazmi and Johan Driesen

3.1 Introduction

There is widespread scientific consensus that greenhouse gas emissions, especially as a result of using fossil fuels to power civilization, are among the primary drivers of anthropogenic climate change [1, 2]. In light of this realization, the overall energy system in general and the built environment in particular are undergoing a profound shift towards decarbonization. Decarbonizing the building sector can be seen as a two-layered process, which comprises of the dual goals of electrifying everything and decarbonizing the electricity mix. Electrification is primarily concerned with the demand side (e.g. electrification of heat, transport, lighting and cooking, etc.), while decarbonizing electricity tackles the supply part (e.g. by replacing fossil fuel plants with renewable energy sources (RES)). The continuous drive towards greater efficiency, resulting from optimization in both design and operation, unifies the supply and demand side. It is obvious therefore that weaning the built environment from fossil fuels will require a cross-sectoral transformation in the energy domain.

3.1.1 Decarbonization

On the supply side, replacing fossil fuel based generation with renewable energy sources offers substantial benefits. The climate change potential for renewable energy sources, such as wind and solar, is much lower than that of their fossil fuel based counterparts, such as coal, oil and even gas [3]. At the same time, the

H. Kazmi (✉) · J. Driesen
KU Leuven, Leuven, Belgium
e-mail: hussainsyed.kazmi@kuleuven.be; johan.driesen@esat.kuleuven.be

M. Sayed-Mouchaweh (ed.), *Artificial Intelligence Techniques for a Scalable Energy Transition*, https://doi.org/10.1007/978-3-030-42726-9_3

economic equation is becoming increasingly more favourable for these renewable sources because of economies of scale: by 2020, IRENA data shows that 77% of onshore wind and 83% of utility scale solar photovoltaic (PV) generated electricity will be cheaper than the most economical fossil fuel based competitors [4]. An interesting development in this context has been the rise of distributed energy resources (DER) such as rooftop solar PV panels. These can act, in practice, as 'negative loads' on the grid, i.e. they reduce the offtake of energy from the grid. When supply does not match demand, they can however also result in substantial problems, including to the stability of the grid. These include the infamous 'duck curve', which identifies very high ramping rates in the evening when peak demand coincides with diminishing solar electricity production. Large scale reverse power flows can also undermine grid stability and equipment longevity [5].

3.1.2 Electrification

After a number of faltering starts, today electrification of demand continues apace. The built environment and transportation sectors can benefit massively in their decarbonization drive by utilizing cleaner energy. These can be distinguished into areas which are cost-effective to electrify today, and those which require technological breakthroughs or economies of scale before they can become feasible. Fully electrifying newly constructed buildings, by focusing on major draws such as heating, cooling and hot water production, are possible today and have been demonstrated in a number of pilot projects. This principle is enshrined in, for instance, the Dutch drive to fully electrify and decarbonize the social housing sector in the coming decades [6]. Carefully drafted building codes aimed at nearly or net-zero energy buildings have also driven the market in this direction in Europe [7], and similar trends can be seen in many other affluent regions such as California. Vehicles which were conventionally not a household electricity draw are also increasingly becoming part of the household demand. While fully electrifying transportation will take decades, regions with favourable policies, such as The Netherlands and Norway, have seen a massive surge in uptake of electric vehicles (EVs) in recent years [8].

3.1.3 Optimization: The Need for Demand Side Management

Demand side management (DSM), as a concept, has existed for a long time. However, it is only in conjunction with increasing electrification of demand and changes in the supply mix that it has become a central figure in the future of energy systems. At its heart, DSM can refer to both demand reduction and demand response, both of which can entail different things for the operative grid [9]. Despite their differences, which we will explore in greater detail, both refer to optimizing energy flows either on a building or a grid level. This paves the way to applying sophisticated artificial intelligence algorithms to achieve the objectives associated with DSM (i.e. either demand reduction or response). The remainder of this section introduces and motivates key concepts from DSM literature, while the next one takes a closer look at the data-driven algorithms which can be used to unlock its potential.

3.1.4 Demand Reduction

Demand reduction, in itself, is a key component of the broader transition to a more sustainable energy mix. It is eventually much more sustainable to reduce overall energy usage, irrespective of the source especially if it can be done while maintaining human comfort and economic productivity. There is also a more subtle reason for focusing on demand reduction. This has to do with the increasing electrification and its multi-level impact on the electricity grids:

1. Increasing electrification in buildings will substantially increase the electrical energy demand in buildings. While monitoring one hundred recently refurbished net-zero energy households in The Netherlands over a year, we discovered that the additional energy demand caused by the heat pump (to provide space heating and hot water) led to a considerable increase in overall electricity demand (sometimes by more than 50%). The problem was especially acute during the winter months due to the concentration of space heating, and the summer months due to the DER generation. This information, along with the power draws associated with the hot water production and space heating, is visualized in Fig. 3.1. It is obvious that while the hot water circuit induces a higher instantaneous power demand, the space heating load causes sustained loading on the grid during the winter months.
2. Increasing electrification in buildings also considerably contributes to higher peak power load on the grid. This is both due to electrification of heat, which causes a higher draw during the winter in many houses simultaneously, and due to distributed energy resources (DERs), such as solar panels which lead to significant reverse power flows. These higher peaks often necessitate grid reinforcement investments to maintain stability—an expensive proposition.

Energy efficiency improvements, either through automation or user engagement, seek to address these issues by reducing the load.

3.1.4.1 Demand Response

As opposed to demand reduction, the objective of demand response is to shift demand in time. This does not necessarily reduce demand and may in fact increase it. However, the objective of demand response programs is frequently to increase local (or self-)consumption of energy generated by DERs to solve local congestion

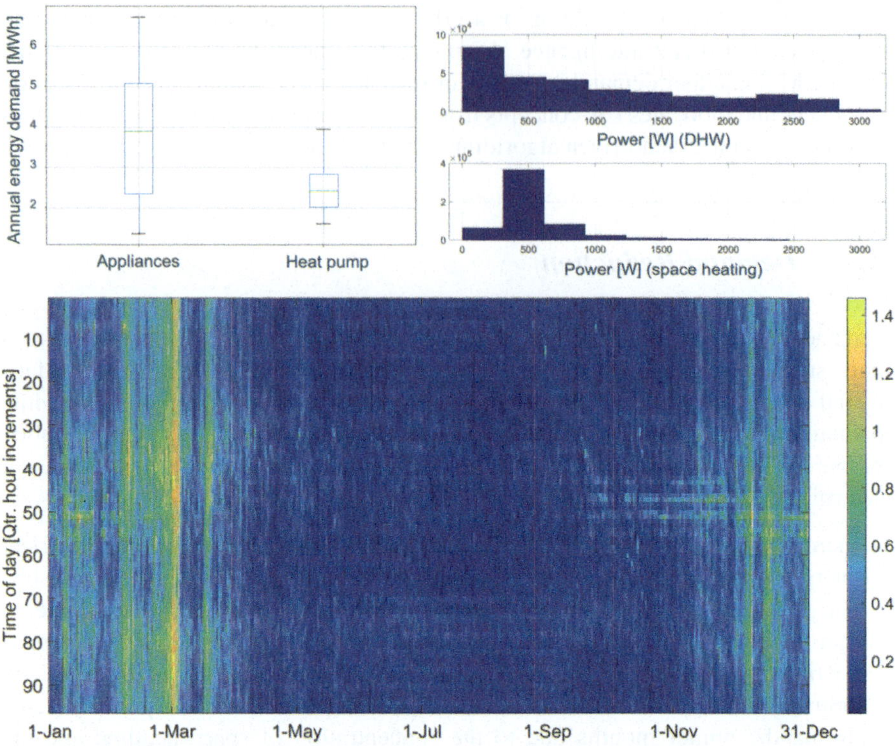

Fig. 3.1 (Top-left): Energy consumed by appliances (e.g. lighting, refrigeration, plug loads) and heat pump (space heating and hot water production); (Top-right): Power consumption by the heat pump for space heating and hot water production; (Bottom): Average additional load [kW] induced by an air source heat pump used for space heating and domestic hot water production: winter consumption spikes due to space heating demand, while summer demand is mostly due to domestic hot water production

or voltage issues, as well as for peak shaving and valley filling [10, 11]. Increasingly, pilot projects are being undertaken with an aim to use the energy flexibility of these buildings to also provide ancillary services to the transmission grid, such as frequency regulation [12, 13]. As with demand reduction, these services can be enabled through automation and user engagement, making use of techniques developed in the artificial intelligence community.

3.2 Artificial Intelligence and DSM

While DSM represents enormous potential in untapped flexibility, there are practical challenges which must be overcome before it can be deployed in practice. This is especially true for the case of residential buildings, which represent extremely limited flexibility individually, but offer tremendous DSM potential collectively. Recent advances in artificial intelligence (AI) can help realize many of these opportunities. This section highlights the current state-of-the-art in using data-driven methods for DSM in buildings. The discussion begins with an overview of use of data-driven methods to engage building users in their energy demand. It then introduces aspects of automated DSM and presents some of the most interesting avenues arising out of contemporary research in the topic.

3.2.1 User Engagement

User engagement, in the context of building energy systems, refers to the interaction between energy companies and building occupants. The energy companies can be utilities, but they can also be other service providers such as software companies. This engagement, which is often enabled by data-driven methods in practice, can be distinguished along two dimensions: (1) getting occupants to make cost-optimal investments in the building and (2) providing feedback to occupants to modify their behaviour in a manner that supports (or does not harm) grid's operation. This section highlights the framework that governs both of these objectives.

3.2.1.1 Engagement on Investments

Service providers or utilities can offer much-needed visibility and information to users to replace less efficient appliances or carbon intensive loads with more efficient or sustainable options. This type of engagement is more of a one-and-done deal, as once the less efficient equipment has been replaced, the engagement cannot continue as-is, and must take on a different form. Common examples of this paradigm include:

1. Lighting, which is arguably the lowest hanging fruit in the efficiency value chain, where both regulatory and market frameworks have come together in recent years to drastically improve efficiency. This has led to remarkable energy efficiency gains when comparing LED technology with older, incandescent bulbs. Another cost-effective example is replacing older appliances such as refrigerators.
2. Thermal loads (for space or water heating and cooling) are among the largest draws for energy in many buildings around the world. Even in high efficiency buildings, thermal loads, as represented by the heat pump in Fig. 3.1, constitute roughly one-third of total electric demand. In older buildings, this figure can be twice as high. By replacing inefficient and carbon intensive heating technologies such as resistance heating and gas boilers with modern heat pumps, it is possible to improve the overall efficiency. Likewise, this information can be used by households to make better informed decision making about the costs and

benefits of a possible refurbishment (this could involve, for instance, investing in improving the insulation of the building facade, etc.).

3. Depending on the ambient conditions and type of the building and the prevailing electricity tariffs, installing solar panels can be an economically viable investment, either with or without electrical storage. This information, in the form of optimally sized solar panels (and potentially battery storage systems), can be communicated directly to users based on their historical consumption patterns.

The widely used cost-optimal strategy can be used to perform this analysis [14]. However, historically, this advice was provided to building occupants in a rather generic manner, i.e. it was not tailored to the individual needs of a household (e.g. informing everyone in a utility's operating area that investing in a more energy efficient heat pump can be more cost-effective than resistance heating). This was mostly done because there was scant data on individual demand and tailored advice could only be obtained by conducting detailed energy audits. With the rise of smart meters, data-driven algorithms have shown great potential in this regard to better understand user behaviour, either through sub-metering where individual loads are monitored or through disaggregation which signifies non-intrusive load monitoring [15]. Disaggregation can help automatically identify individual loads in the building, often through some form of high resolution pattern matching, which can then feed into a recommendation system to provide users tailored feedback. Despite considerable research, this field of work is still very much in its nascency, primarily owing to the existence of multitudes of different devices, all of which can have different electrical signatures. The high frequency sampling required to observe consumption data in real time is also an expensive proposition from both a processing and storage perspective. Future developments in disaggregation are therefore expected to bring these algorithms closer to real world deployment.

3.2.1.2 Engagement on Operation

Besides informing building occupants about the benefits of shifting to more efficient or less carbon intensive appliances, it is also possible to communicate with users about their day-to-day operation of these appliances. Many companies have already trialled such services, with OPower being arguably the most well-known [16]. In such projects, building occupants are informed about their energy usage in the context of other similar households, with the objective of reducing demand. Access to data and meta-data about buildings allows such companies to provide this feedback in a visually appealing and statistically consistent manner. These tests rely on the concept of randomized controlled trials (RCTs) whereby buildings are divided into different test groups to make statistically significant judgements about the effect on demand of providing feedback. Consistent efficiency gains of 2–4% have been shown in the trials by OPower [17].

Other options also exist for operational engagement with building occupants. Utilities or other service providers can engage with building owners or occupants to avoid demand during peak times for a specified time period during a year. This can be done in exchange for reduced tariffs and/or installation of new appliances to the buildings. This is typically realized by messages pushed to end consumers. Alternatively, real time or surge pricing can be activated for these consumers. However research has shown that there is an optimal amount of information exchange between the utility and the end consumers—exceeding this frequently leads to information fatigue and users not paying attention. Furthermore, consumers are typically risk-averse as the savings arising from switching from fixed to dynamic tariffs are limited while the risks can be substantial. These results, especially in the context of demand response, were typified in LINEAR, a large scale project investigating demand flexibility in Belgium [18]. For demand reduction, it is important to keep in mind that the elasticity of electricity usage has been historically known to be quite low, so simple price-based mechanisms are unlikely to lead to substantial energy conservation [19].

3.2.1.3 Discussion on Data-Driven User Engagement

The impact and adoption of the solutions provided by such data-driven algorithms can vary considerably. Deep refurbishment of the building, for instance, can have a high impact on the building energy demand. However, it is a costly investment with a lengthy payback period. On the other hand, replacing lighting in a building is easily adoptable, but its impact on overall energy demand is quite limited. Worryingly, engaging users to modify their behaviour seems to yield only limited benefits. This is further hampered by information overload which can often lead users to churn. Another common problem is the transience of user response change, i.e. it can often revert to the baseline once the engagement program is withdrawn.

3.2.2 Data-Driven Learning and Modelling

While technological advances have made it possible to engage building users for DSM, it has also become increasingly possible to directly control appliances in buildings. The same data required to create models to engage users can typically also be used to create the models necessary to control devices, although real time control places strict constraints on data availability and reliability. This data enables a service provider to create forecasts for both user behaviour and the energy systems under consideration over a specified time horizon. Often, these appliances can be split into controllable and non-controllable loads. These forecasts can then be used to optimize the behaviour of controllable appliances to provide demand side management, while maintaining occupant comfort.

Automation can be seen as the converse of user engagement. While user engagement takes observed (historic) consumption data as input and seeks human intervention to elicit changes in this usage of household appliances, automation modifies device behaviour directly in a way that is cognizant of the human user's behaviour and device limitations. In this way, automation operates in a more confined space than user engagement. However, at the same time, it does away with requiring users to change their behaviour and is therefore often the path of least resistance.

For planning and optimization purposes to track total demand and generation at all times, it is imperative to have good forecasts of user demand. This can be done by analyzing historical consumption and production trends, and using this as the basis for making forecasts for the future. The problem of forecasting requires several high level decisions to be made, including the forecast time horizon, the inputs to the forecast algorithm, whether learning needs to be online or offline, and the choice of the forecast algorithm. This section highlights the differences these choices can make in practice.

3.2.2.1 Creating Forecasts

For forecasting energy demand and production, a number of principles apply from research in data-driven algorithms which can be broadly classified as falling under the umbrella of artificial intelligence. More concretely, there are at least three different, albeit connected, approaches to create forecasts, two of which are directly data-driven. These are frequently termed as white-box, black-box and grey-box models [20]. Here, white-box models refer to the use of domain knowledge (or human expertise) to create a model which explains the dynamics of the process generating data (e.g. the energy usage in a household or electricity production from rooftop solar PV panels). Black-box methods lie on the opposite end of the spectrum, and usually operate in a purely data-driven fashion. Grey-box methods typically combine the two approaches. This information is summarized in Fig. 3.2. In practice, white-box models can be impractical to construct as there are millions of different device types installed in buildings all over the planet, and the dynamics governing them may not be well understood yet in any case. Black-box methods, which rely on data-driven techniques, remove this reliance on a human domain expert, and are consequently much more scalable in practice.

As black-box models require access to observational data to learn a model to create forecasts, the first question that naturally arises is how does learning takes place. This can take one of the following three forms:

1. **Forecasting from raw time series data**: this is arguably the panacea for black-box learning systems, where a function approximation algorithm attempts to learn a dynamics model directly from observational data, with no human involvement whatsoever, shown in Fig. 3.3a. While this can work for some relatively simple tasks, there are nevertheless several practical limitations to this approach. The first is the choice of input training data. The training data dictates the quality of the forecast, however this involves both human judgement (i.e. which data streams are required for a forecast application) and observability (i.e.

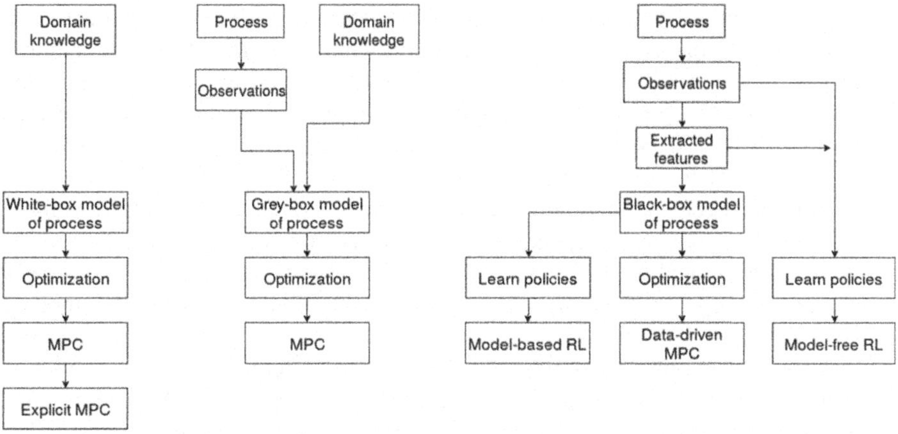

Fig. 3.2 Typical application of different learning and control algorithms to energy systems operating in buildings: a process is a conceptual representation of an energy system in a building; the flowchart shows loosely the white-, grey- and black-box modelling techniques, and also shows the range of different control strategies available to the practitioner. The same models, regardless of their type, can also be used to engage users by providing feedback and to improve design choices. The control component of the figure will be explained in a subsequent section

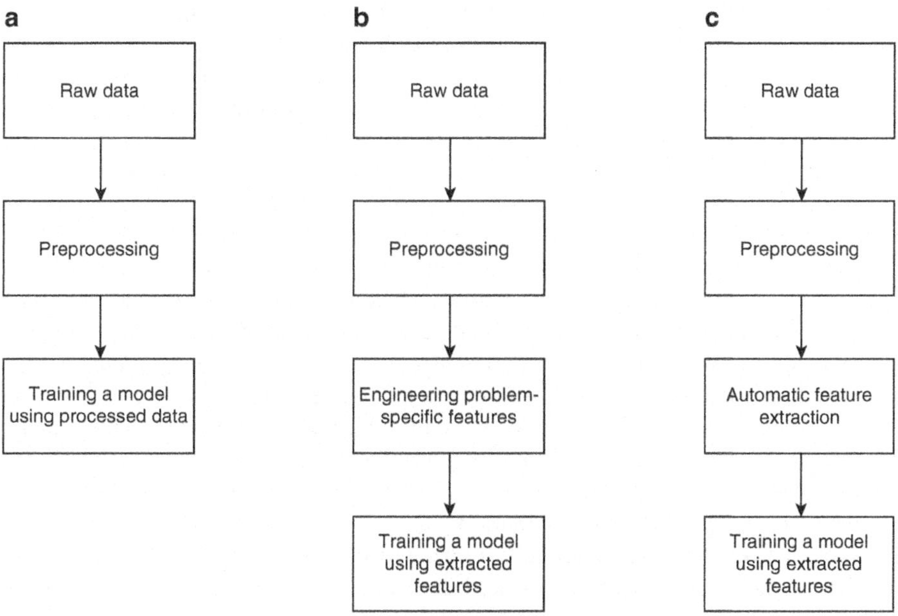

Fig. 3.3 Different techniques used to forecast from raw time series data

if those required data streams are available in a cost-effective manner). This also typically involves choosing a temporal window which defines the amount of past observations that can be used to predict future outcomes. Longer time windows allow a function approximator to learn more complicated relationships (i.e. potentially from further back in time or discover low frequency events). However it also considerably increases the input dimensionality which leads to more stringent requirements on the amount of data required for learning a good model. Likewise, reducing the temporal window length reduces the feature vector dimensionality, but might also make the learning agent unable to properly comprehend system dynamics. The second key limitation of using this method is in the choice of the learning algorithm; the no free lunch theorem states that there is no single learning mechanism which outperforms every other in every situation. This requires testing a handful of staple algorithms in practice to evaluate which one works best. Very recent innovations such as the attention mechanism [21] and recurrent neural networks [22] have considerably expanded the prowess of true sequence models while incorporating explainability into the prediction process. This is however very much an active area of research.

2. **Forecasting using feature engineering**: this is the classical paradigm for machine learning algorithms for tasks where learning directly from raw data is considered impractical, as shown in Fig. 3.3b. In fact, a review of over 50 different forecasting studies of energy systems reveals that most commonly used features used in black-box modelling include historic data, ambient conditions (temperature, wind speed, etc.) and calendar features (time of day, weekend vs. weekday, etc.) [23]. Feature engineering can range from extracting a large number of features which explain some characteristic of the data to making use of detailed domain knowledge to extract specific features. In many cases, extracting meaningful features can be the difference between a machine learning system that converges to a meaningful solution and one that does not. Feature engineering allows direct integration of (some) human knowledge into the problem which can solve many problems such as interpretability (since features are engineered, they can be interpreted), transferability (features can be engineered in a way that they transfer across domains) and performance in low data regimes (feature extraction typically reduces the dimensionality). However, these methods also introduce some new problems. Foremost among these is the choice or selection of features which is, at best, a hazy area in machine learning research. In many instances, this problem is solved by extracting a large number of features and then optimizing over their selection in a way that achieves the best predictive performance through cross-validation.
3. **Forecasting using automatically extracted features**: extracting features from the time series used for forecasting future demand can be particularly useful in poorly explored domains or when training data is sparse. However, while the learner remains black-box in the sense that it is still learning directly from data, some human bias is incorporated into the problem in the selection of features. This can be avoided by using automatic feature extraction algorithms such as principal component analysis [24] or nonlinear autoencoders, etc. [25]. Such

an approach has been investigated in, for instance, [26], albeit in the context of learning value functions used in reinforcement learning to control energy systems. Two potential problems with such automatically extracted features can often be their non-interpretability (i.e. extracted features have no physical meaning) and non-transferability (i.e. extracted features in one domain are not necessarily transferable to a different domain). Arguably the biggest limitation of such techniques is their data requirement; however, in order to learn feature mappings, they require copious amounts of data. In a regime where data might be limited, this creates problems and undesired delays while data is being gathered for automatic feature extraction, rather than learning.

In practice, forecasting generally works best with engineering features, as evidenced by their popularity in published literature. It is important to note however that new developments in machine learning may well change this in the near-future with increasing amount of data and compute power.

3.2.2.2 Online vs. Offline Learning

The model used to create a forecast can be learned either in an online or offline manner.

1. **Offline learning** refers to the fact that the model is learned once, using gathered data, and is not subsequently updated. An example of this appears in [27]. The obvious benefit of this approach is that it leads to models which can be extensively tested before deployment. Where the test data matches the training data, the performance of such offline learning is well-known in advance. This however is also the greatest disadvantage of learning in an offline manner. First, offline learning is often carried out in controlled settings, for instance, in a laboratory which may not represent the many different ways in which building occupants may use energy. This offline availability of devices is also not guaranteed; further, models learned in this manner can often be oblivious to real world changes in operative conditions of the device itself.
2. **Online learning** can be summarized as ‘learning on the job’, whereby the model is continuously updated based on observed interactions between systems and the environment as well as human users. Often, past observations are weighted less heavily to focus on more recent observations, which is useful when the system dynamics underpinning the forecasting problem may change over time. While this reflects the real world use case, in practice it has a significant drawback as well. Since the agent starts off with no prior knowledge about the system, it can drive the system to dangerous regions in the state-space during the exploration phase. This is especially risky when user comfort or grid stability is on the line. An example of such behaviour can be found for space heating systems can be found in [28]. In practice, carefully designed exploration functions such as upper confidence bound [29] should be employed rather than ϵ-greedy ones [30].

3.2.2.3 Choice of Forecasting Algorithm

The choice of algorithm also influences the quality of the resulting forecast, although in the energy domain, where inputs are typically of much lower dimensionality than in computer vision, this is often less of a concern. In practice, a wide array of algorithms are available to the practitioner, and it is not always clear which algorithm will yield the best results. A large scale review of published case studies found that almost three quarters of such projects made use of either support vector machines or neural networks for prediction and modelling [23]. The motivation for focusing solely on these techniques is often suspect however, and it is not at all clear whether other methods would have outperformed them, or achieved the same performance at a lower computational expense. Therefore, while it is certainly possible, and even advisable, to test different algorithms, it is also important to know the strengths and limitations of different algorithm classes to avoid unpleasant surprises during deployment. The following presents an overview of different forecasting algorithm for function approximation in building energy systems, with an emphasis on real world deployability and common pitfalls:

1. **Linear and autoregressive models** are adept at capturing linear dependencies between input and output variables; autoregressive models are a special class of these algorithms which treat historic observations of a time series as input features for the future [31]. While least squares regression with extracted features from observational data is arguably the most popular variant of this class of algorithms, it is possible to fit nonlinear functions with the same models if the input features are converted into a different basis first. This however increases space complexity in a combinatorial manner (as the number of possible combinations of input features grows exponentially). Training a linear least squares model has typically cubic complexity $O(p^3)$ with the number of training features p, while making predictions is linear with p, and therefore extremely fast.
2. **Kernel-based methods**, of which support vector machines and regression are perhaps the most well-known algorithms, are another function approximation technique that has become extremely popular for predicting behaviour of energy systems. These algorithms work by performing the so-called kernel trick, which allows them to calculate higher dimensional feature spaces implicitly, unlike the case of fitting transformed input features to nonlinear basis as described earlier. This makes them more practical from a memory perspective. However, their computational complexity can be much greater, as training is cubic in n, the number of training examples, i.e. it scales poorly with increasing amounts of data. **Gaussian processes**, another powerful kernel-based approximation technique, allow computation of uncertainty bounds and are therefore useful for active and reinforcement learning tasks. However, they also scale poorly with increasing amounts of training data.
3. **Tree-based methods**, such as random forests and gradient boosting, are typically the workhorse in many prediction problems. By employing either bagging or

boosting, they combine a number of weak learners into one strong learner which outperforms the individual models. These methods offer a great balance between computational requirements and predictive performance, and are well capable of fitting both linear and nonlinear functions. Their performance when extrapolating beyond the observed feature space, as opposed to interpolating between observed data points, can often be suspect however. Finally, they scale much better with increasing amounts of training data, and can therefore be more suitable for big data applications than kernel-based methods in their unoptimized form.

4. **Neural networks** are universal function approximators, which means that they can capture any level of complexity in a training dataset, given a high enough model capacity. In fact, the recent boom in AI research has been driven in large part by state-of-the-art results achieved by deep neural networks. While these methods are extremely flexible and increasingly accessible because of open-source tools, they are also prone to overfitting, i.e. memorizing input data as opposed to generalizing well to unseen examples. Their computational complexity too, especially during training, is a huge concern. In fact, recent analysis of large scale deep neural networks trained for natural language processing shows that these models have a massive carbon footprint (ranging from tens of kilograms to tons) [32]. As neural networks trained for energy related tasks grow deeper and ever more pervasive, it is important to keep in mind that training the model should not outweigh the savings it achieves.

In the end, the choice of modelling algorithm should be driven not by hype, but by practical performance and real world constraints on data and computational budget.

3.2.2.4 Forecast Time Horizon

The required forecast time horizon depends on the application and can be distinguished into three classes:

1. **Long-term forecasting**, often focused on creating multi-year horizon forecasts, is typically done for systems planning. An example application of this type of forecasting is forecasting household loads and DER proliferation to aid in the design of the distribution grid.
2. **Day-scale forecasting**, which is often on a day-ahead horizon, is necessary for operational planning and optimization with building loads. This type of forecasting forms the foundation of many classical demand side management programs which require nominating energy demand and production 24 h ahead in time.
3. **Very short-term forecasting**, or nowcasting, is an emerging field of study and typically concerns itself with forecasts on the scale of a few seconds or minutes. This level of forecasting is usually suitable for reactive control, which can react to real time disturbances in planned operation in a more effective manner.

It is important to note that different errors and concerns tend to dominate in all these forecasts, ranging from occupant behaviour to ambient conditions.

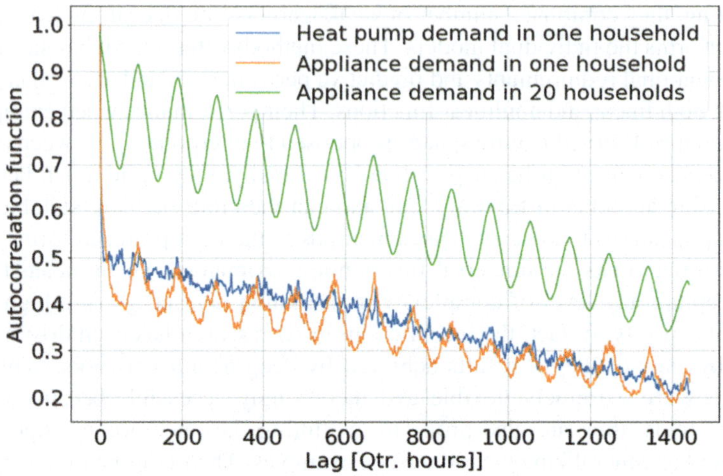

Fig. 3.4 Autocorrelation function calculated for different load types (appliances and heat pump load) and levels of aggregation (single vs. 20 households). A higher level of load aggregation results in a smoother, more periodic autocorrelation function for the aggregated load, hinting at a predictable time series. Single appliance load usage also results in a somewhat periodic autocorrelation function, albeit with lots of noise caused by human behaviour stochasticity. Finally, the heat pump load is not periodic showing that to predict its behaviour exogenous variables (such as ambient conditions and/or a model for the heat pump) are required

3.2.2.5 Aggregation Level

While the forecast time horizon deals with how far ahead we are looking into for planning purposes, aggregation concerns the actual time series that needs to be forecast. This can be distinguished into:

1. The **scale aspect** refers to the particular energy draw being forecast. This could refer to multiple levels: device level, household level or community/neighbourhood level forecasts. Going from device level to community level, the forecast quality generally improves to a certain level before plateauing. This is because human stochasticity becomes less of an issue and the autocorrelation of the time series rises and becomes more smooth, as visualized in Fig. 3.4. The same behaviour can be seen, albeit to a lesser extent with energy production through rooftop solar PV panels. Here the improvement in forecastability stems not from suppression of human stochasticity, but of small variations in ambient conditions.
2. According to the **temporal aspect**, finer grained forecasts tend to be less accurate as the stochasticity of human behaviour is generally reduced by aggregating demand over several time periods. This can be noted, for example, from the fact that it is generally easier and far more accurate to predict energy demand for the whole day as opposed to the demand for any single minute or hour of the day.

On a temporal scale, it is easy to predict that the heating circuit will be activated if the weather will be cold. On the other hand, to know the precise moment when it will be activated requires not just access to how the building has responded to the cold historically, but also the human behaviour modifying it in the moment, for instance, opening or closing windows or occupancy patterns, etc. As the prediction time scale becomes less fine-grained, these random variations caused by individual user behaviour are smoothed out. Furthermore, as seen in Fig. 3.4, it is not always possible to predict the behaviour of an energy draw (e.g. the heat pump) simply looking at the historical time series. However, by incorporating exogenous variables such as temperature forecasts, etc., it might become possible to forecast these. Furthermore, it is possible to create specific forecasts for controllable loads, which take into account human stochasticity which we discuss next.

3.2.2.6 Point and Interval Forecasts

It is often desirable to make stochastic forecasts which capture the variability inherent in the future to some extent. Unlike a point estimate of future energy consumption or production which returns a single value, a stochastic forecast either returns the expected variance in this prediction or multiple scenarios which have a certain probability of occurring. The higher the prediction variance (or the spread between different scenarios), the less likely it is that a single point forecast will be realized. This fact is especially relevant for data which is generated from bimodal (or multimodal) distributions: here, making a point forecast can actually smear the signal to be forecast to values that would never be realized. In the context of energy consumption, consider the case of energy usage during the night. Most of the time, the energy consumption will be very low, but occasionally there will be a load such as a microwave oven, etc. A stochastic forecast would be able to correctly assign these probabilities, while a naively designed point forecast would make misleading predictions.

The utility of stochastic forecasts also arises from the different sources of noise encountered in the prediction problem and the different types of uncertainty they give rise to, including aleatoric and epistemic uncertainty. Aleatoric uncertainty is a function of the process generating the data and is a property that cannot be altered by more sophisticated prediction models or gathering greater amounts of data. Epistemic uncertainty, on the other hand, refers to the uncertainty caused by lack of sufficient data to learn a reliable model for the process. In practice it is possible to squash epistemic uncertainty by driving exploration to improve the quality of the model. At the same time, while it is true that aleatoric uncertainty is irreducible, this too can be quantified by creating stochastic forecasts. These can then be used to make risk-averse (or risk-aware) decisions; for instance, in the framework of stochastic or robust optimization [33]. A number of reinforcement learning algorithms, especially of the model-based variety, can also utilize such forecasts. There are a number of algorithms which can be used in practice to create stochastic forecasts. These range from quantile random forests [34] to, very recently, using dropout as a Bayesian approximation in deep neural networks [35]. This latter is particularly attractive as it can be seamlessly incorporated into many existing forecast workflows which make use of neural networks.

3.2.2.7 Measuring the Quality of Forecasts

In this section, we will elaborate on the different performance metrics used commonly to evaluate different models, and where they may be useful. These methods can be used to compare different methods against one another, but can also be used to combine different models in an ensemble which is designed to outperform all of its constituents.

Model Accuracy
As forecasting is usually posed as a regression problem, one way of defining model accuracy is through the **mean absolute error (MAE)** metric, which measures the average absolute difference between the observations and (model) predictions on a held-out test set. This is given as:

$$MAE = \frac{1}{n}\sum_{i=1}^{n}|y_i - \hat{y}_i| \tag{3.1}$$

Common alternatives to the MAE metric include the **mean squared error (MSE)** or **root mean squared error (RMSE)**. These are also often the primary objective the function approximation method (e.g. a neural network) is trained to minimize, in addition to a regularization term used to minimize overfitting. The MAE is in the same units as the observations $\hat{y}_i$, and is therefore useful to get a quick idea about the efficacy of the forecast algorithm. The related **mean absolute percentage error (MAPE)** scales the error relative to the observations. It is important to note here that in practice, squared error terms as the ones used in MSE and RMSE metrics weigh large prediction errors much more severely than MAE. This can be either desirable or undesirable depending on the learning problem, the quality of observation data and the probability of anomalies.

Another performance metric used extensively in literature is the **R-squared** (R^2), also known as the coefficient of determination, a statistical measure of how close the predictions and observations are. Its value lies between 0 and 1, where a value closer to 0 indicates that the model explains very little variability of the response data around its mean, while a value of 1 indicates that the model explains all of this variability. Higher values of R^2 usually indicate a better fit.

It is important to note here that both the MAE (and its variants) and the R^2 metrics measure only the magnitude of error in the prediction (and not its direction). This has implications for learned model quality, as residuals must be analyzed to determine if they are unbiased. One way to do this is by simply dropping the absolute term in the MAE formulation: this gives an idea about potential bias in the forecast. Thus while MAE measures how far the predictions vary from observations in general, the **mean error (ME)** is a measure of whether the model systematically under- or over-predicts observation data. Another way of doing this is by carrying out **tests for normality** on the residuals (i.e. the residuals must be Gaussian distributed about zero). This can identify if there is a systematic prediction error. Finally, a more sophisticated alternative is to calculate the **autocorrelation function of the residuals**. If no significant autocorrelations exist at any time besides lag zero, this means that the prediction model has been able to capture all the trends and seasonalities in the input data, so the error term is completely random. Where this is not the case, either more data is required (when the model is overfitting) or a larger capacity model needs to be used to create the predictions (when the model is underfitting).

The Temporal Dimension

The model forecast accuracy is not a fixed term, especially for the online case discussed above. Rather, it evolves over time as more data is gathered about the system. This characteristic is captured by considering three dimensions for whichever error statistic (MAE, MAPE, etc.) is considered.

1. **Initial performance.** Model performance at the beginning of the learning process (possibly after an initial stabilization period) is an important consideration as it identifies how well the system behaves in the absence of much data. A model which has a much higher accuracy than another model is therefore considered to exhibit jumpstart performance.
2. **Learning rate.** Learning rate reflects the rate at which the model performance improves over time, i.e. with the acquisition of additional data points.
3. **Asymptotic performance.** Model performance in the limit is an important concern, especially for control and simulation purposes. If the model never converges to a reasonable accuracy, it should not be used for active control or user engagement purposes. Additional data should help the model improve its performance until it plateaus. This plateau should not be induced by limitations of the learning algorithm, and should be at the level dictated by the noise inherent to the system.

3.2.3 Advanced Topics in Modelling

Two key limitations hinder the real world applicability of data-driven approaches. These include (1) huge dataset requirements and (2) data privacy concerns. This section provides an overview of some concepts which have arisen to address these issues.

3.2.3.1 Transfer Learning

In many real world situations, it is possible to use previously collected data or trained models to bootstrap a new forecasting model, i.e. providing a jumpstart to the predictive performance, as shown in Fig. 3.6. The objective of doing this is to learn a highly accurate model of the system under consideration in as little time and at as little cost as possible. This transcends the traditional limitations of data inefficiency associated with black-box techniques. It can also considerably lower the computational budget required to train the model from scratch.

Transfer learning, which allows this accelerated black-box modelling in a principled manner, refers to the case where knowledge—in some form—is transferred from a source to a target [36]. In general, transfer learning makes sense when plenty of data has been collected for a source domain or task, which can be utilized to improve the predictive accuracy of forecasts made for a different target domain or task. Figure 3.5 shows a visualization of the main concept behind transfer learning.

If black-box modelling is to go beyond the performance levels we have already witnessed in the past few years, transfer learning is likely to play a role in this, even though adoption in the energy domain has been fairly limited. Very recently, some research has appeared which addresses this shortcoming [20, 37–39]. Practically, transfer learning involves two key concepts: a domain and a task [36, 39] which can be defined as:

1. The **domain** $\mathcal{D}$ consists of a feature space $\mathcal{X}$ and a marginal probability distribution $P(X)$ over the feature space where $X = \{x_1, x_2, \ldots, x_n\} \in \mathcal{X}$. Here $\mathcal{X}$ includes the space of all possible feature vectors, whereas x_i is a particular feature vector corresponding to some input and X is a particular learning sample.

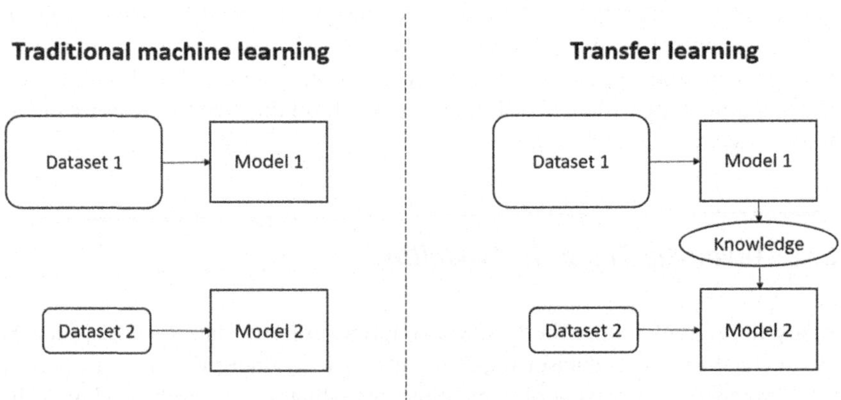

Fig. 3.5 A visual overview of traditional machine learning workflows, as opposed to those incorporating transfer learning; transfer is especially useful in settings where there is limited amount of annotated data (represented by dataset two in this case). A typical machine learning algorithm would have poor performance for this generalization, while with transfer learning it is possible to bootstrap its performance using knowledge gained from learning with dataset one

2. Given a domain, $\mathcal{D} = \{\mathcal{X}, P(X)\}$, a **task** $\mathcal{T}$ consists of a label space $\mathcal{Y}$ and a conditional probability distribution $P(Y|X)$, which is typically to be learned from the training data in the form of pairs $x_i \in X$ and $y_i \in \mathcal{Y}$. The task $\mathcal{T}$ is then given by $\{\mathcal{Y}, P(Y|X)\}$.

Thus, in the context of creating load forecasts, an example of the input feature space $\mathcal{X}$ can be all possible combinations of historical smart meter data as well as other relevant sensor readings such as temperature sensors, etc. The marginal distribution $P(X)$ over this feature space then quantifies the probability of observing a specific feature vector (e.g. a specific energy consumption pattern occurring in a household). This depends on consumer behaviour, device characteristics and ambient conditions. $\mathcal{Y}$ is the set of all possible labels, which in this case would be the set of real numbers, in case energy can be both generated and consumed locally. The conditional distribution $P(Y|X)$ is the specific mapping between historical user behaviour, device specifications and ambient conditions, and future energy demand that we are interested in learning.

There are four possibilities in which transfer learning can be applied (where either of source domain and task can differ from the target domain and task). The most practical ones among these are when either the marginal or conditional probability distribution are different across the source and target, i.e. $P(X_s) \neq P(X_t)$ and $P(Y_s|X_s) \neq P(Y_t|X_t)$, respectively. The former refers to the case of homogeneous populations, where household behavioural patterns or ambient conditions cause the state-space to be explored differently, but the underlying data generating process (e.g. appliance type) remains the same. This can be contrasted with the latter where conditional probabilities differ, i.e. heterogeneous populations are considered. This can lead to transfer across devices which do not share identical dynamics. Following the terminology in [36], we refer to the former as **transductive transfer** and the latter as **inductive transfer**. While similar techniques may be applied to achieve both, a differentiation can be made between:

1. **Feature sharing** is the form of transfer learning where features from a different but related source domain or task are directly shared with a target to improve learning performance. Both raw observations and extracted features can be used for this purpose, as long as the training input and output feature vectors share the same dimensionality across source and target. Which features to share for transfer is still an open area of research, and different studies have shown different results employing various metrics such as similarity, diversity, etc. [40].
2. **Parameter sharing** usually involves the training of a function approximator such as a neural network with a large amount of source data. The weights of this neural network are then used as initialization for the target domain or task which fine-tunes the network weights with observed data in the target domain. The fine-tuning is usually done with a much smaller learning rate to retain the representations already learned by the network while nudging it to adapt to the new learning problem [41].

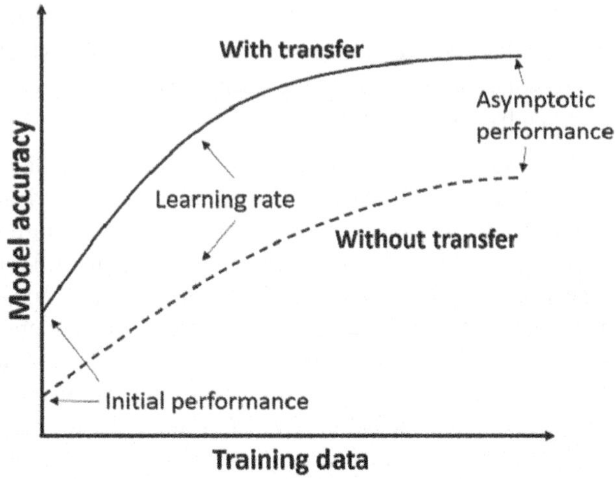

Fig. 3.6 A demonstration of different ways in which transfer learning can accelerate learning; the jumpstart is illustrated by the higher starting point, the difference in learning rate is captured by the higher slope, and the improved asymptotic performance is illustrated by the higher asymptote, adapted from [42]

On an abstract level, feature sharing corresponds to sharing experiences across learners, while parameter sharing encourages transfer of knowledge. The choice of which one is more suitable depends on the nature of the problem and the disparity between the source and target in the transfer learning problem. Transfer through parameter sharing is generally associated with deep learning, while sharing raw features can be seen as a naive alternative which is not guaranteed to work in heterogeneous settings, and can sometimes even lead to negative transfer. The benefits encapsulated by transfer learning are summarized in Fig. 3.6. It is straightforward to see why this technique can help with reducing the data intensity of learning algorithms.

3.2.3.2 Privacy-Preserving and Privacy-Aware Learning

Privacy concerns underpin most contemporary innovations in data-driven learning and forecasting. This is definitely the case in creating forecasts and models for energy usage and production in buildings, which require gathering and sharing smart meter or other sensor data with a central server. While a number of different solutions have been proposed which address parts of the problem including differential privacy [43], homomorphic encryption [44] and secure multi-party communication [45], federated learning is an end-to-end solution which allows learning in a centralized manner with distributed data gathering mechanisms [46]. This is especially desirable when shared representations are being learned which can make use of transfer learning as described above. The central idea behind federated learning arises from performing decentralized model updates on individual devices (or nodes). These model updates (or a subset thereof) are then pushed to a central server where they are aggregated across multiple devices. This helps with not just smoothing out individual anomalous behaviour, but also further anonymises user data, as it is might still be possible to infer user behaviour from individual model updates. The final model at the central server is then pushed back to individual nodes for consumption. This ensures that user data never leaves their premises, and addresses a large part of the data privacy debate.

3.2.4 Control

Learning is only one part of the larger problem space, as shown in the pipeline in Fig. 3.2. The learning agent also needs to make decisions and execute actions based on the model it has learned so far, or it needs to provide actionable input to the building occupants. This allows it to gather even more data to improve its representation of the system. Here, we briefly present some objective functions which agents controlling buildings, or clusters of buildings, can optimize towards. This is followed by a discussion of the algorithms which practitioners and researchers can make use of to achieve this optimization in practice.

3.2.4.1 Objective Functions

The structure of optimizing energy usage in buildings for enabling DSM derives from the interactions between the energy loads, the building occupant, any DERs and storage elements and the energy grid. The control agent in this case, which is also learning and creating forecasts for the future, controls the flexible (or controllable) devices in a building following a policy, π_n. This policy is designed to maximize a specified objective function such as energy efficiency, subject to certain constraints which can depend on both physical aspects of the energy system and consumer preferences. These objectives include both local optimization objectives where buildings are treated on their own, and global optimization objectives where multiple buildings are aggregated together to achieve a shared objective such as grid stability. Typically, it is easier to optimize towards local objectives as the sensing and coordination costs are more limited. Some of the most commonly employed objectives for DSM programs include:

1. **Energy efficiency**, or demand reduction, refers to the goal whereby the overall energy used by a building or a cluster of buildings is minimized while still respecting constraints based on user and system preferences. A number of practical examples of such optimization exist in literature for a number of different loads including building space heating and hot water systems [20, 47, 48]. Typically, the focus of these studies is thermal draws as these represent untapped flexibility potential in the system due to use of naive control mechanisms. Optimizing energy efficiency is usually a local optimization problem in the sense that the energy demand reduction of a building does not influence the energy demand of its neighbours.

2. **Price optimization** is a generalization of the energy efficiency case, where the price of electricity is allowed to vary according to the time of day and/or year. In this case, the objective of the control agent is to minimize energy costs. The simplest case of a variable tariff is the dual pricing scheme offered to households in many countries [28, 49]. However, in many places such as The Netherlands, it is becoming increasingly possible for households to directly access real time pricing. One drawback of this mechanism means the exposure of households to large scale price fluctuations which will likely increase with growing variable renewable energy proliferation. This can be mitigated by placing price caps, but also by smart agents controlling flexible loads in the building.
3. **Self-consumption** of local DER energy production is a specialized objective which is designed to promote consumption of locally produced resources such as solar energy. Primarily, this targets maximizing the usage of local electricity generation from rooftop solar panels in households to minimize grid interaction. Large scale injection of electricity during the summer months can easily overload the local low-voltage grid, especially when coupled with low consumption—as is frequently the case in many Northern European countries during summer. Local consumption, either at the household or neighbourhood level, is meant to alleviate this problem [50].
4. **Providing ancillary services** to the energy grid can work on two different levels: distribution and transmission. On the distribution side, load (or production) can be modulated to alter the power flow or voltage in the distribution grid. This problem typically requires at least a few controllable appliances (or solar panels) in a neighbourhood [6, 51]. On the other hand, ancillary services for the transmission grid are typically centred on frequency regulation. When frequency drops below the standard (50 Hz in Europe), demand should be reduced or production should be increased instantaneously. When the frequency rises above the standard, demand must be ramped up or production must go down. As we transition to more variable renewable energy sources, where supply side controls will not necessarily be sufficient, demand needs to be made flexible too. Both heating and electric vehicles will be the ideal fit for this [52]. However, to ensure stability, the market rules for ancillary services require strict compliance and, as the resources required to participate in these services are typically in the megawatt range, exclude demand side management in buildings. This will slowly change in the coming years as the energy grid undergoes rapid decarbonization and electrification.

In addition to these objective functions, there are usually constraints placed on the optimization problem as well. These include device limitations and occupant comfort bounds. Sometimes, when these constraints need not be strictly enforceable, they can also be made part of the objective function that the agent optimizes. While historically control of appliances in most buildings took a naive rule-based form, there are two main alternatives which have developed over time to improve its performance. There are two different ways to look at this. The more common view is to distinguish them as model predictive control and reinforcement learning. This is rather disingenuous as it lumps the model-based variants of reinforcement learning, some of which are actually closer to model predictive control, with the model-free reinforcement learning algorithms. Therefore, in this section, we distinguish these classes of control along model-based and model-free controllers, while also highlighting the conceptual differences and similarities between model predictive control and model-based reinforcement learning.

3.2.4.2 Model-Based Algorithms

Unlike rule-based controllers which do not possess any anticipatory prowess, model-based controllers make use of a model of the system dynamics to look into the future and explicitly optimize operation. This model can be obtained by making use of the techniques discussed earlier in this chapter, whereas the actual optimization can be based either on a passive formulation such as conventional model predictive control, or one based on active learning such as reinforcement learning.

Model Predictive Control (MPC)
MPC combines a model of the system with conventional solvers making use of convex or non-convex optimization algorithms to arrive at optimal control actions for the future [47]. The discussion of convexity derives from the structure of the problem, although in many practical cases convex relaxations have been introduced to solve non-convex problems. This is, in many practical situations, a trade-off between computational complexity and the actual performance of the controller. Regardless of the optimization method chosen, the objective is to improve the operational performance of the device compared to some baseline, usually obtained via naive rule-based control.

A distinctive feature of MPC is that the objective (or cost) function to be minimized or maximized has to be explicitly stated. Economic MPC (EMPC), where the objective is to reduce operational costs (e.g. energy prices as described above), is arguably the most popular formulation. Where the cost is constant, this reduces to the standard form of demand reduction (or improving energy efficiency). This formulation has the following cost function [53]:

$$J_e = \sum_i p_{el}(i)\dot{W}_{el}(i) \tag{3.2}$$

where $\dot{W}_{el}(i)$ is the power consumed and p_{el} is the electricity price, which varies in time according to the tariff structure. By minimizing J_e over a receding time horizon, it is possible to obtain monetary savings. This process fundamentally pushes energy usage from times of higher energy prices to when it is cheaper. p_{el} can be the actual price, or it can be a forecast of the price. Other formulations seeking to optimize user thermal comfort, by minimizing temperature variations outside some predefined criterion [54], or to provide ancillary services to the grid by demand response exist as well.

In addition to defining the objective function, a model predictive controller typically also explicitly defines the constraints placed upon the optimization problem. These can take on many forms, but are mostly derived from physical limitations of the device. For instance, a controller might be forced to make binary decisions while controlling a heat pump, because it can only turn the system on or off. Likewise, while a modulatable heat pump can theoretically consume continuous-valued power at any given time, the amount of this consumption is bounded above by its nominal capacity, and below by zero.

Typically model predictive control has been deployed in a deterministic setting. This assumes that future consumption and production of energy, as well as energy prices, is forecast as point estimates. As we have discussed previously, this is seldom the case. Especially with the proliferation of distributed generation and greater electrification of loads, this forecast error is a source of considerable problems in optimal control. A popular alternative is to employ a variant of stochastic programming for the actual optimization. This paradigm makes use of a stochastic forecast, reduced to a set of scenarios each with a distinct probability. These scenarios are then used by the optimizing agent to reduce the possibility of a worst case scenario occurring. In many services, such as the provision of heat to end users during winter or ancillary services to the grid, it is important to plan keeping this uncertainty in mind.

Despite widespread usage, there are some key limitations of MPC in its conventional form:

1. Defining an objective function and constraints in a rigid manner can lead to issues during the operational phase, especially if different components of the system are replaced or upgraded over time. A common practical example is when the constraints originally defined in the optimization problem do not hold due to a system upgrade.
2. The requirement for a model to use with the controller and the potential computational cost. The limitation surrounding the existence of a prior model can be side-stepped by learning the model online, an approach we refer to as data-driven MPC. It is important to keep in mind here that the only learning in this paradigm is of system dynamics (i.e. how the system behaves over time). In other words, the optimization problem is still solved at every time step.
3. The MPC formulation is especially wasteful in domains where the controller is expected to solve the same (or similar) optimization problem on a recurring basis (e.g. when it revisits similar states on a recurring basis). In other words, it is a memory-less optimization framework, which can exacerbate its computational complexity problems. This renders it unsuitable for problems which require very fast response times such as frequency regulation. This concern has been addressed partly by formulations such as explicit MPC, which create a tabular mapping of optimal control actions from all possible states of the system.

However, this is only feasible when the problem dimensionality is very low. As the state-action space grows, this quickly becomes infeasible.

Model-Based Reinforcement Learning
Model-based reinforcement learning (RL) lies somewhere between conventional data-driven MPC formulations and model-free RL algorithms, which we discuss in the next section. While the lines are continuously shifting, there are two fundamental conceptual differences between MPC and model-based RL, as most practitioners employ it:

1. **Exploration** ensures that there is a principled way for the treatment of uncertainty inherent to data-driven models. When the agent is uncertain about its representation or the optimal course of action to adopt, it takes more exploratory steps in the hopes of discovering something useful about the environment (or its own dynamics). When it has learned an accurate representation and/or control policy, it turns its attention to exploiting this knowledge, rather than exploring needlessly. There are many ways of ensuring exploring agents which have already been mentioned in the section on point and interval forecasts.
2. **Policy-side learning** ensures that, over time, the controller learns optimal policies in addition to a model for the system. This means that, when faced with similar optimization problems, it does not need to perform expensive computations repeatedly, but can just run inference on a function approximation algorithm. This learning can therefore considerably speed-up real world performance and, in doing so, it addresses a major shortcoming of model predictive controllers.

3.2.4.3 Model-Free Methods

At the opposite end of the control spectrum lie model-free reinforcement learning based algorithms, which learn (optimal) policies directly from acting in and interacting with the environment [30]. This can be, in many instances, an attractive proposition because it does away with the need to learn a model for the system altogether. In recent years, a number of increasingly complex formulations have been proposed to solve specific problems with basic formulations such as Q-learning or SARSA. These algorithms can be especially effective in domains where the input space is high dimensional and learning an appropriate model might be impractical because of the task's complexity [55].

A note on the terminology used in reinforcement learning is necessary here. Markov decision processes (MDPs) form the foundation of many RL formulations which allow us to interleave modelling and control of energy systems. An MDP is a special type of a stochastic process, where the state transition probabilities (i.e. the conditional probability distribution of future stures) follow the Markov property, and are influenced not just by the stochasticity of the environment, but also by the actions of a decision making agent. An MDP is typically defined by the tuple: $\{S, A, P, R\}$. Here, the control agent is in some state $s \in S$, whereupon it can choose an (allowed) action $a \in A$. Given this decision and the stochasticity inherent to the MDP, the agent moves to a new state s', based on the state transition function $P_a(s, s')$, and receives a reward based on $R_a(s, s')$. In MDPs, the state transition function is conditionally independent of all previous states and actions, except for the current state s and action a. This ensures that an MDP satisfies the Markov property [30]. Based on the MDP, the agent learns a policy, π, to perform control actions given its present state, given as:

$$\pi : S \times A \rightarrow [0, 1] \tag{3.3}$$

$$\pi(a \mid s) = P(a_t = a \mid s_t = s) \tag{3.4}$$

which defines the probability of taking control action a in state s, and marks a probabilistic policy. This does not however say anything about how good this certain policy is at maximizing rewards. Therefore, to determine the optimal policy, a value function is defined for each state as the expected return the agent will receive (and reflects the *goodness* of the state). The value of a policy can therefore be defined as:

$$V^{\pi}(s) = \mathbb{E}[R \mid s, \pi] \tag{3.5}$$

A policy, π, is then considered to be optimal, π^*, if it achieves the highest expected return, starting from any initial state, and is given as:

$$V^*(s) = \max_{\pi} V^{\pi}(s) \tag{3.6}$$

It is also possible to take a step further, and define the state-action values rather than simply the state-values. This is typically given by the Q-value of a state-action pair, and is defined as:

$$Q^{\pi}(s, a) = \mathbb{E}[R \mid s, a, \pi] \tag{3.7}$$

Once a Q-value has been learned for each state-action pair, it is straightforward to obtain the optimal policy. A number of algorithms have been devised to learn either the Q-value or the optimal policy directly from data, including Q-learning, SARSA, Fitted Q-iteration and Deep Q-Networks, besides many others [30]. One key distinguishing factor in modern RL algorithms is the use of function approximation to replace tabular Q-values, which allows for much better generalization, especially in high dimensional spaces.

Despite impressive advances, it is well established in control circles in general that model-free controllers cannot compete with their model-based counterparts on most realistic control tasks [56]. Similar findings have been reported for the energy domain [28]. One reason for this is the fact that model-based algorithms can use the learned models to simulate many trajectories of possible futures, thereby further accelerating the discovery of good policies. Model-free controllers are constrained to only make use of observations which have been gathered in the past and cannot usually generate their own data.

3.3 Practical Challenges

This section highlights some of the most important technical challenges still plaguing many smart grid projects of the type discussed in this chapter. Besides technical challenges, economic and social issues can often derail such projects. These include the problem of diffuse rewards, high installation and maintenance costs of sensors, as well as data privacy and security issues. Often, there are promising technical solutions to these challenges that are yet to be commercialized, but for brevity these are not discussed further. This section instead focuses on the technical challenges that still need to be addressed while implementing data-driven algorithms to perform demand side management in buildings.

3.3.1 Challenges with Function Approximation

3.3.1.1 Data Requirements

The lack of availability of data can often hinder adoption of smart solutions. This need for data can manifest both in terms of requirements for extensive sensing (to perform state estimation) and historical data (to train models for system identification). Where the input state-space has been inadequately explored, the modelling algorithm will typically make incorrect predictions. Furthermore, as the dimensionality of the input state-space grows, the amount of training data required to learn a reliable model can quickly become unmanageable. Transfer learning is generally intended to serve as a general-purpose solution for this practical challenge of generalization from limited data. However, it does not address the very real possibility of physical changes to an operative device which can fundamentally alter the data generating process. Likewise, the question of what the model actually knows and what it does not know is an important one, and can be addressed by making use of Bayesian or ensemble based methods. Getting reliable uncertainty estimates with deep learning remains however an active area of research. Finally, using existing models to explain the behaviour of previously unseen devices remains an open research problem.

3.3.1.2 Practical Costs of Deep Learning

Deep learning has, for better or worse, become the method of choice for most function approximation tasks in the energy domain. Often where simpler methods would suffice, practitioners employ deep neural networks because of the hype surrounding these methods. While highly versatile machine learning tools, there are however considerable hidden costs associated with this choice. Deep neural networks can easily learn the noise in the training data and thus fail to generalize to unseen examples. Even when ample training data has been gathered, deep neural networks require a lot of computational power to train, which is not always available, and is in any case counterproductive because of the associated environmental footprint. Finally, neural networks, or any data-driven method for that matter, need to be periodically retrained as more data becomes available. This represents further computational expense. At the same time, retraining a neural network repeatedly with different training data introduces an additional point of failure in the system. It is entirely possible for the neural network to fail to converge to a suitable minima during training due to their stochastic nature. This can be avoided by careful test-driven development, a practice often eschewed in practice. This issue can only be addressed as development tools and processes in the data pipeline mature.

3.3.2 Challenges with Control

3.3.2.1 Response Times

As already mentioned, a number of ancillary services require extremely fast response times. This is especially true for the frequency reserve, where some services require even sub-second response times, a feat almost certainly impossible to achieve with a central controller which involves both network latency and planning time. Instead, it is much more practical to perform control at a local level in this case, directly in response to fluctuations in frequency measurements, even though it may increase local sensing costs. Learning optimal control policies over time offers a practical means to achieve this goal. Similarly, response times may be constrained by the device itself. For instance, many heat pumps on the market nowadays have a rather lengthy response time (after being switched on). This means that these devices may typically not be suitable for providing very fast frequency response to the grid on their own. Finally, there is also a substantial asymmetry in the amount of flexibility different devices can offer to the grid, i.e. the potential upward modulation of energy demand for a flexible load is not the same as how much its energy can be reduced at any given time. In general, aggregating multiple different sources of flexibility can increase the robustness of demand side management. The less correlated these sources of flexibility are, the greater their impact on making the aggregation portfolio resilient to random variations caused by user behaviour and ambient conditions.

3.3.2.2 Distributed vs. Centralized Control

The discussion about centralized vs. distributed controllers goes beyond the device response time. Centralized approaches involve a single agent gathering data from multiple agents (buildings), performing an optimization or learning task, and communicating results to the agents (typically in the form of actions to take). This allows the central agent to directly optimize towards a single, global objective, and reduces complexity if information about the individual agents is readily available (in real time). On the other hand, this approach has high computational complexity, incurs substantial communication costs and suffers from network latency. Decentralized algorithms, on the other hand, operate on an individual agent level and can remove the effects of network latency. These approaches have the potential to be more resilient as they do not have a single point of failure. However, they often suffer from several other limitations, such as requiring different agents to learn how to cooperate and compete to obtain shared objectives, which can be both data and compute intensive. Furthermore, for critical services, testing and debugging decentralized agents can be highly challenging due to the wide variety of scenarios that can be experienced in practice.

3.3.2.3 Demonstrability

Demonstrability refers to the fact that a provider of ancillary services is contractually obligated to provide the service that has been committed. This can take the form of providing a certain amount of upwards or downwards flexibility, and can invoke both capacity and activation based mechanisms [52, 57]. Failure to comply might lead to penalties, and even exclusion from the pool. However, demonstrability is far from straightforward in many cases with DSM. Occasionally, this can be due to failures in sensing and/or communication networks, but more frequently it is due to the requirement for a model which defines a reference baseline to which the service provider's response is compared against. Human stochasticity and unexpected ambient condition changes (and their interaction) can adversely affect how this plays out in the real world, and there is no universally accepted solution to handling this at this point in time.

3.4 Conclusion

Despite the challenges highlighted in this chapter, demand side management powered by data-driven methods will enable the next generation of smart buildings and grids. This is a timely development that will enable societal level electrification and decarbonization of the building stock. Technological innovations in machine learning research, such as large scale transfer learning across disparate domains and tasks, federated learning which allows model-learning without access to user data and reinforcement learning which avoids the pitfalls of model predictive control, seem particularly promising. At the same time, practitioners must ensure that these algorithms are implemented in a way that is cognizant of not just their strengths, but also their limitations.

References

1. E.A. Rosa, T. Dietz, Human drivers of national greenhouse-gas emissions. Nat. Clim. Chang. **2**(8), 581 (2012)
2. E. Wolff, Climate change: evidence and causes. Sch. Sci. Rev. **96**(354), 17–23 (2014)
3. IPCC, *Special Report on Renewable Energy Sources and Climate Change Mitigation* (Cambridge University Press, Cambridge, 2011)
4. IRENA, *Renewable Power Generation Costs in 2018* (IRENA, 2019)
5. M. Obi, R. Bass, Trends and challenges of grid-connected photovoltaic systems—A review. Renew. Sust. Energ. Rev. **58**, 1082–1094 (2016)
6. D. van Goch, et al., Rennovates, flexibility activated zero energy districts, H2020. Impact **2017**(5), 29–31 (2017)
7. E. Annunziata, M. Frey, F. Rizzi, Towards nearly zero-energy buildings: the state-of-art of national regulations in Europe. Energy **57**, 125–133 (2013)
8. IEA, *Global EV Outlook* (IEA, Paris, 2019)
9. P. Palensky, D. Dietrich, Demand side management: demand response, intelligent energy systems, and smart loads. IEEE Trans. Ind. Inf. **7**(3), 381–388 (2011)
10. W. Liu, et al., Day-ahead congestion management in distribution systems through household demand response and distribution congestion prices. IEEE Trans. Smart Grid **5**(6), 2739–2747 (2014)
11. D. Pudjianto, et al., Smart control for minimizing distribution network reinforcement cost due to electrification. Energy Policy **52**, 76–84 (2013)
12. S.A. Pourmousavi, M. Hashem Nehrir, Real-time central demand response for primary frequency regulation in microgrids. IEEE Trans. Smart Grid **3**(4), 1988–1996 (2012)
13. Y. Lin, et al., Experimental evaluation of frequency regulation from commercial building HVAC systems. IEEE Trans. Smart Grid **6**(2), 776–783 (2015)
14. M. Ferrara, V. Monetti, E. Fabrizio, Cost-optimal analysis for nearly zero energy buildings design and optimization: a critical review. Energies **11**(6), 1478 (2018)
15. J.Z. Kolter, M.J. Johnson, REDD: a public data set for energy disaggregation research, in *Workshop on Data Mining Applications in Sustainability (SIGKDD), San Diego, CA.*, vol. 25. No. Citeseer (2011)
16. A.J.C. Cuddy, K.T. Doherty, M.W. Bos, *OPOWER: Increasing Energy Efficiency through Normative Influence (A)* (Harvard Business School Case Collection, 2010)
17. A. Laskey, O. Kavazovic, Opower. XRDS Crossroads ACM Mag. Stud. **17**(4), 47–51 (2011)
18. B. Dupont, et al., LINEAR breakthrough project: large-scale implementation of smart grid technologies in distribution grids, in *2012 3rd IEEE PES Innovative Smart Grid Technologies Europe (ISGT Europe)* (IEEE, Piscataway, 2012)
19. M.G. Lijesen, The real-time price elasticity of electricity. Energy Econ. **29**(2), 249–258 (2007)
20. H. Kazmi, et al., Multi-agent reinforcement learning for modeling and control of thermostatically controlled loads. Appl. Energy **238**, 1022–1035 (2019)
21. A. Vaswani, et al., Attention is all you need, in *Advances in Neural Information Processing Systems* (2017)
22. I. Sutskever, O. Vinyals, Q.V. Le, Sequence to sequence learning with neural networks, in *Advances in Neural Information Processing Systems* (2014)

23. K. Amasyali, N.M. El-Gohary, A review of data-driven building energy consumption prediction studies. Renew. Sust. Energ. Rev. **81**, 1192–1205 (2018)
24. J. Shlens, A tutorial on principal component analysis (arXiv: 14041100), pp. 1–12 (2014)
25. I. Goodfellow, Y. Bengio, A. Courville, *Deep Learning* (MIT, Cambridge, 2016)
26. B. Claessens, P. Vrancx, F. Ruelens, Convolutional neural networks for automatic state-time feature extraction in reinforcement learning applied to residential load control. IEEE Trans. Smart Grid **9**(4), 3259–3269 (2016)
27. H. Kazmi, et al., Generalizable occupant-driven optimization model for domestic hot water production in NZEB. Appl. Energy **175**, 1–15 (2016)
28. A. Nagy, et al., Deep reinforcement learning for optimal control of space heating. arXiv preprint:1805.03777 (2018)
29. A. Carpentier, et al., Upper-confidence-bound algorithms for active learning in multi-armed bandits, in *International Conference on Algorithmic Learning Theory* (Springer, Berlin, 2011)
30. R.S. Sutton, A.G. Barto, *Reinforcement Learning: An Introduction* (MIT, Cambridge, 2018)
31. H. Akaike, Fitting autoregressive models for prediction. Ann. Inst. Stat. Math. **21**(1), 243–247 (1969)
32. E. Strubell, A. Ganesh, A. McCallum, Energy and Policy Considerations for Deep Learning in NLP. arXiv preprint:1906.02243 (2019)
33. M.V.F. Pereira, L.M.V.G. Pinto, Multi-stage stochastic optimization applied to energy planning. Math. Program. **52**(1–3), 359–375 (1991)
34. N. Meinshausen, Quantile regression forests. J. Mach. Learn. Res. **7**(Jun), 983–999 (2006)
35. Y. Gal, Z. Ghahramani, Dropout as a Bayesian approximation: representing model uncertainty in deep learning, in *International Conference on Machine Learning* (2016)
36. S.J. Pan, Q. Yang, A survey on transfer learning. IEEE Trans. Knowl. Data Eng. **22**(10), 1345–1359 (2009)
37. E. Mocanu, et al., Unsupervised energy prediction in a smart grid context using reinforcement cross-building transfer learning. Energy Build. **116**, 646–655 (2016)
38. Q. Hu, R. Zhang, Y. Zhou, Transfer learning for short-term wind speed prediction with deep neural networks. Renew. Energy **85**, 83–95 (2016)
39. H. Kazmi, J. Suykens, J. Driesen, Large-scale transfer learning for data-driven modelling of hot water systems, in *Building Simulation* (2019)
40. S. Ruder, B. Plank. Learning to select data for transfer learning with Bayesian Optimization. arXiv preprint:1707.05246 (2017)
41. J. Yosinski, et al., How transferable are features in deep neural networks?, in *Advances in Neural Information Processing Systems* (2014)
42. L. Torrey, J. Shavlik, Transfer learning, in *Handbook of Research on Machine Learning Applications and Trends: Algorithms, Methods, and Techniques* (IGI Global, Hershey, 2010), pp. 242–264
43. J. Zhao, et al., Achieving differential privacy of data disclosure in the smart grid, in *IEEE INFOCOM 2014-IEEE Conference on Computer Communications* (IEEE, Piscataway, 2014)
44. F.D. Garcia, B. Jacobs, Privacy-friendly energy-metering via homomorphic encryption, in *International Workshop on Security and Trust Management* (Springer, Berlin, 2010)
45. C. Thoma, T. Cui, F. Franchetti, Secure multiparty computation based privacy preserving smart metering system, in *2012 North American Power Symposium (NAPS)* (IEEE, Piscataway, 2012)
46. J. Konecny, et al., Federated learning: strategies for improving communication efficiency. arXiv preprint:1610.05492 (2016)
47. F. Oldewurtel, et al., Use of model predictive control and weather forecasts for energy efficient building climate control. Energy Build. **45**, 15–27 (2012)
48. J.E. Braun, Reducing energy costs and peak electrical demand through optimal control of building thermal storage. ASHRAE Trans. **96**(2), 876–888 (1990)
49. F. Ruelens, *Residential Demand Response Using Reinforcement Learning: From Theory to Practice*. PhD Thesis, KU Leuven (2016)

50. O. De Somer, et al., Using reinforcement learning for demand response of domestic hot water buffers: a real-life demonstration, in *2017 IEEE PES Innovative Smart Grid Technologies Conference Europe (ISGT-Europe)* (IEEE, Piscataway, 2017)
51. O. Sundstrom, C. Binding, Flexible charging optimization for electric vehicles considering distribution grid constraints. IEEE Trans. Smart Grid **3**(1), 26–37 (2011)
52. D. Fischer, H. Madani, On heat pumps in smart grids: a review. Renew. Sust. Energ. Rev. **70**, 342–357 (2017)
53. G. Masy, et al., Smart grid energy flexible buildings through the use of heat pumps and building thermal mass as energy storage in the Belgian context. Sci. Technol. Built Environ. **21**(6), 800–811 (2015). masy2015smart
54. R. De Coninck, L. Helsen, Practical implementation and evaluation of model predictive control for an office building in Brussels. Energy Build. **111**, 290–298 (2016)
55. V. Mnih, et al., Human-level control through deep reinforcement learning. Nature **518**(7540), 529 (2015)
56. B. Recht, A tour of reinforcement learning: the view from continuous control. Annu. Rev. Control Robot. Auton. Syst. **2**, 253–279 (2019)
57. A. Balint, H. Kazmi, Determinants of energy flexibility in residential hot water systems. Energy Build. **188**, 286–296 (2019)

Chapter 4
A Multi-Agent Approach to Energy Optimisation for Demand-Response Ready Buildings

Oudom Kem and Feirouz Ksontini

4.1 Introduction

For decades, electric grids have been employed to deliver electricity from utilities to customers. This classical approach, in which power generation is centralised, provides a uni-lateral communication between the utilities and the customers subscribing to the "generation follows loads" principle. The increasing penetration of renewable energy sources has led the transition towards the next generation of grids, the smart grids, which allows a bilateral communication between the utilities and their customers. This new paradigm empowers the demand side, providing the possibility for the customers to offer electricity and flexibility. From the system point of view, benefits such as reduced peak demands, more efficient transmission and distribution, and improved quality of services can be achieved. From the demand side perspective, customers are more in charge of managing their electricity bills through flexible consumption as well as additional revenue streams from the network, and they are less dependent on the utilities.

To reap the aforementioned benefits, customers need to be ready for this paradigm-shift. The challenge addressed in this work is enabling customers to adopt flexible consumption patterns and to support the demands from the system. Our purpose is to render the customers capable of providing the flexibility and participating in demand-response schemes. To this end, we aim at optimising the energy consumption and generation on the demand side and at the building level. The objectives of the optimisation are threefold: optimise the consumption and injection to grids of locally generated power, support demand-response signals from

O. Kem (✉) · F. Ksontini
CEA, LIST, Gif-sur-Yvette, France
e-mail: oudom.kem@cea.fr; feirouz.ksontini@cea.fr

M. Sayed-Mouchaweh (ed.), *Artificial Intelligence Techniques for a Scalable Energy Transition*, https://doi.org/10.1007/978-3-030-42726-9_4

external parties, and exploit the flexibility to reduce energy bills considering the revenues from subscribing to demand-response schemes.

In the optimisation, we consider a wide range of dynamic devices providing different types of flexibility such as shiftable/deferrable devices, sheddable devices, and curtailable generators. Each device possesses its own dynamic constraints and objectives. The optimisation needs to take into account the constraints of each device and the demand-response requests, while ensuring that the constraints of the customers are satisfied. Performing such an optimisation over a time horizon entails dealing with a large number of variables, making it computationally impractical to solve in a centralised manner. To solve this optimisation problem in a distributed fashion, thereby ensuring efficiency, scalability, and privacy, we propose a multi-agent optimisation approach that is based on the Alternating Direction Method of Multipliers (ADMM) [9, 12, 15]. For each type of devices, we model its objective function and constraints incorporating user constraints and demand-response incentives, when applicable. The result of the optimisation is the optimal energy flow (i.e., consumption and generation profiles) that takes into account the incentives from demand-responses schemes, while respecting user constraints.

The remainder of this chapter is organised as follows. Section 4.2 presents the context and background of this work providing details on the flexibility and the demand-response schemes considered in the optimisation. Section 4.3 reviews related work from the literature. Section 4.4 describes the proposed models for different types of demand-response requests. Section 4.5 presents a conceptual model of the building, the optimisation problem, and the algorithm upon which our system is based to solve the problem. Section 4.6 describes our main proposal, the multi-agent system for energy management as well as the optimisation models to incorporate demand response in the optimisation. Section 4.7 details the experiments conducted to validate the proposal and the analysis of the empirical results. Section 4.8 discusses various aspects of the current state of the proposal. Section 4.9 concludes the chapter and presents the directions for our future work.

4.2 Context

4.2.1 Building Flexibility

In this work, we address energy management in buildings that are equipped with a wide range of energy-consuming devices and energy-producing devices such as household appliances in residential buildings, photovoltaic, and local generators. Building energy consumption normally consists of fixed consumption and flexible consumption. Fixed consumption refers to the consumption that cannot be changed in any manner. Flexible consumption, in contrast, may be modified to a certain extent, thus providing some sort of flexibility. In [10], flexibility is defined as "the modification of a power pattern of an energy resource at various time scales in reaction to an external signal in order to provide a technical service for an energy system and/or an economical service for a system stakeholder". In other words, flexibility represents the aggregated consumption and generation of flexible devices (i.e., devices whose consumption may be changed in some way). In the context of this work, the following types of devices, distinguished based on the flexibility of their consumption or production, are considered:

- fixed loads: devices whose consumption cannot be altered in any way,
- shiftable loads: devices whose consumption can be shifted within a specified time interval,
- sheddable loads: devices whose consumption can be shed to a certain extent at a specified time interval,
- and curtailable generator: energy-producing devices whose production may be curtailed at a specified time interval.

The flexibility provided by each device is determined by its user. For instance, lights and TVs are usually considered as fixed loads by their users as they are not to be switched off by the system when the users need them. However, in some cases, users may explicitly indicate that their lights are sheddable (on/off or shed to a lower level of brightness), in which case the lights are considered as sheddable loads. Typical examples of shiftable loads are washing machines and dish washers as their consumption usually can be scheduled within a time interval without affecting user comfort.

Each type of devices possesses its own constraints. Users of the devices may also impose their preferences on how the devices should operate. User preferences for each device should be private to each device as to respect user privacy, a criterion particularly important in a multi-user environment. Device constraints, user preferences, and privacy concern need to be taken into account in building energy management.

4.2.2 Demand-Response Schemes

As a part of demand side management, demand response is designed to motivate changes in consumption patterns of customers by providing financial incentives in the form of changes in energy prices over time or as direct payments [28]. Unlike energy reduction, which is also a demand side management program, demand response is concerned with shifting consumption to different points in time rather than reducing the overall energy consumption [10]. Demand response can be categorised into price-based and incentive-based. Price-based demand response or implicit demand response provides customers with time-varying energy tariffs. Incentive-based programs offer direct payments to customers to change their consumption patterns upon request.

The ability to participate in demand-response programs depends on the flexibility voluntarily provided by the consumers. Such flexibility can be exploited to adapt the consumption and/or generation to benefit from the revenues from demand-response participation. Each customer subscribing to a demand-response program receives a set of demand-response requests customised based on their consumption and generation. In this work, both price-based and incentive-based demand-response schemes are considered in the optimisation to align with our objective to render buildings ready for demand response.

4.3 Related Work

Energy optimisation addressed in this work involves various types of dynamic devices connected by both alternating current (AC) and direct current (DC) lines, forming a network of devices. The aim is to minimise a network objective, which is the sum of the objective functions of the devices, subject to the constraints of devices and lines. This problem is considered as a dynamic optimal power flow problem as mentioned in [20]. In the literature, many algorithms and techniques have been proposed for building energy optimisation such as artificial neural network-based model [14, 23], ant colony optimisation algorithm [16], reinforcement learning [24], and genetic programming with fuzzy logic [1], just to name a few. However, each device possesses its own dynamic constraints, objectives, and preferences provided by its user. Performing an optimisation with such devices over a time horizon entails dealing with a large number of variables, making it computationally impractical to solve in a centralised manner [20].

Distributed optimisation techniques are naturally applied to power networks considering the inherent graph-structured nature of the distribution networks. Solving complex problems using a distributed technique has become prevalent in the literature. Distributed optimisation has been shown to be applicable for the optimal power flow problem [3]. In [11], the authors propose some strategies for distributed problem decomposition to achieve desirable properties such as scalability. Dual decomposition [4] is an example of distributed optimisation methods. It preserves the privacy of the cost function and local constraints of each solver agent. However, it is not robust in the sense that it requires many conditions such as strict convexity of all local cost functions for convergence to optimality. An advance in decomposition methods, ADMM provides robustness and possesses the privacy-preserving feature. Its applicability for solving the dynamic optimal power flow is shown in [20].

Various optimisation models [18, 20] have been proposed for different types of devices. They are generic models that incorporate device cost function and constraints. One of the main objectives of our approach is to take into account also the revenues from participating in demand-response programs, and thus incorporating demand-response incentives in the models. Many works have been done to address demand response in building energy optimisation such as [21, 27, 31], but they focus mainly on price-based demand response such as real-time pricing or time-of-use pricing. Optimisation considering the participation in different types of incentive-based demand-response programs, however, is much less investigated. Existing works addressing incentive-based demand response such as [25] and [29] propose models that take into account an incentive offered to the consumer during peak times. This method motivates consumers to modify their behaviour the same manner as the energy tariff is used in price-based demand response. It is incompatible for more elaborated demand-response requests such as load shedding and load shifting that require a specific amount of consumption to be shed or shifted in some specified time interval. In [25], the authors go further to include also the notion of consumers' inconvenience level in the model to represent the disparity between the base consumption and the deviated consumption as a result of modifying the consumption in response to the incentive. The inconvenience is an important notion that we will also consider in our approach.

Multi-agent systems (MAS) is a powerful tool for modelling and developing complex and distributed systems due to the properties offered by the agents, namely autonomy, social ability, reactivity, and pro-activity [13]. The distributed nature of problem solving inherent in MAS introduces reliability in the event of agent failure as tasks can be reassigned to other agents [8]. MAS has been proven an efficient solution to tackle problems in the context of power systems as shown in various works [2, 5–7, 17, 22, 30]. Owing to the aforementioned properties of MAS, in our approach, MAS is used as an abstraction for modelling, designing, and implementing the distributed optimisation algorithm, namely ADMM. Agents' ability to autonomously operate and to interact or communicate with one another, enabling privacy for each decision-making entity and coordination among the agents, renders MAS compatible for our approach. Each agent performs its local optimisation based on their model that incorporates demand-response incentives as well as consumers' inconvenience when applicable.

4.4 Demand-Response: Models, Incentives, and Scales

In this section, we provide the conceptual models for different types of demand-response requests encapsulating the parameters necessary for a customer to adapt their consumption pattern to respond to the requests. An analysis on demand-response incentives, their values, and scales are provided.

4.4.1 Price-Based Demand-Response Model

A price-based demand response is essentially a tariff (e.g., time-of-use or real-time pricing). The tariff is modelled according to the optimisation time horizon and sampling time. For instance, in a case where the optimisation is done for the next 24-h horizon with 15-min sampling time, the tariff consists of energy prices for the next 24 h discretised into 15 min.

4.4.2 Incentive-Based Demand Response

Incentive-based demand-response requests are defined with respect to the base consumption, which is the expected imported energy of a customer (i.e., predicted energy import). They provide incentive payments to motivate customers to modify their consumption patterns.

4.4.2.1 Load Shedding Request Model

A load shedding request is characterised by the following parameters:

- deviation period: the time period in which the requested shedding is to be executed,
- deviated consumption: the target consumption after shedding. This parameter is expressed with respect to the base consumption (e.g., shed 10% from the base consumption),
- and incentive: the payments rewarded to the customer provided that the load shedding request is fulfilled.

4.4.2.2 Load Shifting Request Model

A load shifting procedure consists of two phases: deviation phase and recovery phase. In the deviation phase, a specified amount of consumption is to be reduced from a given time interval defined as the deviation period. In the recovery phase, the amount of consumption reduced from the deviation period is supposed to be consumed within a given time interval of the recovery phase, called the recovery period. A load shifting request is characterised by the following parameters:

- deviation period: the time period during which the consumption is to be shifted,
- deviated consumption: the target consumption after shifting,
- recovery period: the time period during in which the shifted consumption is recovered (i.e., consumed in addition to the base consumption),
- and incentive: the payments rewarded to the customer provided that the load shifting request is fulfilled.

4.4.2.3 Incentives, Inconvenience, and Scales

Customers may experience inconvenience when adapting their consumption to respect demand-response requests, especially so in the case of load shedding. Inconvenience may be defined as the disparity between the base consumption and the consumption following demand-response requests, which reflects consumers' discomfort. To make an optimal decision whether to shift or shed the consumption, we factor in the inconvenience as well as the incentives.

Load shedding is carried out by reducing the consumption of sheddable loads. The model for sheddable loads factoring in the incentive and the inconvenience are provided later in Sect. 4.6.2.2. Similarly, load shifting is executed by shifting the consumption of shiftable loads. The model incorporating the incentive for load shifting is provided in Sect. 4.6.2.6. In the proposed models, we require that the values of the incentives and inconvenience are expressed in the same unit as the energy price since the system currently optimises the consumption to reduce energy bills. However, it is noteworthy that the values may be expressed in other units depending on the objective of the optimisation (e.g., reducing carbon footprint).

4.5 Distributed Building Energy Optimisation

In this section, we provide a conceptual model of the building that captures the integral aspects considered in the optimisation. Then, we describe concisely the optimisation problem and the algorithm upon which we base our system to solve the problem.

4.5.1 Conceptual Model of the Building

To capture the elements of the building that are integral in energy management, we model a given building as follows.

Definition 1 (Model of the Building) Let U be a set of users in a given building E and D a set of devices consuming or producing energy in the building. Let further C be a set of device constraints and *UPref* the preferences of users on how each device should operate (e.g., interval in which a device is allowed to operate). Then, a building defined as follows:

$$E = (U, D, C, UPref, S) \tag{4.1}$$

where

- U is a set of users in E;
- D is a set of devices located in E and considered in the optimisation. We make a separation between static devices $SD \subset D$, whose state is either constant or has no impact on the optimisation, and dynamic devices $DD \subset D$, whose state may evolve over time;
- C is a set of constraints of the devices in E. A device $d \in D$ has a set of constraints $C_d \subset C$;

- *UPref* represents a set of user preferences on the devices in E. A user $u \in U$ owns a set of devices $D_u \subset D$, and has a set of preferences on their devices $UPref_u \subset UPref$. For each device they own $d_u \in D_u$, the user may configure d_u based on their preferences $UPref_u^d \subset UPref_u$;
- S represents the external sources supplying energy for the building.

4.5.2 Dynamic Optimal Power Flow Problem

The devices, whose energy consumption or production profile is to be optimised, are connected at the building level, forming a network of devices. Figure 4.1 depicts an example of the network. Optimising the energy flow in such a network is to minimise the network objective function subject to the constraints of each device in the network.

We model the network as an energy coordination network [20] composed of a set of terminals T, a set of devices D, and a set of nets N. A terminal models a transfer point through which the energy flows between a device and a net. A net represents an exchange zone that constrains the energy schedules of its associated devices. Each device and each net is associated with a set of terminals.

Each terminal $t \in T$ has an associated energy flow schedule $p_t = ((p_t(1), \ldots, p_t(H)) \in \mathbb{R}^H$ over a time horizon $H \in \mathbb{N}^+$. A time horizon is a set of time periods. For instance, a time horizon of 24 h consist of 96 time periods, with the interval of 15 min between each time period. Then, $p_t(\tau)$ where $\tau \in [1, H]$

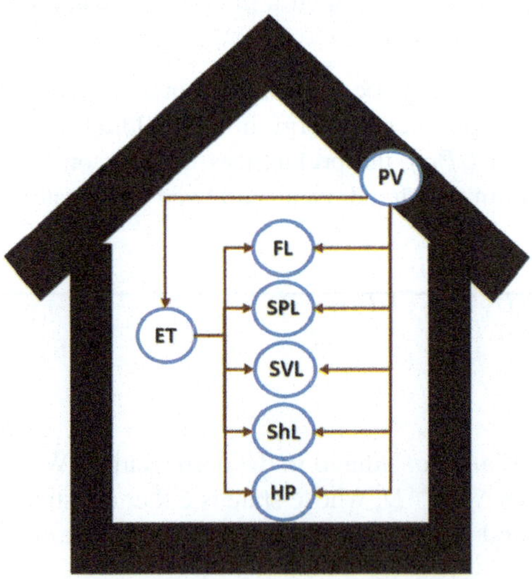

PV	Photovoltaic
ET	External tie
FL	Fixed load
SPL	Shiftable-profile load
SVL	Shiftable-volume load
ShL	Sheddable load
HP	Heat pump

Fig. 4.1 Example of energy-consuming and energy-producing devices in a building

is the amount of energy consumed ($p_t(\tau) > 0$) or generated ($p_t(\tau) < 0$) by device $d \in D$ in time period τ through terminal t, where t is associated with d.

Each device $d \in D$ has a set of energy schedules denoted by $p_d = \{p_t | t \in d\}$, which can be associated with a $|d| \times H$ matrix. The set of all energy schedules associated with a net $n \in N$ is denoted by $p_n = \{p_t | t \in n\}$. Using the same notation, we denote the set of all terminal energy schedules by $p = \{p_t | t \in T\}$, which we can associate with a $|T| \times H$ matrix.

For each device $d \in D$, we use $'d'$ to refer to both the devices and the set of terminals associated with the device. Each device d possesses a set of $|d|$ terminals and has an objective function $f_d : \mathbb{R}^{|d| \times H} \to \mathbb{R}$. Then, $f_d(p_d)$ extends over the time horizon H and encodes the cost (e.g., energy cost, fuel consumption, or carbon emission) of operating device d according to the schedule p_d. Furthermore, every device has a set of constraints C_d which p_d must satisfy. Similarly, each net $n \in N$ has a set of $|n|$ terminals, an objective function $f_n : \mathbb{R}^{|n| \times H} \to \mathbb{R}$, and a set of constraints C_n to satisfy. Provided an energy coordination network, we define the optimisation problem as follows:

$$
\begin{aligned}
& min_{p \in \mathbb{R}^{H \times |T|}} \sum_{d \in D} f_d(p_d) + \sum_{n \in N} f_n(p_n) \\
& subject\ to\ p_d \in C_d, \quad \forall d \in D \\
& p_n \in C_n, \quad \forall n \in N
\end{aligned}
\tag{4.2}
$$

The previous example of a building energy network can be transformed as an energy coordination network as shown in Fig. 4.2. The devices consuming or producing electrical power are connected to the electricity net (*E-Net*). Heat pump

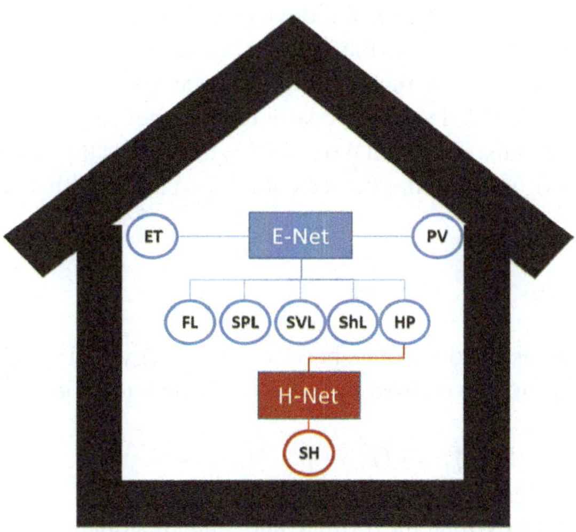

E-Net	Electricity net
H-Net	Heat net
SH	Space heating

Fig. 4.2 Example of a building energy coordination network

(*HP*) converts electrical power to heat and supplies the heat to space heating (*SH*) via a heat net (*H-Net*).

4.5.3 Alternating Direction Method of Multipliers

To solve the optimisation problem specified in Eq. (4.2), we implemented a solution based on ADMM. ADMM iteratively solves the problem until the convergence is reached. In each iteration, ADMM performs the following steps:

Step 1 Device-minimisation step (executed in parallel by each device)

$$\forall d \in D,\ p_d^{k+1} = argmin_{p_d \in C_d}(f_d(p_d) + \frac{\rho}{2}||p_d - \dot{p}_d^k + v_d^k||_2^2) \tag{4.3}$$

Step 2 Net-minimisation step (executed in parallel by each net)

$$\forall n \in N,\ \dot{p}_n^{k+1} = argmin_{p_n \in C_n}(f_n(p_n) + \frac{\rho}{2}||p_n - p_n^k - v_n^k||_2^2) \tag{4.4}$$

Step 3 Price update (i.e., scaled dual variables) (executed in parallel by each net)

$$\forall n \in N,\ v_n^{k+1} = v_n^k + (p_n^{k+1} - \dot{p}_n^{k+1}) \tag{4.5}$$

In the first step, each device computes in parallel its best response to the price and energy requested by nets. At any iteration $k+1$, p_d^{k+1} represents the device's response to the request $<\dot{p}_d^k + v_d^k>$. Upon receiving the offers from all the devices connected to it, each net checks if the convergence has been reached. If there is no convergence, nets compute a new request for the devices considering the devices' previous offers and send the new request to the devices, which corresponds to the second step. In the third step, nets update the scaled dual variables.

The convergence criteria for ADMM applied locally for each net are based on the net's local primal and dual residuals. The primal residual is defined as the difference between the energy schedule offered by the devices and those required by the net, formally:

$$\forall n \in N,\ r_n^k = p_n^k - \dot{p}_n^k \tag{4.6}$$

The dual residual is defined as the difference between the energy required by the net in the two consecutive iterations weighted by a scaling parameter, formally:

$$\forall n \in N,\ s_n^k = \rho(\dot{p}_n^k - \dot{p}_n^{k-1}) \tag{4.7}$$

The stopping criterion requires that the primal and dual residuals are small, formally:

$$||r_n^k||_2 \leq \epsilon_n^{pri} \; and \; ||s_n^k||_2 \leq \epsilon_n^{dual} \tag{4.8}$$

where $\epsilon_n^{pri} > 0$ and $\epsilon_n^{dual} > 0$ are, respectively, the feasibility tolerance for the primal and dual feasibility conditions. These tolerances can be chosen using an absolute and relative criteria such as

$$\begin{aligned} \epsilon_n^{pri} &= \sqrt{|r^k|\epsilon^{abs}} + \epsilon^{rel} max\{||p_n^k||_2, ||\dot{p}_n^k||_2\} \\ \epsilon_n^{dual} &= \sqrt{|s^k|\epsilon^{abs}} + \epsilon^{rel} ||\rho v_n^k||_2 \end{aligned} \tag{4.9}$$

where $|x|$ returns the cardinality of vector x. The choice of the absolute stopping criterion (i.e., ϵ^{abs}) depends on the scale of the typical variable values. Semantically, ϵ^{abs} stands for the maximum error allowed for each constraint when assuming that errors are distributed uniformly among nets. For the relative stopping criteria, a reasonable value might be $\epsilon^{rel} = 10^{-3}$ or 10^{-4}, depending on the application. The cost functions and constraints of the devices and nets differ according to their types, and are presented in the following section.

4.6 Multi-Agent System for Energy Management

The distributed energy optimisation is modelled as a multi-agent system, enabling the coordination and the execution of the tasks necessary to perform the optimisation. This section presents the conceptual model of the system, how the system takes into account demand-response requests in the optimisation, and how it addresses device constraints and the privacy concerns.

4.6.1 System Conceptual Model

The multi-agent energy management system (MEM) is composed of agents assuming different roles required to carry out the optimisation. The conceptual model of the system is a tuple:

$$MEM = (NA, DA) \tag{4.10}$$

where

- *NA* is the model of net agents assuming the role of the nets in the energy coordination network;
- *DA* is the model of device agents representing the devices in the network.

The role of an agent determines the knowledge it possesses, the actions it can perform, and the behaviour it comports. Each of the roles in the system is described in the following subsections.

4.6.1.1 Net Agent Model

A net agent assumes the role of a net modelled in the energy coordination network. Its role is to ensure that, in each time period, there is a balance between the energy flowing into and out of its terminals. To enable net agents to perform their role, each net agent is modelled as follows:

$$NA = (Opt_{net}, C_{net}, Comm_{net}) \tag{4.11}$$

where

- Opt_{net} is the optimisation steps, namely net-minimisation (Eq. 4.4) and price update (Eq. 4.5), executed locally by a net agent;
- C_{net} represents the energy balance constraint upheld by a net. Nets are lossless energy carriers. They require that in each time period there is a balance between energy flowing into and out of their terminals. Such a constraint is expressed formally as below:

$$\sum_{t \in n} p_t(\tau) = 0, \tau = 1, \ldots, H \tag{4.12}$$

- $Comm_{net}$ is the communication ability of the agent, enabling it to send and receive messages with other agents. Concretely, this corresponds to the implementation of the functions required for sending, receiving, and processing messages. In some agent-based platforms, such functions are already available and ready to be used when developing the agents.

Each net agent iteratively computes the optimisation steps Opt_{net} until a convergence is reached, while ascertaining that C_{net} is respected.

4.6.1.2 Device Agent Model

The role of a device agent, which represents a device in the energy coordination network, is to perform a local optimisation, while respecting the constraints of the device and its user(s). The optimisation result is communicated to its associated net agent(s). Formally, the model of a device agent is as follows:

$$DA = (Opt_{dev}, C_{dev}, Pref, I_{dev}, Comm_{dev}) \tag{4.13}$$

where

- Opt_{dev} is the device-minimisation step (Eq. 4.3) executed by the device agent. The cost function considered in this step varies according to the device type (e.g., fixed load, shiftable load, or sheddable load). The cost function for each type is presented in the following subsection;
- C_{dev} represents the constraints of the device to be satisfied. The constraints are device-specific and presented in the following subsection;
- *Pref* consists of the preferences of the device's user(s) regarding the manner in which the device may operate (e.g., the interval in which the device is allowed to be switched on or off);
- I_{dev} refers to additional information required by the device agent to perform the optimisation (e.g., energy prices);
- $Comm_{dev}$ is the communication capacity of the agent required for communicating with its associated net agent(s).

Each device agent executes its device-minimisation step iteratively considering the updates from its associated net(s).

4.6.2 Optimisation Models for Demand-Respond Ready Agents

In this section, we provide a detailed description of the device optimisation models for different types of device flexibility, which include the cost functions and the constraints of the devices as well as demand-response incentives and inconvenience when applicable.

4.6.2.1 Fixed Load

A fixed load (FL) [20] models a device whose power consumption profile must be satisfied, thus providing no flexibility. Fixed loads have a zero cost function. Their sole constraint $c_{FL} \in C_{dev}$ ensures that their required consumption p_{req} is satisfied in each time period, formally:

$$p_{FL}(\tau) = p_{req}(\tau), \tau = 1, \ldots, H \tag{4.14}$$

where p_{FL} is the actual consumption of FL.

4.6.2.2 Sheddable Load

A sheddable load (ShL) [20] models a device whose consumption can be shed, to a certain extent and at a cost, when there is a need to reduce the demands (e.g., lighting system - on/off or selecting a lower level of brightness). The cost function of sheddable loads considers the inconvenience $p_{inc} \in Pref$ resulting from the shedding and the demand-response incentive p_{shed}, formally:

$$f_{ShL}(\tau) = (p_{inc}(\tau) - p_{shed}(\tau)) * (p_{req}(\tau) - p_{ShL}(\tau)) \tag{4.15}$$

where $p_{shed}(\tau) \geq 0$, p_{req} is the required consumption, and p_{ShL} the actual consumption after shedding. In this model, for a shedding to be carried out at time period τ, the value $p_{shed}(\tau)$ needs to be big enough to compensate for the inconvenience such that:

$$p_{inc}(\tau) - p_{shed}(\tau) < energy\ price(\tau) \tag{4.16}$$

In this way, the algorithm chooses to shed the consumption as the shedding cost is inferior to the price for importing energy to fulfill the load's consumption if it were not shed. ShL's constraint $c_{ShL} \in C_{dev}$ ensures that the consumption after shedding does not surpass the required consumption, formally:

$$0 \leq p_{ShL}(\tau) \leq p_{req}(\tau), \tau = 1, \ldots, H \tag{4.17}$$

4.6.2.3 Shiftable-Volume Load

A shiftable-volume load (SVL) [20] represents a device that must consume a certain amount of energy (i.e., the volume) within a given time interval. A distinct property of shiftable-volume loads is that there is no constraint on the consumption in each time period as long as the total consumption required is met during the given interval. For example, a battery-powered vacuum cleaner can be considered as a shiftable-volume load as it requires a certain amount of consumption to charge the battery but can be spread over a time interval.

The time interval allowed to activate the device is a user preference $pref_{SVL} \in Pref$ and imparted to the device agent as part of its knowledge. Shiftable-volume loads have a zero cost function. First, they encode a hard constraint $c_{SVL-1} \in C_{dev}$ mandating that the required amount of consumption V is satisfied within a time interval between a given earliest time period A and a latest time period D, formally:

$$\sum_{\tau=A}^{D} p_{SVL}(\tau) = V \tag{4.18}$$

where p_{SVL} is the actual consumption of SVL. The second constraint $c_{SVL-2} \in C_{dev}$ limits the consumption in each time period by a specified amount L^{max}, expressed as follows:

$$0 \leq p_{SVL}(\tau) \leq L^{max}, \tau = A, \ldots, D \tag{4.19}$$

4.6.2.4 Shiftable-Profile Load

A shiftable-profile load (SPL) models a device whose consumption can be shifted within a given time interval. The difference from shiftable-volume loads is that shiftable-profile loads have a constraint on the amount of energy consumed in each time period. For instance, washing machines are a typical example of shiftable-profile loads. The activation of the machines can be scheduled within an interval, but once activated, it requires a specific amount of consumption in each time period throughout its running time.

As in the case of SVL, the time interval is provided by the device's user $pref_{SPL} \in Pref$ and passed to the device agent as part of its knowledge. Shiftable-profile loads have a zero cost function. They encode a hard constraint $c_{SPL-1} \in C_{dev}$ requiring the consumption p_{SPL} to be within a given time interval between the earliest time period A and the latest time period D, formally:

$$\begin{aligned} p_{SPL}(\tau) &= 0, \tau = 1, \ldots, (A-1) \\ p_{SPL}(\tau) &= 0, \tau = (D+1), \ldots, H \end{aligned} \tag{4.20}$$

where p_{SPL} is the actual consumption of SPL. The second constraint $c_{SPL-2} \in C_{dev}$ ensures that the actual consumption matches the required profile p_{req}, formally:

$$\begin{aligned} \bigcup_{\tau=A}^{D-\delta} &\Big(p_{SPL}(\tau) = p_{req}(1)\Big) \cap \Big(p_{SPL}(\tau+1) = p_{req}(2)\Big) \cap \ldots \\ &\cap \Big(p_{SPL}(\tau+\delta-1) = p_{req}(\delta)\Big) \end{aligned} \tag{4.21}$$

where $\delta = |p_{req}|$.

4.6.2.5 (Curtailable) Generator

A generator represents a source of locally produced energy (e.g., photovoltaics). A curtailable generator (CG) refers to a source of energy whose production can be curtailed. The cost function of curtailable generators incorporates a curtailment cost $p_{curt} \in Pref$, formally defined as follows:

$$f_{CG}(\tau) = p_{curt}(\tau) * (p_{gen}(\tau) - p_{CG}(\tau)) \tag{4.22}$$

where p_{gen} is the expected generation and p_{CG} the generation after curtailment. CG encodes the following constraint $c_{CG} \in C_{dev}$:

$$p_{CG}(\tau) \leq p_{gen}(\tau), \tau = 1, \ldots, H \tag{4.23}$$

A non-curtailable generator has a zero cost function. Its sole constraint ensures that the actual energy flow from the generator p_G matches the energy produced by the generator p_{gen}, formally described as follows:

$$p_G(\tau) = p_{gen}(\tau), \tau = 1, \ldots, H \tag{4.24}$$

4.6.2.6 External Tie

An external tie (ET) [20] is not an actual physical device. It is an abstract notion used to represent a connection of the building to an external source of energy. Transactions with ET consist in pulling energy from the source or injecting energy to it. Its cost function factors in the prices of importing $P^{imp} \in I_{dev}$ and exporting $P^{exp} \in I_{dev}$ energy defined as follows:

$$\begin{aligned} p_{ET}(\tau) < 0, \quad & f_{ET}(\tau) = P^{exp}(\tau) * p_{ET}(\tau) \\ & p_{ET}(\tau) = 0, \quad f_{ET}(\tau) = 0 \\ p_{ET}(\tau) > 0, \quad & f_{ET}(\tau) = (P^{imp}(\tau) + p_{shift}) * p_{ET}(\tau) \end{aligned} \tag{4.25}$$

where $p_{ET}(\tau)$ is the amount of pulled (positive value) or injected (negative value) energy at a given time period τ. Price-based demand-response (i.e., energy tariffs) is considered in the model as P^{imp}.

The incentive for load shifting p_{shift} is also incorporated in the model. A positive value of p_{shift} incentivises the reduction of consumption which can be used to shift the consumption for load shifting, while a negative value encourages consumption, applicable for recovering the shifted load in the recovery period. The constraint $c_{ET} \in C_{dev}$ encoded by ET restricts importing and exporting energy by some specified limit P^{max}, expressed formally as follows:

$$|p_{ET}(\tau)| \leq P^{max}, \tau = 1, \ldots, H \tag{4.26}$$

4.6.2.7 Converter

A converter (CON) [20] models the device that transforms energy from one form A to another form B. Some instances of converters are gas boiler and heat pump. Therefore, converters are connected to two different energy networks. Converters have a zero cost function. The first constraint $c_{CON-1} \in C_{dev}$ to be satisfied is the conversion equation based on the conversion efficiency k of the converter. The constraint is formally expressed as follows:

$$-p_B(\tau) = k \cdot p_A(\tau), \tau = 1, \ldots, H \tag{4.27}$$

where $p_A(\tau)$ is the input energy profile and $p_B(\tau)$ the output energy profile at time period τ. Note that the value of $p_B(\tau)$ is negative as it models the generated energy, and not the consumption. The second constraint $c_{CON-2} \in C_{dev}$ assures that the amount of energy input at each time period of the transaction does not exceed some specified limit, which is described as follows:

$$C^{min}(\tau) \leq p_A(\tau) \leq C^{max}(\tau), \tau = 1, \ldots, H \tag{4.28}$$

where $C^{max}(\tau)$ is the maximum energy input at a given time period τ and $C^{min}(\tau)$ the minimum energy input.

4.6.2.8 Space Heating

A space heating or a thermal load [20] models a physical space (e.g., room, building) with temperature profile T and whose temperature must be kept within minimum T^{min} and maximum T^{max} temperature limits. The temperature of the space evolves as follows:

$$T(\tau+1) = T(\tau) + \left(\frac{\mu}{c}\right)\left(T^{amb}(\tau) - T(\tau)\right) - \left(\frac{\eta}{c}\right)p_{therm}(\tau), \tau = 1, \ldots, H-1 \tag{4.29}$$

where $T(1)$ is the initial temperature of the space, T^{amb} the outdoor temperature, μ the ambient conduction coefficient, η the heating or cooling efficiency, c the heat capacity of indoor air, and p_{therm} the heating or cooling power consumption profile.

Thermal loads have a zero cost function. First, a thermal power input has power schedule p_{therm} that is constrained as follows:

$$0 \leq p_{therm}(\tau) \leq Q^{max}, \tau = 1, \ldots, H \tag{4.30}$$

where $Q^{max} \in \mathbb{R}^H$ is the maximum heating power. Second, its temperature profile $T \in \mathbb{R}^H$ must be kept between a minimum temperature $T^{min} \in \mathbb{R}^H$ and a maximum temperature $T^{max} \in \mathbb{R}^H$, formally described as follows:

$$T^{min}(\tau) \leq T(\tau) \leq T^{max}(\tau), \tau = 1, \ldots, H \tag{4.31}$$

4.6.3 *Analysis on Constraints and Privacy*

4.6.3.1 Constraints and User Preferences

In the proposed multi-agent model, device constraints, defined previously in the device agent model as C_{dev}, are expressed as hard constraints to be satisfied in the optimisation. Such constraints are imparted to each corresponding device agent as a part of its knowledge when the agent is instantiated. Thus, while conducting its local optimisation, each agent can ensure that the constraints are not violated. Similarly, user's preferences *Pref* for each device are encoded as hard constraints in the device agent model, enabling the agents to ensure its upholding.

4.6.3.2 Privacy

User preferences and constraints for a device are known only to the device agent representing the device. Such data are stored locally and used in the agent's local optimisation process. They are not shared with neither other device agents nor net agents. Moreover, employing ADMM enables the coordination exchanges in which the preferences, the constraints, and the cost structures of each device are reflected in its private cost function. More precisely, as described in Sect. 4.5.3, the coordinated messages p_d^{k+1} sent to the associated net agent are simply the representation of the device agent's response to the net agent's request $< \dot{p}_d^k + v_d^k >$.

4.7 Validation of the Approach

We implement the proposed multi-agent energy optimisation system using JADE (Java Agent DEvelopment),[1] a framework for developing agent-based applications. JADE provides a simple yet complete agent abstraction, supports asynchronous peer-to-peer agent communication, and facilitates the overall development of a distributed system. Tasks to be carried out by an agent are implemented as *behaviours* to be executed by the agent. Agent knowledge can be passed as *arguments* to an agent when it is instantiated. All the experiments are executed on an Intel Core i7 CPU at 1.90 GHz computer with a 16 GB RAM. Various experiments are conducted to validate the system with regard to the following criteria: reduction of energy bills and the participation to demand-response programs, while ensuring that user preferences and device constraints are satisfied and privacy preserved.

4.7.1 Case Study

The evaluations are conducted using a case study which represents a prosumer building with a connection to an energy supplier (i.e., external tie) and equipped with a photovoltaic (PV) for local uses. The building has a fixed amount of consumption (i.e., fixed load) that must be satisfied. It also provides a certain amount of flexible consumption (i.e., shiftable load and sheddable load) that can be shifted or shed to some extent over a given time frame. Various scenarios, with different levels of

[1] https://jade.tilab.com/.

flexibility, are used in the experiments. The configurations of the experiments for each scenario are as follows:

- Time horizon: the time horizon for the experiments is 24 h divided into 96 time periods (TP) of 15-min interval;
- Building consumption: for the next 24-h predicted consumption of the building, we use, as sample data, the consumption data from UK Elexon's non-domestic unrestricted consumers[2];
- Fixed load: in each scenario, the fixed consumption represents a certain percentage of the overall consumption. It varies depending on the scenario;
- Flexible load: for each scenario, a certain percentage of the overall consumption is considered flexible. The percentage varies depending on the scenario;
- PV production: the production of the PV is based on the data from one of our projects;
- Energy tariffs: we use the two-banded Time-Of-Use tariffs from EDF (i.e., an energy supplier in France) consisting of a peak price (from Monday to Friday between 7:00 and 23:00): €0.158/kWh and an off-peak price (from Monday to Friday between 23:00 and 7:00 and weekends starting from Friday 23:00 until Monday 7:00): €0.11/kWh.

In the experiments, flexibility is concretely modelled by means of shiftable devices and sheddable devices, whose consumption accounts for the amount of flexibility provided in each scenario. User preferences such as the earliest and the latest TPs to activate each shiftable device and device constraints, based on the technical specifications of the devices, are encoded in the corresponding device agents. It is important to note that each device agent is instantiated with the knowledge of its user preferences and device constraints, unknown to other agents, in order to preserve the privacy. The output of the optimisation is a schedule for all the devices over the time horizon, essentially indicating whether to turn on or turn off each device at the beginning of each 15 min.

4.7.2 Reduction of Energy Bills Through Flexibility and Price-Based Demand-Response

In the first validation, we experiment with four different scenarios, with varying percentages of flexibility proportional to the total consumption of the building. The first scenario has no flexibility at all, second with 20% flexibility, third with 50% flexibility, and fourth with 100% flexibility. Figure 4.3 depicts the energy imported for each of the scenarios. The imported energy is the energy pulled from the supplier via the external tie. Since the building also produces energy, the imported energy is the amount required after self-consumption using the energy from PV.

[2]https://www.elexon.co.uk/knowledgebase/profile-classes/.

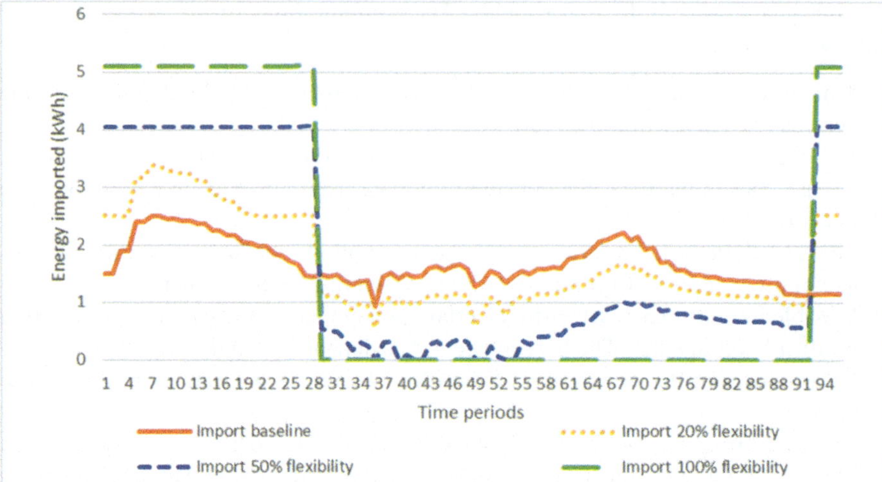

Fig. 4.3 Comparison of energy imported in function of different levels of flexibility

In the first scenario (shown as Import baseline), we suppose that all the consumption is fixed, providing no flexibility. Therefore, the system is not able to optimise the consumption based on the energy tariffs. By introducing flexibility, the system can optimise the flexible consumption to benefit the off-peak price, the result of which can be seen in Fig. 4.3. The amount of imported energy during the peak period between TP 29 (i.e., 7:00) and TP 91 (i.e., 23:00) decreases as the flexibility increases, with no import when the flexibility reaches 100% (i.e., no fixed loads and all consumption considered flexible). The reason behind this result is that the optimisation algorithm schedules the flexible consumption in the off-peak periods.

Figure 4.4 illustrates the reduction of energy bills as the amount of flexibility increases. The reduction is nearly linear to the increment of flexibility. In the best-case scenario, in which all the consumption is flexible and can be shifted, the reduction attains over 20%. In a more realistic scenario where approximately 20%[3] of the consumption is shiftable, the energy bill decreases roughly 5%. Energy bills are the consequence of importing energy. The system exploits the provided flexibility to schedule the flexible consumption to benefit from both off-peak periods and self-generated energy.

It is noteworthy that, in the aforementioned results, all the flexible consumption can be scheduled freely by the system without any constraints. In reality, it is common for users to specify the interval in which the shiftable consumption is allowed to be scheduled. To validate the system's capacity to respect such preferences from users, in the second validation, we simulate user preferences on shiftable devices. For this experiment, a set of shiftable devices are only allowed to be shifted within an interval between TP 41 and 61 (i.e., 10:00–15:00).

[3]Different studies have shown that around 10–20% of demand can be time-shifted [26].

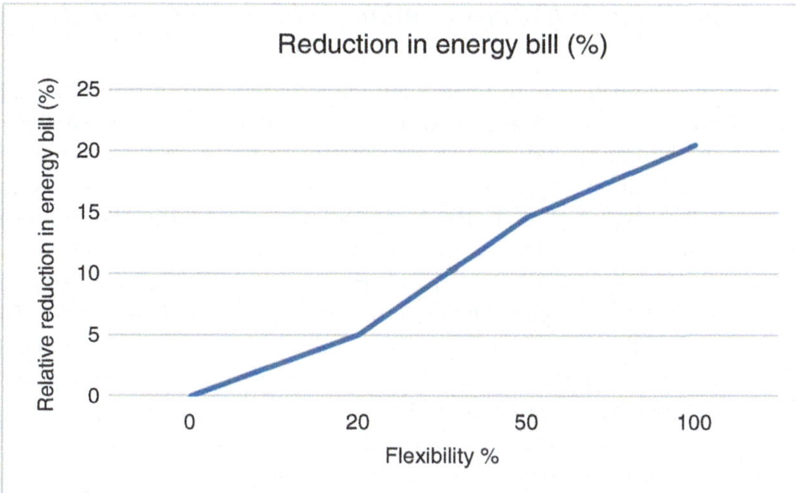

Fig. 4.4 Reduction of energy bills in function of different levels of flexibility

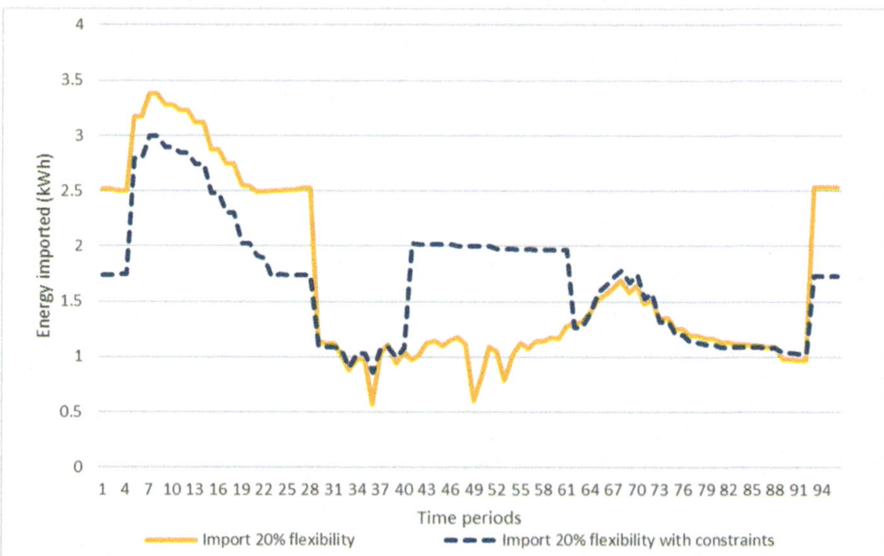

Fig. 4.5 Satisfying user preferences on shiftable loads

Figure 4.5 demonstrates the comparison of imported energy between the scenario of 20% flexibility without any constraint and that with the previously mentioned constraint. Without constraints, all the shiftable devices are scheduled to operate in the off-peak period, while, with the constraint, they are set to execute as specified in the constraint regardless of the energy price.

4.7.3 Prioritising Self-Consumption and Injection to Grids

The aim of this validation is to investigate the system's ability in optimising the consumption of the energy self-produced by the building (i.e., generated by PV installed in the building). We conduct an experiment based on the same use case as in the first validation. In this experiment, we consider the worst-case scenario in which all the consumption is fixed, providing no flexibility. The motivation behind this choice is to evaluate self-consumption independently of the flexibility. As demonstrated in Fig. 4.6, the energy imported between TP 22 and TP 70 is lower than the consumption, as depicted by the Import baseline curve in the figure. This is due to the fact that the optimisation prioritises the use of energy from PV and only imports the rest. It is important to note that in this scenario all the energy generated by PV is consumed even without flexibility as the amount of consumption is much more significant than the amount of self-produced energy throughout the whole 24 h. Therefore, introducing flexibility has no impact on the result.

Flexibility becomes essential for self-consumption when the self-produced energy is more than the consumption during certain period of the horizon. In such a case, the optimal decision is to shift the flexible consumption from other periods to the period with residual self-produced energy. To study this particular case, we conduct an experiment with the same configuration as the previous one, except that we use a lower amount of consumption which is based on the consumption

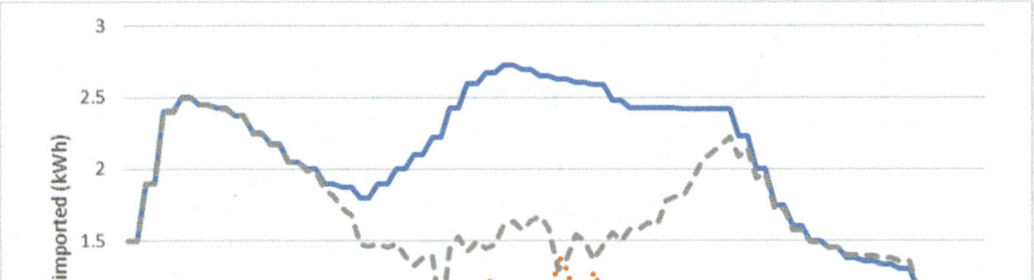

Fig. 4.6 Self-consumption without flexibility

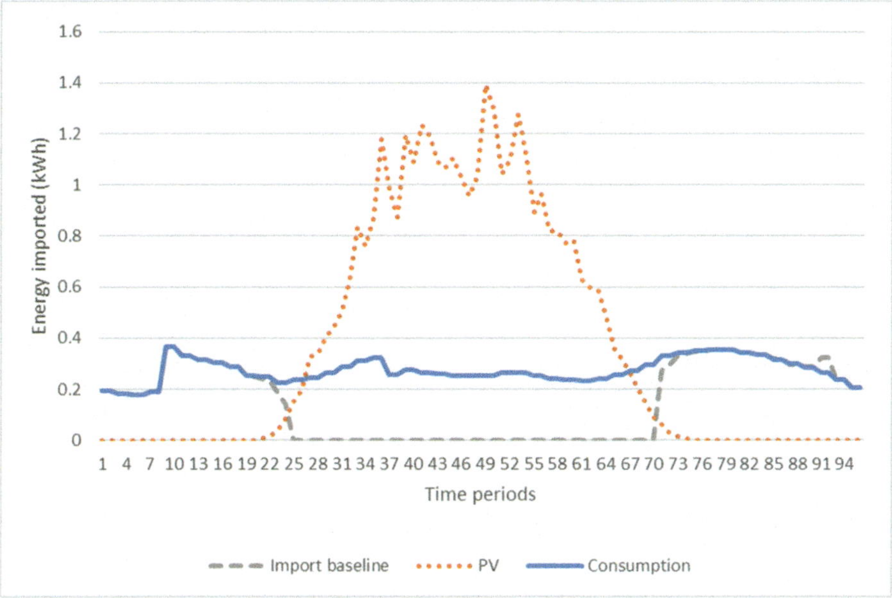

Fig. 4.7 Domestic self-consumption without flexibility

data from UK Elexon's domestic economy 7 consumer[4] for the autumn season. The results shown in Fig. 4.7 show the consumption, PV production, and import without any flexibility. It is worth noticing that between TP 25 and TP 70, there is no power imported. The reason is that the power produced by PV exceeds the required consumption.

By introducing flexibility, the system is able to schedule the flexible consumption in the period where there is more production from PV to benefit from the locally generated power as shown in Fig. 4.8. The percentage of self-consumption after optimisation in function of the flexibility is demonstrated in Fig. 4.9. The explanation for such a result is that when there is more flexible consumption, more loads can be shifted to benefit the self-produced power.

In the same scenario, the energy produced from PV and left over from the local consumption is injected to the grid. The amount of exported energy for different levels of flexibility is depicted in Figs. 4.10 and 4.11. The results show that with more flexibility the optimisation leads to more self-consumption, and thus less energy exported to the grid. A particularity in this scenario is that there is approximately 9 kWh of energy exported even though the flexibility is 100%. This is owing to the fact that PV produces more power than the total consumption over the 24-h horizon.

[4] https://www.elexon.co.uk/knowledgebase/profile-classes/.

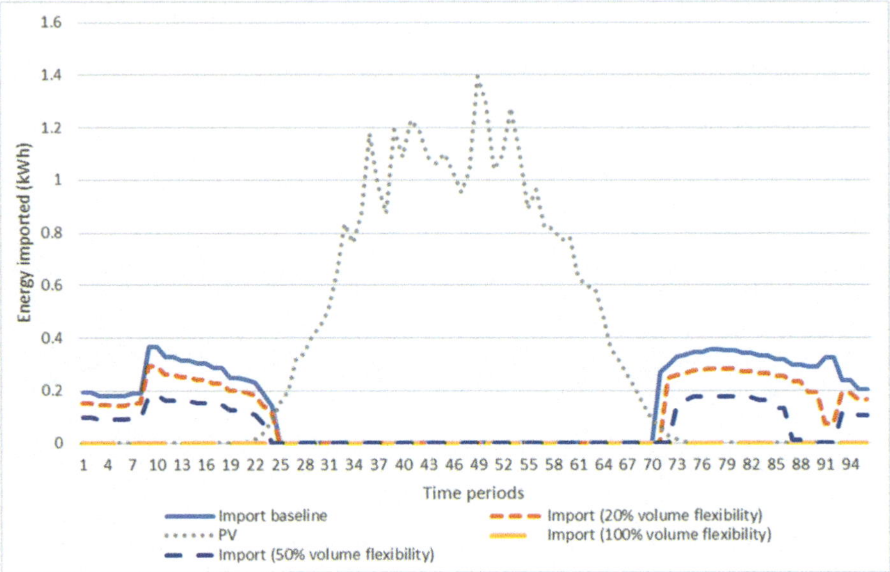

Fig. 4.8 Domestic self-consumption with flexibility

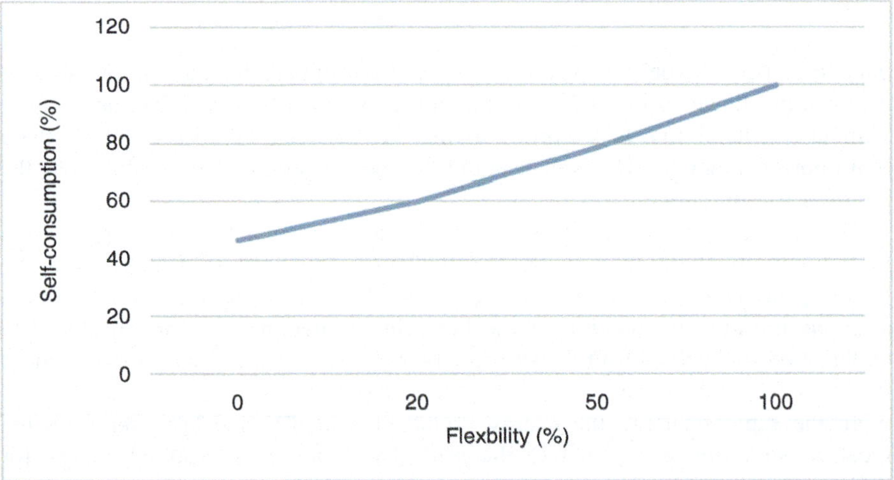

Fig. 4.9 Percentage of self-consumption after optimisation in function of different levels of flexibility

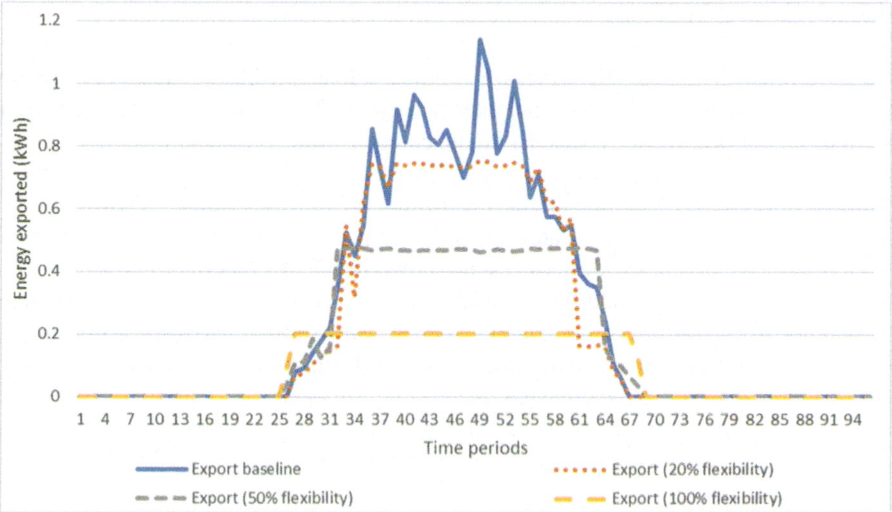

Fig. 4.10 Energy injected to grid over a 24 h horizon in function of different amounts of flexibility

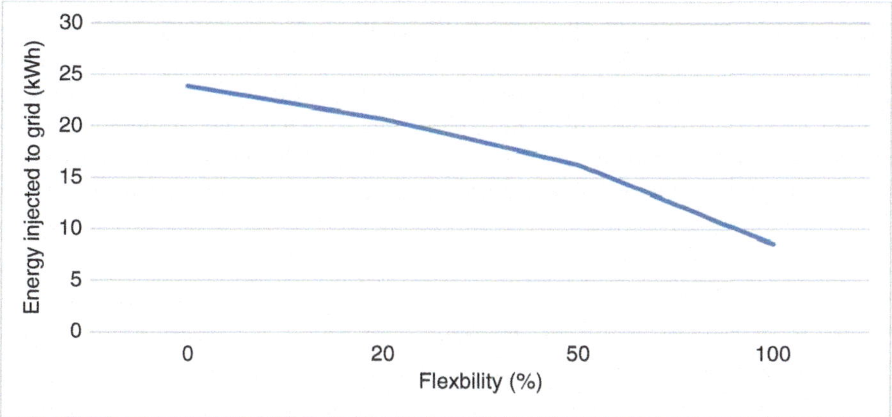

Fig. 4.11 Total energy injected to grid in function of different levels of flexibility

4.7.4 Participation to Incentive-Based Demand-Response

This validation investigates the ability of the system in exploiting the flexibility to participate in incentive-based demand-response programs including load shedding and load shifting requests.

4.7.4.1 Load Shedding

To experiment with load shedding, we simulate a load shedding request to shed the consumption such that the imported energy between TP 5 and TP 14 (i.e., deviation period) is reduced to 3 kW. Various inconvenience values p_{inc} and incentive values for load shedding p_{shed} were generated in an empirical manner for the validation purpose. Figure 4.12 demonstrates the results of the experiment. Providing the incentive p_{shed} that compensates the inconvenience value during the deviation period p_{inc}, we obtain the desired result (shown in Fig. 4.12 as *Import shed*).

In reality, users may find load shedding at a certain time more acceptable than at another time. Therefore, the inconvenience level should also vary accordingly. To simulate this case, we experiment the load shedding request with varying inconvenience levels by increasing p_{inc} value between TP 8 and TP 10. Since the incentive cannot compensate the inconvenience between TP 8 and TP 10, no load shedding is carried out in that period (shown in Fig. 4.12 as *Import varying inconveniences*).

4.7.4.2 Load Shifting

This experiment aims at validating the feasibility of carrying out load shifting requests. We simulate a request to shift 10% of the imported energy between TP 1 and TP 28 (i.e., deviation period) to between TP 56 and TP 75 (i.e., recovery period). Figure 4.13 depicts the results of the experiment. To reduce the consumption in the deviation period, we provide an incentive p_{shift} that compensates the energy price during that period. This actually encourages more consumption during the deviation

Fig. 4.12 Carrying out a load shedding request

Fig. 4.13 Carrying out a load shifting request

period, which is contradictory to the objective of shifting. However, we set the import limit P^{max} in the deviation period to *Import baseline −10%*. Therefore, the consumption, though encouraged, will not surpass the limit, but attains the desired limit. The same mechanism is applied in the recovery period, except with the import limit of *Import baseline +10%*, to encourage consumption of the shifted amount from the deviation period.

4.8 Discussion

4.8.1 Handling Inapplicable Demand-Response Requests

The rewards for customers' participation to price-based demand response are directly reflected in their energy bills as the scheme is embodied in the energy tariffs. For incentive-based demand response, however, the extent to which a consumer is rewarded is more complex to determine. The remuneration can be done for consumers individually, for instance, via a contract between DSO and the consumers, or through a contract with an aggregator. Demand-response requests meant for a consumer could be generated with the knowledge of the consumer's aggregated consumption or even the overall flexibility, but not the flexibility provided at different time periods by different flexible devices. In consequence, there is a possibility that a request demands shedding or shifting of loads more than the actual consumption of sheddable or shiftable loads, respectively.

The solution to this situation depends on the contract between the consumer and the aggregator or DSO. If participation to a demand-response program means following the request in its entirety, then, in this situation, the consumer is unable to change their consumption as request, and thus is not rewarded. In a more compromising case, the consumer should try to adjust their consumption as close as possible to the request, and they are remunerated accordingly. Both of the cases are suitable with our approach. For the first case, the request (e.g., load to shed or shift) can be considered as a hard constraint, so the optimiser can either realise it, in which case the consumer gets the total payment, or not at all. For the second case, the optimisation models already incorporate the incentive, so the optimiser can find the optimal consumption considering the incentive.

4.8.2 Validation on Real Test Sites

In the framework of this research, further validation of the system will be conducted on real test sites. As mentioned in the previous section, the current validation of the approach is based on a case study, and the experiments were conducted using simulated data. The case study used for the validation is intentionally designed to bear resemblance to the real test sites from the types of devices to the consumption and production. The values for demand-response requests, incentives, and inconvenience used in the experiments were generated in an empirical manner. We are actively working with partners who are responsible for generating demand-response requests in order to calibrate the values to reflect the real cases in which the solution will eventually be deployed. It is noteworthy that there are no empirical comparisons of our system to existing solutions. The main reason is that, to the best of our knowledge, there are no solutions in the literature that take into account both price-based demand response as well as incentive-based demand response to such an elaborated extent and that share the same set of validation criteria, which can be used for the comparison.

4.9 Conclusions and Future Work

In this paper, we propose a multi-agent energy management that optimises energy consumption in a distributed and efficient manner. The aims of the system are to reduce energy bills and enable participation to various demand-response schemes, while respecting user preferences as well as device constraints, and preserving privacy. The architecture of the system modelled as a multi-agent system is proposed. A set of device optimisation models are provided to address different types of devices based on their flexibility and to incorporate demand response. Concrete specifications of the both price-based and incentive-based demand-response requests encapsulating the necessary parameters required in the optimisation are proposed. It has been shown that the system is able to reduce energy bills up to over 20% in the best-case scenario by exploiting the flexibility of the consumption and the production offered by the building to benefit the off-peak price of the time-of-use tariffs. The system is able to carry out the simulated load shifting and load shedding requests. In all the experiments, user preferences and device constraints are satisfied.

The results demonstrated in this chapter focus primarily on the building energy consumers. The benefits gained from using the approach are from consumers' perspective. The impacts of demand response on the energy management also benefit other actors at the higher levels of the chain such as aggregators and utilities. The study of the impacts of our approach in enabling buildings to participate in demand-response schemes on a larger scale such as at the district level consisting of multiple buildings is one of the directions worthy of our future investigation. Furthermore, energy storage devices such as battery and hot water storage have not been considered in the experiments. All excess energy is directly injected to the grid. However, the storage devices present an interesting potential for cost saving, especially in the situation where there is a lot of PV production, which is to be investigated in the next step of our work.

These days, the shifting towards smart buildings with the support of Internet of Things (IoT), while offering numerous advantages, poses necessary challenges that need to be addressed. Smart buildings are not limited to residential buildings, but cover a wide range of building types from connected industrial buildings to smart transit stations and airports. Such buildings are dynamic and open in the sense that devices may enter or exit the environments at any time. Furthermore, there may be multiple users in an environment, each of whom possesses a set of devices in the environment and is able to configure their devices based on their preferences. On top of this, certain devices may be activated or deactivated at any moment by their owners, requiring them to be included in or excluded from the optimisation process. The energy management system should be designed to support dynamic changes of their components and to allow dynamic additions of new components. The system should be able to detect and handle such dynamics and openness of the environment. Building upon our multi-agent infrastructure, agents could be equipped with the capability to acquire information concerning the devices (e.g., via sensors) to update their knowledge of the devices and to detect the activation or deactivation, thus allowing them to inform their associated net agent of the events to include or exclude certain devices in the next optimisation iteration. In [19], the authors present an attempt at addressing energy management in such complex and multi-user environments, handling the dynamics and openness.

Acknowledgements This research was supported by the European project TABEDE[5] (funded by European Union's Horizon 2020 research and innovation programme) and by the Commissariat à l'Énergie Atomique et aux Énergies Alternatives (CEA).

---

[5]http://www.tabede.eu/.

References

1. S. Ali, D. Kim, Building power control and comfort management using genetic programming and fuzzy logic. J. Energy South. Afr. **26**(2), 94–102 (2015)
2. K. Butler-Purry, N. Sarma, I. Hicks, Service restoration in naval shipboard power systems. IEE Proc. Gener. Transm. Distrib. **151**(1), 95–102 (2004)
3. G. Cohen, D.L. Zhu, Decomposition coordination methods in large scale optimization problems: the nondifferentiable case and the use of augmented Lagrangians, in *Advances in Large Scale Systems* (JAI Press, Greenwich, 1984), pp. 203–266
4. G.B. Dantzig, P. Wolfe, Decomposition principle for linear programs. Oper. Res. **8**(1), 101–111 (1960)
5. A. Dimeas, N. Hatziargyriou, A multi-agent system for microgrids, in *Hellenic Conference on Artificial Intelligence* (Springer, Berlin, 2004), pp. 447–455
6. A. Dimeas, N. Hatziargyriou, A MAS architecture for microgrids control, in *Proceedings of the 13th International Conference on Intelligent Systems Application to Power Systems* (IEEE, Piscataway, 2005), p. 5
7. A.L Dimeas, N.D. Hatziargyriou, Operation of a multiagent system for microgrid control. IEEE Trans. Power Syst. **20**(3), 1447–1455 (2005)
8. A. Dorri, S.S. Kanhere, R. Jurdak, Multi-agent systems: a survey. IEEE Access **6**, 28573–28593 (2018)
9. J. Eckstein, Parallel alternating direction multiplier decomposition of convex programs. J. Optim. Theory Appl. **80**(1), 39–62 (1994)
10. EURELECTRIC, Everything you always wanted to know about Demand response. Tech. rep., EURELECTRIC (2015)
11. D.A. Fisher, H.F. Lipson, Emergent algorithms-a new method for enhancing survivability in unbounded systems, in *Proceedings of the 32nd Annual Hawaii International Conference on Systems Sciences, HICSS-32*. Abstracts and CD-ROM of Full Papers (IEEE, Piscataway, 1999), p. 10
12. M. Fortin, R. Glowinski, Chapter III on decomposition-coordination methods using an augmented Lagrangian, in *Studies in Mathematics and Its Applications*, vol. 15 (Elsevier, Amsterdam, 1983), pp. 97–146
13. S. Franklin, A. Graesser, Is it an agent, or just a program? A taxonomy for autonomous agents, in *International Workshop on Agent Theories, Architectures, and Languages* (Springer, Berlin, 1996), pp. 21–35
14. D. Fuselli, F. De Angelis, M. Boaro, S. Squartini, Q. Wei, D. Liu, F. Piazza, Action dependent heuristic dynamic programming for home energy resource scheduling. Int. J. Electr. Power Energy Syst. **48**, 148–160 (2013)
15. D. Gabay, Chapter ix applications of the method of multipliers to variational inequalities, in *Studies in Mathematics and Its Applications*, vol. 15 (Elsevier, Amsterdam, 1983), pp. 299–331
16. A. Galbazar, S. Ali, D. Kim, Optimization approach for energy saving and comfortable space using ACO in building. Int. J. Smart Home **10**(4), 47–56 (2016)
17. J. Hossack, S. McArthur, J. Mcdonald, J. Stokoe, T. Cumming, A multi-agent approach to power system disturbance diagnosis, in *2002 Fifth International Conference on Power System Management and Control* (IEEE, Piscataway, 2002)
18. T.G. Hovgaard, S. Boyd, L.F. Larsen, J.B. Jørgensen, Nonconvex model predictive control for commercial refrigeration. Int. J. Control **86**(8), 1349–1366 (2013)
19. O. Kem, F. Ksontini, A multi-agent system for energy management in a dynamic and open environment: architecture and optimisation, in *IEEE/WIC/ACM International Conference on Web Intelligence, WI '19* (ACM, New York, 2019), pp. 348–352
20. M. Kraning, E. Chu, J. Lavaei, S. Boyd, et al., Dynamic network energy management via proximal message passing. Found. Trends® Optim. **1**(2), 73–126 (2014)

21. N. Li, L. Chen, S.H. Low, Optimal demand response based on utility maximization in power networks, in *2011 IEEE Power and Energy Society General Meeting* (IEEE, Piscataway, 2011), pp. 1–8
22. S.D. McArthur, E.M. Davidson, J.A. Hossack, J.R. McDonald, Automating power system fault diagnosis through multi-agent system technology, in Proceedings of the 37th Annual Hawaii International Conference on System Sciences (IEEE, Piscataway, 2004), p. 8
23. J.W. Moon, J.J. Kim, Ann-based thermal control models for residential buildings. Build. Environ. **45**(7), 1612–1625 (2010)
24. J.Y. Park, T. Dougherty, H, Fritz, Z. Nagy, LightLearn: an adaptive and occupant centered controller for lighting based on reinforcement learning. Build. Environ. **147**, 397–414 (2019)
25. D. Setlhaolo, X. Xia, J. Zhang, Optimal scheduling of household appliances for demand response. Electr. Power Syst. Res. **116**, 24–28 (2014)
26. Smart Grid Program Team, The journey to Green Energy or a Quest for Flexibility. Tech. rep., Eandis, 2015. https://www.edsoforsmartgrids.eu/wp-content/uploads/Vision-Paper_Journey-to-Green-Energy_Eandis.pdf
27. K.M. Tsui, S.C. Chan, Demand response optimization for smart home scheduling under real-time pricing. IEEE Trans. Smart Grid **3**(4), 1812–1821 (2012)
28. U.S Department of Energy, Benefits of demand response in electricity markets and recommendations for achieving them. Tech. rep., U.S Department of Energy, 2006
29. C. Wang, Y. Zhou, J. Wang, P. Peng, A novel traversal-and-pruning algorithm for household load scheduling. Appl. Energy **102**, 1430–1438 (2013)
30. H.F. Wang, Multi-agent co-ordination for the secondary voltage control in power-system contingencies. IEE Proc. Gener. Transm. Distrib. **148**(1), 61–66 (2001)
31. Z. Zhao, W.C. Lee, Y. Shin, K.B. Song, An optimal power scheduling method for demand response in home energy management system. IEEE Trans. Smart Grid **4**(3), 1391–1400 (2013)

Chapter 5
A Review on Non-intrusive Load Monitoring Approaches Based on Machine Learning

Hajer Salem, Moamar Sayed-Mouchaweh, and Moncef Tagina

5.1 Introduction

The energy consumption is increasing. Electricity consumption in the average EU-25 household has been increasing by about 2% per year during the past decade. Moreover, the demand in peak periods (especially in winter) is increasing dramatically compared to the baseline. The peak power in France increases by 1600 MW per year, which is equal to the activation of two nuclear reactors of 900 MW per year [4]. As the main purpose of the electric grid is to ensure the balance between the energy supply and demand, this is hard to reach at peak periods when demand is higher. The traditional method at peak times, called direct control, consists of increasing energy production by activating thermal power stations which produces a large part of man-made CO_2 emissions to the atmosphere. Henceforth new solutions are proposed to overcome these shortcomings. Namely, these strategies are called demand side management. It is defined as the actions that influence the way consumers use electricity to achieve savings and higher efficiency in energy use. The aim is to reduce the stress of the electrical grid and decrease congestion situations and load shedding. These actions result in reducing the consumer electrical bill and better manage the peak loads by the electrical utilities.

Non-Intrusive Load Monitoring is an area in computational sustainability that aims at determining which appliances are operating from the aggregated load reported by a smart meter. Non-Intrusive Load Monitoring may include com-

H. Salem (✉) · M. Tagina
University of Manouba, Manouba, Tunisia
e-mail: hajer.salem@imt-lille-douai.fr; moncef.tagina@ensi-uma.tn

M. Sayed-Mouchaweh
Institute Mines-Telecom Lille Douai, Douai, France
e-mail: moamar.sayed-mouchaweh@mines-douai.fr

M. Sayed-Mouchaweh (ed.), *Artificial Intelligence Techniques for a Scalable Energy Transition*, https://doi.org/10.1007/978-3-030-42726-9_5

munities, industrial sites, commercial sites, office towers, buildings, and homes. In this review, we focus on residential NILM which consists of recognizing the appliances that are active from the aggregated load in houses. Residential homes consume within 20–30% of the total power consumption in Europe [32] which means consumers will be charged higher prices and more supply is produced using unclean sources.

An interesting point highlighted in recent studies [28] is that occupant presence and behavior in buildings have been shown to have large impacts on space heating, cooling and ventilation demand, the energy consumption of lighting and space appliances, and building controls. Careless behavior can add one-third to a building's designed energy performance, while conservation behavior can save a third [26] (see Fig. 5.1). One particularly interesting example is an experiment regularly performed by the company 3M at their headquarters in Minnesota: office workers are asked to switch off all office devices, lights, etc. not in use during peak-price periods. The results of such that experiment were profound: the building's electricity consumption dropped from 15 MW to 13 MW in 15 min and further to 11 MW over 2 h [26]. For this purpose making residents aware of their consumption in detail, by bias of NILM systems, is important for energy sustainability.

An example of Non-Intrusive load Monitoring system input and the desired output is illustrated in Figs. 5.2 and 5.3. The main question is how to disaggregate this load especially in cases where appliances consume in different states? Usually, a NILM system is composed of a smart meter that performs data acquisition, a model that models appliances or household power profiles, a disaggregator that infers the

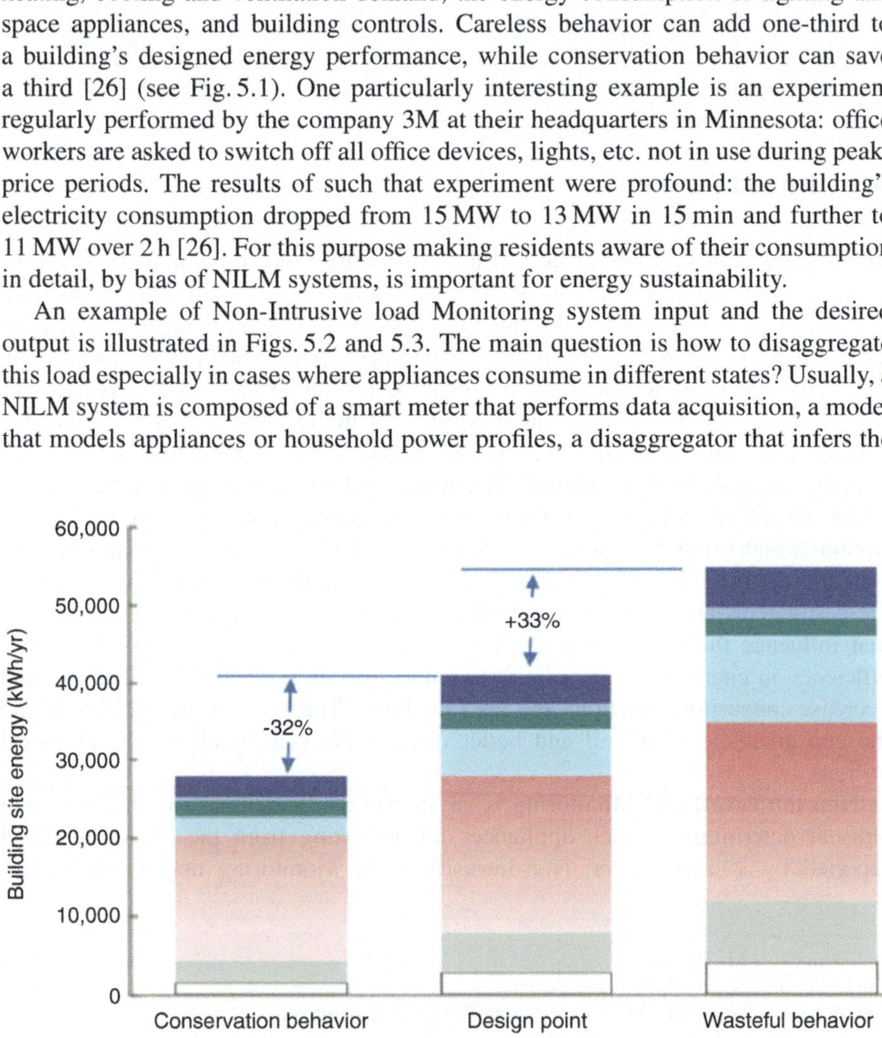

Fig. 5.1 Energy-unaware behavior uses twice as much energy as the minimum that can be achieved [35]

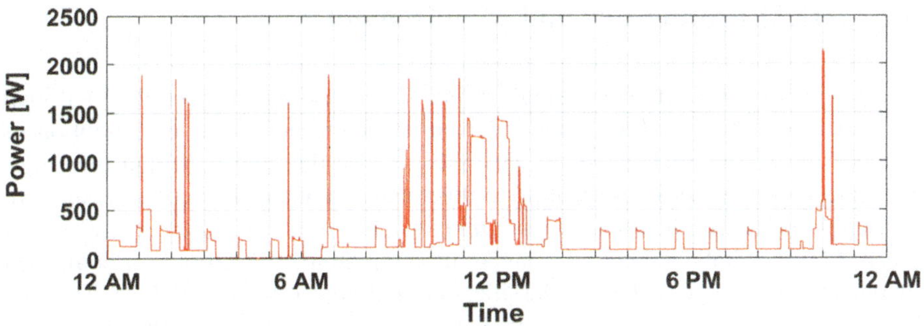

Fig. 5.2 Aggregate load data example: Input to a NILM system

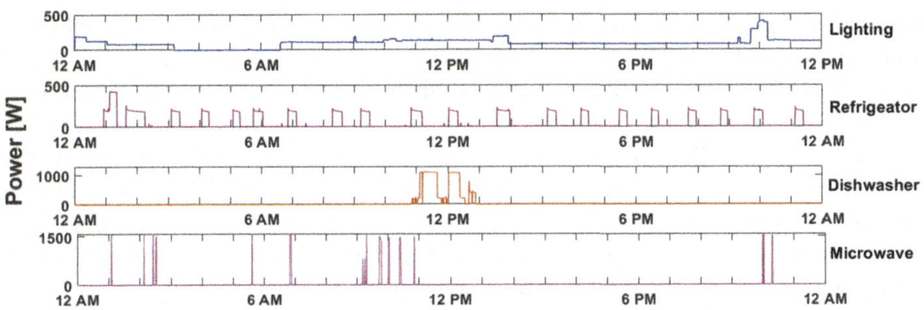

Fig. 5.3 Disaggregated loads of active appliances: Desired NILM system output

sequence states of active appliances, and an estimator that maps these states to a consumption profile.

Many approaches in the literature tackled this problem but many challenges still have to be addressed. In the sequel of this paper, we focus on NILM state of the art. We extend NILM requirements introduced in [40] to present two new requirements that have been claimed by residents. Furthermore, we highlight the challenges facing NILM giving deep explanations. We also discuss NILM open problems. An overview of different NILM state-of-the-art approaches is presented. Approaches are categorized according to the sampling rate of data used, the learning type and we distinguish between event-based and model-based approaches. This review focuses on machine learning and data mining techniques adopted. It also highlights NILM challenges from a data processing perspective.

The remainder of this paper is organized as follows: Sect. 5.2 introduces the NILM problem, a NILM framework, and challenges facing NILM. Section 5.3 reviews approaches based on machine learning and focuses mainly on approaches based on hidden Markov models (HMM) in Sect. 5.3.3. Finally, the paper is concluded in Sect. 5.3.5 and a comparative study is discussed.

5.2 NILM Framework and Requirements

The problem of disaggregating single appliances consumption from the total load in a house can be stated as follows: Given the aggregate power consumption of a house for T time instants, $Y = \{y_1, y_2, \ldots, y_T\}$, the goal is to find the set of appliances' active states contributing to this total load and estimate their respective consumption. Hence the total load at a time instant t is $y_t = \{y_t^1, y_t^2, \ldots, y_t^m., y_t^M\}$ where M is the total number of appliances in a household. This problem is composed of two stages: operational state classification and power estimation [43]. The first stage aims at detecting the operational state of an appliance from the aggregated load composed of several appliances with different active operational states. The second stage looks to convert an appliance's operational state to an estimation of its power demand. Hart [8] and Zoha [43] identified four appliance types: ON/OFF, finite number of active states, constantly on, and continuously variable. Simple on/off appliances are the easiest to detect. Such appliances are lamps, toaster, or a kettle. Finite-state appliances are more complex and can be modeled using a finite-state machine [8]. They are also called multi-state appliances. More recent works modeled these appliances based on probabilistic graphical models such as hidden Markov models. Appliances like dishwashers, clothes dryers, washing machines have several electrical components within them that may be turned ON and OFF at different times. For instance, a dishwasher has many wash cycles that start and stop, a water pump that drains the water between wash cycles, and a heating element that heats water (during wash cycles), dries the dishes, and can keep plates warm [18]. The power drill is an example of a continuously variable device because it has a variable speed. Constantly ON devices remain active throughout days consuming a fixed rate of electricity such as smoke detectors. Almost appliances in households fall under the category of multi-state appliances.

NILM has many practical applications in the smart grid. For instance, it helps to reduce a consumer electricity bill by providing details to consumers about the consumption of each used appliance. Furthermore, NILM has a direct application to propose new consumption plans that avoid peak periods where electricity prices are the most expensive. The use of highly consuming appliances (e.g., a clothes dryer) can be deferred until off peak hours and consequently, ensuring power grid stability.

5.2.1 *NILM Framework*

A NILM framework is composed of three main modules: data acquisition, feature extraction, inference and learning as depicted in Fig. 5.4. In the following, we explain the different sampling rates (low and high frequency) of consumption data. Then we discuss the different feature categories that could be extracted from data and their dependence on the sampling rate. Finally, we discuss the different methods for inference and learning.

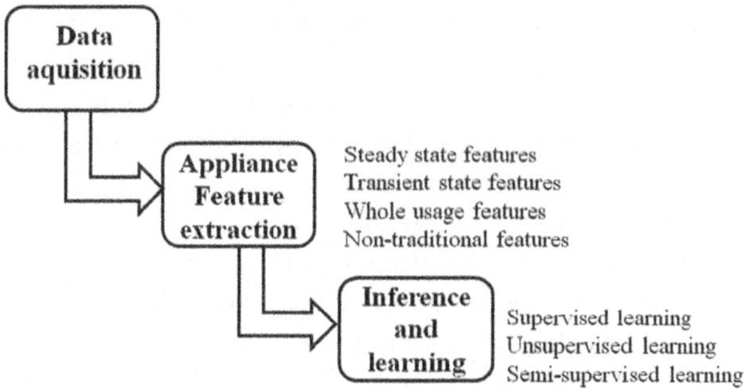

Fig. 5.4 NILM framework example

5.2.1.1 Data Acquisition

The data acquisition module refers to the hardware used to collect the consumption data. We distinguish two frequency sampling rates; High-frequency rate that is around 10–100 MHz and low-frequency rate that is up to 1 Khz. The sampling rate determines the type of information that can be extracted from the electrical signals [43].

- **Low frequency:** Traditional power metrics such as real power, reactive power, root mean square (RMS) voltage and current values can be computed at a low sampling rate (e.g., 120 Hz). The computed metrics are either reported to the backend server via a network interface card (NIC) or processed inside the meter [43]. Although these smart meters are unable to capture the transient-state features, they are more suited for residential deployment because of the return of investment issues.
- **High frequency:** To capture the transient events or the electrical noise generated by the electrical signals, the waveforms must be sampled at a much higher frequency in a range of 10–100 MHz. These types of high-frequency energy meters are often custom-built and expensive due to sophisticated hardware and are tailored to the type of features that needs to be extracted from the signal. They are often deployed for commercial or industrial NILM [27].

5.2.1.2 Feature Extraction

NILM approaches use the information provided by smart meters to identify single appliances consumption from the total load. Many smart meters with different sampling frequencies are available to measure the aggregated load of the house. The smart meter frequency rate influences the type of information collected and hence the features extracted and the machine learning and data mining techniques that could be deployed. Approaches for feature extraction could be classified into (1) steady-state analysis, (2) transient-state analysis and (3) whole usage analysis as the combination of both and (4) non-traditional features. In the following, we discuss these four categories in brief.

- **Steady-state features:** Steady state refers to the stable operational state where features remain constant. Steady-state analysis uses typically the active power where values remain constant and the difference between two power readings is different than a threshold. Approaches in the literature based on steady state demonstrated their effectiveness to disaggregate a small number of appliances from the total load. Nevertheless, these approaches fail to distinguish between two appliances that draw a similar level of power. Indeed, different appliances could have similar power level consumption in several states. Considering only the active power to distinguish between these states is not enough discriminant. To overcome this shortcoming many approaches proposed the use of more features. The combination of active and reactive power is proposed to differentiate between appliances drawing a similar level of active power. Besides, features like RMS current, RMS voltage, active and reactive power, and power factor are proposed to differentiate between similar appliances with similar power consumption at steady states. However, the acquisition of such data requires high-frequency sampling rates that could not be provided by residential smart meters. Indeed smart meters with high-frequency sampling rates are expensive to be deployed in the residential sector and for a large scale deployment [34]. Another solution is to consider non-traditional features like the ones discussed below.
- **Transient-state analysis:** It uses short periods between appliance steady state to disaggregate a household total load. However, the transient state may occur in a small lapse of time that is less than 1 s and hence requires a high sampling frequency rate to be captured. Moreover, transient state does not reflect the appliance real consumption.
- **Whole usage analysis:** An alternative approach to the two aforementioned methods was proposed by Hart [8]. It is a combination of features extracted by steady and transient-state analysis. Indeed he modeled the appliance operation using finite-state machine where states represent the steady state and arcs represent transitions. This approach succeeds to represent multi-state appliances but fails to represent continuously variable appliances. Many approaches have extended finite-state machine models using probabilistic approaches and more precisely variants of hidden Markov models. Indeed, the Markov assumption that a current state depends on the previous state describes perfectly appliances' behavior. To sum up, hidden Markov models represent a suitable trade-off between the representation of the physical structure of appliances (finite state of operational states) and the representation of conditionally dependent events.
- **Non-traditional features:** Non-traditional features refer to contextual and behavioral-based features. They have been considered as additional features to steady-state regime features to distinguish between appliances with similar consumption profile. Time of usage, duration of usage, and seasonal context patterns have been used in recent works and improved accuracy results. For instance, cooling appliances (i.e., fridge, freezer) have discriminating duration patterns. Figures 5.5 and 5.6 show an example of two fridge states duration that are concentrated, respectively, around 12 and 23 min. Besides, some appliances

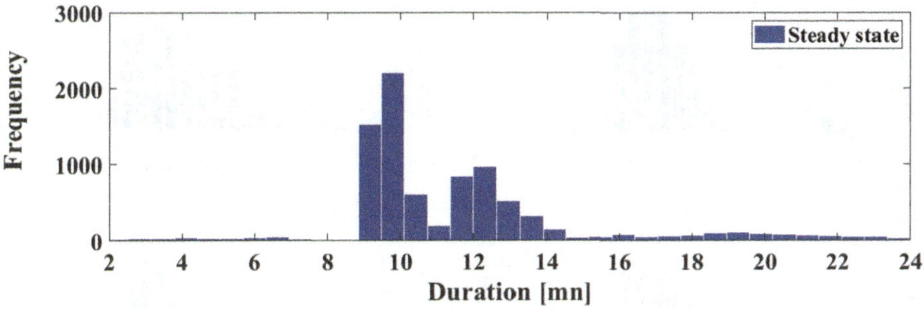

Fig. 5.5 Histogram of a steady-state fridge dwelling time

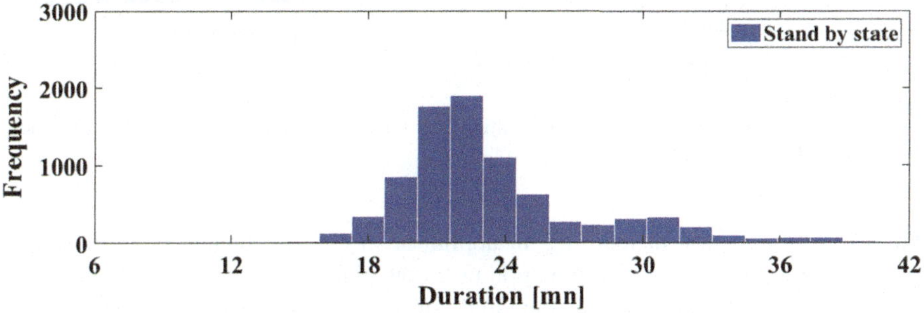

Fig. 5.6 Histogram of a stand by state fridge dwelling time

have a discriminating time of usage. Figure 5.7 illustrates THE time of usage histograms of a lamp, a television, a stereo, and an entertainment appliance.

The data acquisition for NILM can further be categorized into whole-house and circuit-level data [43]. As almost NILM systems use whole-house data, the approach proposed in [23] suggests using circuit-level data to identify low-power appliances (stand by states of several appliances) in presence of high-power appliances (more than 150 W). The approach seems to alleviate the complexity of the problem but on the extent of installation cost.

5.2.1.3 Inference and Learning

Inference in NILM refers to the problem of recognizing the set of active states from the aggregated load in real-time. Two main approaches tackle the inference problem in NILM: optimization and machine learning [25, 43]. Optimization approaches show limits facing computational complexity [1, 9]. Machine learning approaches may be divided into two types: approaches based on events and approaches based on models. Approaches based on events aim at detecting a change in the aggregated load and classify this change to match an appliance state consumption. Approaches

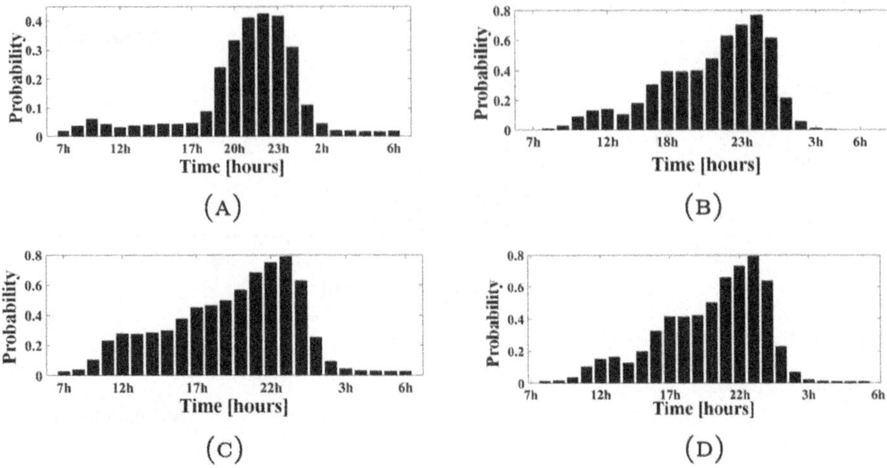

Fig. 5.7 Histograms of usage time probabilities over hours of the day: (**a**) Lamp, (**b**) Television, (**c**) Stereo, (**d**) Entertainment

based on models aim at learning an appliance model parameters or a household power profile, then recognize the appliance consumption in real-time. Approaches differ according to the available data and consequently according to their learning type: supervised, semi-supervised, and unsupervised learning. Machine learning based approaches are the core of this paper and are discussed in detail in Sect. 5.3.

5.2.2 *Requirements*

To attain a successful implementation of a NILM approach, Zeifman [40] presented in 2012 some requirements that should be fulfilled. Many of these requirements have been met in different state-of-the-art approaches. However, the installation of smart meters in residential and new scientific studies points out the need to meet new requirements. We extend the requirements presented in [40] to the following ones:

- **Non-intrusiveness:** The approach must only require consumption data to be collected from a single point of measurement to avoid the cost of installing sensors for each appliance.
- **Adaptability to low sampling frequency data:** The approach must be able to monitor appliances given aggregate power measurements at 1 min intervals since it represents available residential smart meters frequency sampling in Europe. This allows for avoiding the cost of additional hardware (storage, acquisition, etc.) installation.

- **Unsupervised disaggregation:** The training phase should be automatic and independent of the user. The user does not have to label the consumption measurements of appliances, their number, or to switch them ON/OFF to obtain data about each active state of each appliance.
- **Ability to generalize to unseen houses:** It is very unlikely to have available ground truth consumption data for each appliance because sub-metered data are very expensive to be collected. Generalization of the built model to identify appliances used in unseen households is a popular property in machine learning but few NILM approaches considered this aspect to date [31].
- **Protecting consumer privacy:** Recently, many smart meters have been deployed in Europe. However, resident feedback following the smart meter installation in households points out the need to protect privacy requirement. A big issue facing NILM deployment is privacy loss because residents usually complain about sharing their personal data with utility companies. A potential solution is to perform data processing at the household level. Henceforth, personal data are not shared with external parts.
- **Near real-time capabilities:** The approach must be able to infer active appliances, their states, and consumption in real-time. This requirement is fulfilled in many NILM approaches. However, we notice a lack of approaches able to learn online the parameters of a specific appliance from a generic model.
- **Adaptability to new appliances:** The learning approach has to be able to be extended to recognize new appliances when they are added to the load.

5.2.3 Difficult Scenarios for Disaggregation

In addition to the aforementioned requirements, NILM faces some challenges. Indeed, disaggregation is particularly difficult in some scenarios as follows:

- **States that have a small power level consumption:** The energy consumption of certain appliances active states can be less than the event detection threshold or could be considered as noise. These states cannot be detected from the aggregated load. For instance, tablet or smartphone consumes during charging less than 5 W which is within the consumption variance of several appliances.
- **Appliances that have states with similar consumption:** Different appliances could have similar energy consumption in several states. As illustrated in Figs. 5.8, 5.9, and 5.10, the circled states of the laptop, stereo, and freezer have often equivalent consumption readings. Considering only the active power to distinguish between these states may not be enough (Fig. 5.11).
- **Unknown appliances:** A sequence of consumption data may be generated by a new appliance which does not have a prior model. A challenge is to detect this new appliance and distinguish its consumption from an aggregation of two or more of already known states that are active.

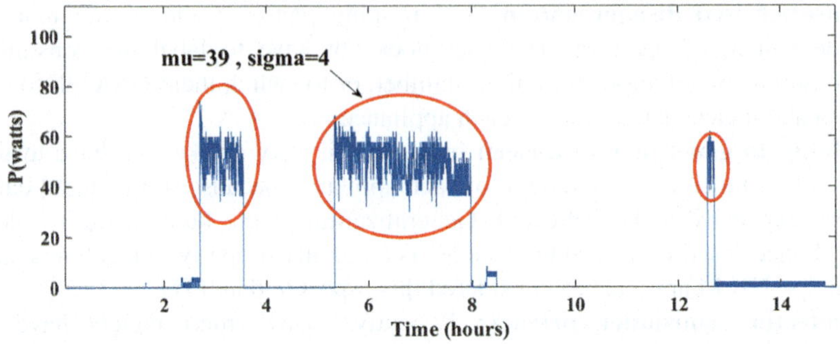

Fig. 5.8 Laptop consumption

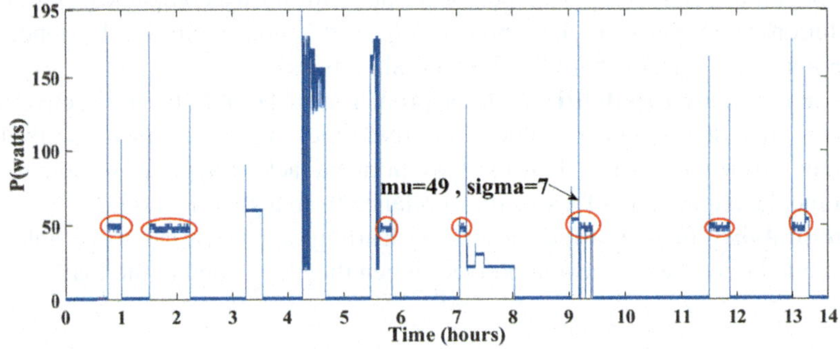

Fig. 5.9 Stereo consumption

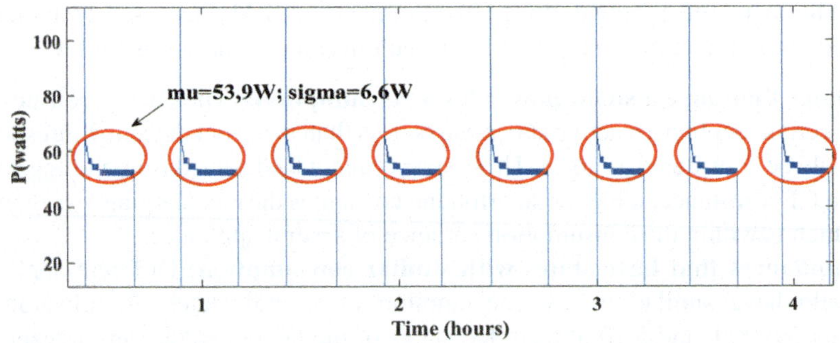

Fig. 5.10 Freezer consumption

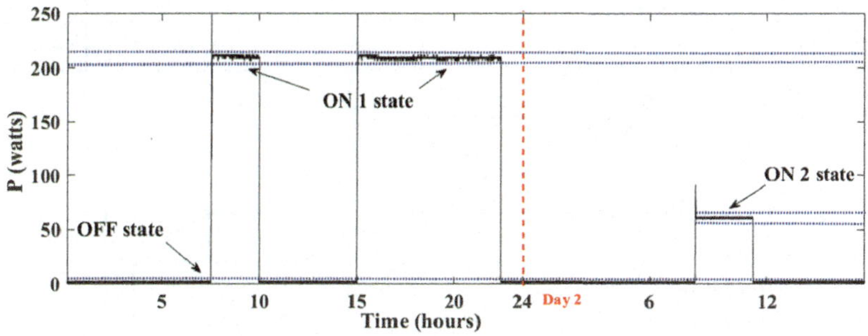

Fig. 5.11 Example of multi-state appliances

5.3 Review of Machine Learning Based Approaches for NILM

Machine learning approaches could be divided into two main categories: approaches based on events and approaches based on models. These approaches differ according to the granularity of data used and the learning type. Different models and classification algorithms are used in the literature. Figure 5.12 illustrates a taxonomy of the discussed approaches. In the sequel, we discuss these two families of approaches with a main focus on the one based on models.

5.3.1 Approaches Based on Event Detection

Event-based approaches aim at classifying appliances based on detected switch events in the aggregated load. These approaches follow usually the steps depicted in Fig. 5.14. First, data are collected from a meter in a household. A pre-processing is performed that depends on the frequency of collected data. Second, edge change is detected and classified as an appliance state power profile. Edge change detection consists in detecting if an appliance changes its state from one to another (from a power profile to another).

A big issue facing this type of approach is the used change detection mechanism. Indeed, it is highly dependent on the used threshold. On the one hand, a small threshold increases the rate of false detection of state changes (see Fig. 5.13a). On the other hand, a big threshold may miss the detection of appliances with small consumption profiles (see Fig. 5.13a). A naive approach deployed in almost power disaggregation approaches [14, 38] is to monitor power consumption readings and to flag an event when the power change deviates beyond a fixed threshold. For instance, a change in the active power of 5 W between two successive readings (y_{t-1} and y_t) is considered as an indication of a transition between two states according to the

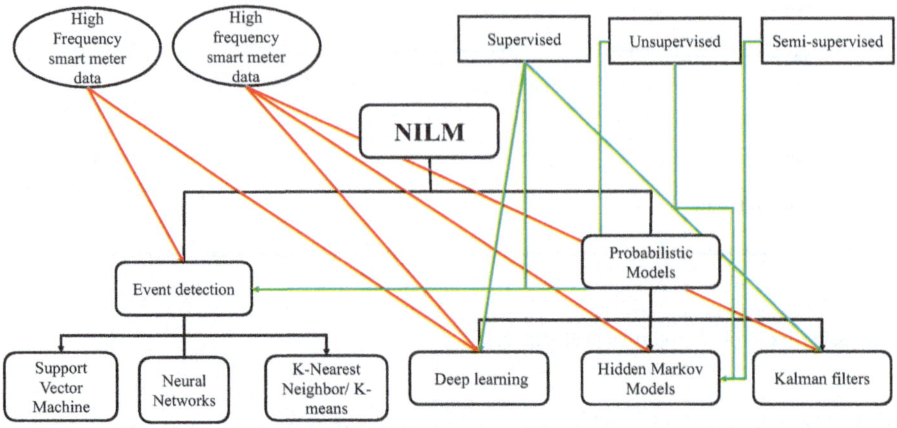

Fig. 5.12 NILM state-of-the-art approaches taxonomy

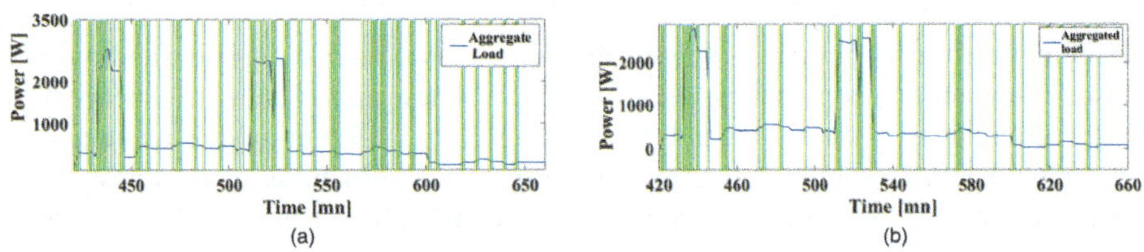

Fig. 5.13 Event detection based on a threshold, (**a**): $\delta = 5$ W, (**b**): $\delta = 20$ W

switch continuity principle (SCP) in NILM [19].[1] A method for adaptive threshold selection which is based on the generalized likelihood ratio (GLR) has been proposed [3]. This method succeeds to properly distinguish between state transition events, noise, and consumption variance within an appliance state consumption. However, computing the likelihood at each observation point leads to computational complexity and is not appropriate for real-time application (Fig. 5.14).

An unsupervised NILM event detector based on a sliding window kernel Fisher discriminant analysis (KFDA) has been proposed [2]. It determines accurately start and end times of transient states instead of returning only a change point and thus provides a good segmentation into steady states and transient states. However, this approach requires data sampled at high rates (12 KHz) to extract the needed features.

[1]The SCP assumes that only one appliance ever changes state at any given point in time. This assumption holds if the sampling time is reasonably short.

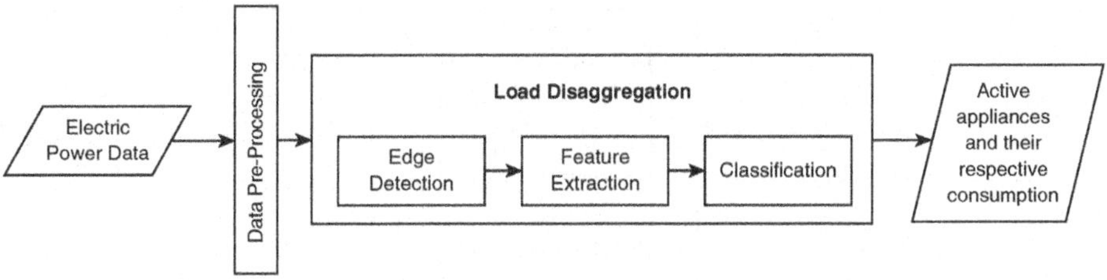

Fig. 5.14 Global dynamic of event-based approaches for NILM

Third, feature extraction is performed to select the most discriminating features to distinguish between appliances. The final task is to classify which appliance has triggered an event-based on the features extracted in the previous step and the labeled data.

Several machine learning algorithms were proposed recently for NILM to perform the classification including supervised methods such as neural networks [33], support vector machine (SVM) [17]. Event-based detection approaches are usually supervised [38]. Indeed, they require appliances annotation in each house which shows an intrusive behavior and a burden on users' dailies. For instance, occupants have to switch on and off devices to allow the system to learn appliances power profiles [38].

Some works proposed unsupervised approaches that are mainly based on a signature database to classify an appliance [3] based on its signature (or the appliance transient profile). However, it is impossible and very expensive to collect all appliances' signatures in a database because many models (brands/instances) exist for each appliance type. More recently, an event detection method based on signal processing [42] has been proposed. The approach uses median filtering and window processing to detect appliances in active states. Classification is performed based on graph filtering. However, the approach is not totally event-based, since the disaggregation is performed based on inference algorithms such as the one proposed for Factorial HMM [14]. Finally, we have to point out that one prominent drawback of event-based methods is that they assume all appliances' switch events are independent. This is not the case for multi-state appliances because each appliance state depends on its previous one.

5.3.2 *Approaches Based on Probabilistic Models for NILM*

The problem of power disaggregation is time-dependent by nature. NILM researchers have always thought models of sequential data and time as possible solution. For this reason, most of the modern approaches for NILM are based on

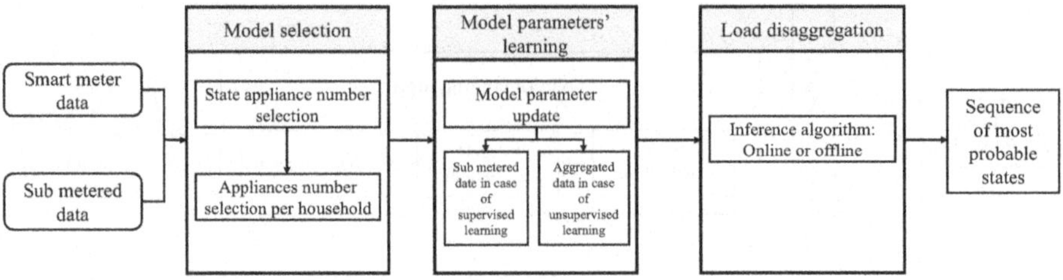

Fig. 5.15 Global diagram of probabilistic based approaches for NILM

hidden Markov models and deep learning. Several deep learning based approaches [12, 16, 24, 41] have been proposed for NILM. They showed very promising results regarding generalization to unseen appliances and presented major improvements in terms of complexity. However this performance is still prone to: the need to high-frequency data which is impractical in the residential sector due to the return of investment issues; the difficult process of training; and sometimes dependency on labeled data. Deep learning training is computationally expensive. Indeed, training takes days of processing on a fast GPU which cannot be an embedded process inside a smart meter.

Probabilistic models based approaches follow the steps depicted in Fig. 5.15. First, data acquisition and feature extraction are performed. The most discriminating features are used to build the model structure. The model parameters are learned during the training phase using available training data. Finally, the updated model is used for disaggregation. We distinguish three main different learning and inference algorithms categories for NILM: supervised ones which usually use sub-metered data for training, unsupervised algorithms which are free of training, and semi-supervised algorithms which use an aggregated load for training. The most adopted probabilistic graphical model for NILM is hidden Markov model and its extensions. In the following, we review the state-of-the-art approaches based HMM for NILM discussing their strengths, shortcomings, and limitations.

5.3.3 Review of Hidden Markov Models (HMM) for NILM

The most popular disaggregation approaches are based on hidden Markov models (HMMs) solutions [7, 11, 13–15, 20–22, 30, 31, 39, 40] because they succeed to correctly model dependence between event transitions and steady-states consumption. Several versions of HMMs have been deployed and have shown interesting results [25]. Despite the arithmetic complexity of inference algorithms, these models still present the best compromise in terms of meeting NILM requirements. In the sequel, we discuss the state-of-the-art methods based on HMMs. We categorize these approaches according to supervised, unsupervised, semi-supervised learning, and non-parametric Bayesian learning and we discuss HMM-based approaches using additional features to active power. A comparative discussion is proposed in the end of this section underlining the limits of the state of the art.

5.3.3.1 Supervised HMM-Based Approaches

Supervised learning algorithms are trained based on data collected from sub-metered data [7, 13, 15, 20, 39, 40]. Accuracy results are very promising and reach 80% [20]. However, these approaches are impractical for large scale applications because spreading sensors in each circuit level is expensive. A sparse HMM is proposed [20] where each HMM state is considered as the composition of all appliances' states and called superstate HMM. In other words, the HMM state represents the state of the household (all the appliances states in the house). This model differs from FHMM representation for NILM by adding the representability of loads dependence. Indeed, the transition matrix of such a model incorporates the probability of switching from one appliance state to another state from a different appliance. An exact inference algorithm has been proposed for the super state HMM. It can recognize more than twenty appliances in real-time and shows promising disaggregation accuracy results. Unfortunately, according to our opinion, this approach violates the main requirement of NILM which is "unsupervised disaggregation." Indeed, the disaggregator is based on a prior model that is built from sub-metered data.

To tackle continuously consuming appliances, particle filter has been used to approximate the PDFs of HMM states with the nonlinear appliance (drill, dimmer, ...) behavior. The approach called "Paldi" [7] considers also non-Gaussian noise that cannot be handled by HMM and FHMM. In the first step, HMMs are built based on sub-metered data to model each appliance consumption behavior. The HMM structure, the transition matrix, and the observation distributions are learned offline. Then, a FHMM is constructed as a combination of the built HMMs to model a household total consumption. Particle filters are used to estimate the posterior density of the FHMM. Paldi succeeds to identify nonlinear appliance behavior such as the drill. However, it is dependent on the choice of an appliance power demand specified offline and cannot work properly if appliances with similar power profiles exist in the same household.

5.3.3.2 Unsupervised HMM-Based Approaches

A bench of unsupervised approaches based on variants of factorial hidden Markov models (FHMM) have been proposed [5, 14, 29]. These approaches share almost the same scalability issues. Indeed, they are not applicable for a number of appliances greater than eighteen [25]. The first unsupervised approach without the use of any prior knowledge has been proposed for power disaggregation in [14]. The proposed approach selects the number of appliances, the number of states per appliance and extracts training samples for each appliance from the aggregated load. It uses a snippet heuristic that looks for periods (samples in the aggregated load) where an appliance is turning ON then OFF. However, this is indecipherable for almost appliances because appliance states that consume for long-duration could not be captured consuming consequently in the aggregated load and are interrupted by the activation of other appliances. The proposed inference algorithm called AFAMAP for additive factorial approximate MAP is a variational approximate inference based on integer programming which overcomes the expensive computation procedure of exact inference. On the one hand, the algorithm constrains the posterior to allow only one Markov chain to change its state at any given time. This constraint is called the one at a time assumption and it follows the SCP principle. On the other hand, it performs inference using both an additive factorial HMM (AFHMM) and a difference factorial HMM (DFHMM) but constrains the posterior over the hidden variables in both models to agree. This constraint is used to alleviate the sensibility of the model to noise. AFAMAP improved significantly accuracy results and showed robustness to noise in factorial models for NILM and inference for FHMMs in general. However, it is intractable for an important number of appliances (parallel Markov chains). Finally, this approach still needs an expert to label disaggregated appliances in the house.

A more recent work [5] extends the AFAMAP algorithm proposed in [14] to a bivariate Gaussian observation model considering both the active and reactive power. The approach proposes a model training method that is different from the one deployed in [14]. A method for extracting an appliance footprint from an aggregated power signal is proposed. Appliances activation is detected based on a fixed threshold. However, temporal information has to be specified a priori to detect ON and OFF states. Several footprints have to be detected for complex appliances such as washing machines. Clustering is performed to group states belonging to the same appliance and approximate states power consumption profiles where bivariate Gaussian distributions are approximated to each appliance state active and reactive power. The extended version of AFAMAP algorithm proposed in [14] improves the state-of-the-art results. However, it shares the same complexity issues as his predecessors.

To sum up, recent unsupervised approaches improved disaggregation accuracy; however, they still suffer from scalability issues and the need for expert annotation after the disaggregation phase.

5.3.3.3 Semi-Supervised HMM-Based Approaches

Approaches for NILM based on machine learning have been either supervised or unsupervised. Both learning approaches have their advantages and limits. Parson et al. [30] proposed a semi-supervised approach to combine their respective strengths. The aim is to take advantage of available appliances data extracted from public data sets and to avoid sensors intrusiveness. Therefore, the approach main idea is to create generic appliances models for all houses and to update these models during the training phase using the aggregate load from each specific house. Authors demonstrate that using prior knowledge for creating generic models gives the same results as using sub-metered data [31] and outperforms the state of the art using factorial hidden Markov model (FHMM). However, a major drawback of this approach is the adopted training phase. Indeed, the approach aims at looking for periods where only one appliance is operating in the house to extract its signature. This entails the storage of all training samples which has a spatial complexity of $O(T * D)$ where T is the training data size and D is the data dimension. Besides, the training samples selection method has to iterate on all training data for each appliance. Such training data cannot be stored or processed at the level of the smart meter because the deployed training algorithm needs several days of household consumption multiplied by the sampling frequency which involves the need for high storage and processing capabilities. For instance, training in the context of [30] approach is performed on the cloud. Unfortunately, this solution presents privacy issues regarding sharing consumers' personal data. We believe that this is impractical because of the following reasons: First, almost appliances are rarely deployed alone. The approach is more adapted for cooling appliances. Second, waiting for long periods to find training samples for a single appliance pushes occupants to lose confidence in the proposed system; finally, the proposed method for generic prior models creation is not the most appropriate because expert information cannot handle operation behavior of all appliances' types.

5.3.3.4 HMM-Based Approaches with Non-traditional Features

Integrating non-traditional based features seems to alleviate scalability issues [29] in FHMM. Non-traditional features refer to contextual based and behavioral based features. Hour of the day, day of the week, and duration have been proposed as additional features for FHMM [13]. Variants of FHMM using contextual features are proposed such as factorial hidden semi-Markov model (FHSMM) where the duration is approximated to a Gamma distribution. Moreover, a conditional FHMM (CFHMM) is developed where the transition probabilities are not constant but conditioned on the time of the day and day of the week. Furthermore, a combination of the aforementioned models is developed and called CFHSMM for conditional factorial hidden semi-Markov model. A modified expectation-maximization algorithm is developed for this model to estimate the model parameters and a simulated annealing algorithm is deployed to infer the sequence of hidden states because

Viterbi algorithm is intractable for the CFHSMM. The proposed model improved the disaggregation accuracy which may be explained by the discriminating behavior of contextual features. However, the effect of each feature on the disaggregation accuracy has not been investigated through the paper. Besides, the accuracy improvement has been at the expense of increasing the model complexity. Indeed, the proposed model is not tractable for a number of appliances greater than ten [22]. Besides, the approach is claimed to be unsupervised; however, training and evaluation are performed on the same data which makes us unable to evaluate the generalization capabilities of the approach.

Time of the day and seasonal context-based patterns have been incorporated into a recent NILM approach [6] to discriminate between appliances. A whole year of usage data has been used for training to build usage patterns which is impractical for real-world applications. Besides, generalization to unseen houses has not been tested. Power consumption patterns of appliances, user presence, and time usage have been investigated in [29]. Data analysis has been performed to highlight the discriminating power of these additional features. The adopted inference algorithm is an extension of the AFAMAP algorithm proposed in [14] with contextual features. This extension shows promising results where the precision increased by approximately 15%. Nevertheless, the approach is considered as supervised because the same appliances are used for training and testing and its performance in the case of unseen appliances cannot be evaluated. Besides, non-intrusiveness is violated because user presence is detected by installing presence sensors in each room in the house.

5.3.4 Non-parametric HMM-Based Approaches

The aforementioned approaches are not fully unsupervised because they still need a prior on the number of appliances in the household. Non-parametric Bayesian models have been proposed to tackle the issue of setting a fixed number of appliances in NILM [10, 11, 36, 37]. A first attempt has been proposed in [11]. Semi-Markov chain is combined with hierarchical Dirichlet process and a Bayesian non-parametric hidden semi-Markov models (HSMM) has been proposed. The model explicitly models the duration distributions and can model multi-state appliances. The approach does not rely on training data but prior information and learns the model parameters during inference. Non-parametric Bayesian models are used to fulfill generalization requirement; however, the proposed Bayesian non-parametric HSMM has been evaluated on only one data set and only some data segments have been chosen which makes us enable to evaluate its generalization capabilities.

A more recent approach proposed in [37] can be extended to an infinite number of appliances and an unbounded number of states per appliance called (IUFHMM) for infinite unbounded factorial hidden Markov model. The approach is generic for all households because it is independent of prior knowledge. Only parameters' prior distribution hyper have been set a priori. All the model parameters, including the number of appliances per house and the number of states per appliance, are learned from data. Experiments showed that inferring the number of appliances from data gives better results than FHMM. However, inferring the number of states from data has not improved significantly the obtained results. Indeed, almost appliances have a number of states less than four. Hence, setting the state's number at four gives approximately the same results than inferring this number from data. One shortcoming of this work remains in the fact that it is unable to determine the label of chains (appliances). Besides, experiments with a high number of devices have not been reported which makes us unable to evaluate the inference algorithm tractability.

5.3.5 Conclusion and Discussion

In this paper, Non-Intrusive Load Monitoring (NILM) has been introduced and the NILM framework has been detailed. Existing approaches for NILM have been discussed. Then, machine learning based approaches for NILM have been categorized according to learning type and the used features. Table 5.1 presents our classification of NILM approaches according to NILM requirements introduced in Sect. 5.2.

The goal behind this comparative analysis is to point out the state-of-the-art limits. Firstly, the conducted analysis showed the importance of prior information for the model robustness. Indeed, unsupervised learning can identify up to ten appliances but shows inaccuracy for twenty or more appliances. Moreover, none of the works proposed in the literature proposed an online training approach which makes training heavy processing that cannot be performed on the level of the smart meter. Furthermore, meeting consumers privacy is requested by users in particular due to smart meter deployments in real households at large scale Europe. Finally, almost proposed approaches in the literature are unable to propose a generic approach that can be adaptive to the number of appliances and can label appliances after disaggregation without the need for expert annotation.

Table 5.1 A comparative analysis of machine learning based methods for NILM

Category	Learning type	Methods	Requirements						
			Non-intrusiveness	Low sampling	Unsupervised disaggregation	Generalization	Privacy	Real-time capabilities	Tractability
Event-based methods	Unsupervised subject to signature database	[3]	+	+	+	–	–	+	–
	Supervised	[38]	–	+	–	–	–	–	–
	Unsupervised	[2]	+	+	+	–	–	+	–
	Unsupervised	[42]	–	+	+	–	–	–	–
Deep learning based methods	Supervised	NeuralNILM[12]	–	–	–	+	–	–	–
	Supervised	[24]	–	+	–	+	–	–	–
	Supervised	[16]	–	–	–	+	–	–	–
	Supervised	Sequence-to-point [41]	–	–	–	+	–	+	–
HMM and FHMM based methods	Supervised	Paldi [7]	–	+	–	–	–	–	+
	Supervised	Super state HMM [20]	–	+	–	–	–	+	+
	Unsupervised	FHSMM, CFHMM, CFHSMM [13]	+	+	+	–	–	–	–
	Unsupervised	AFHMM, DFHMM [14]	+	–	+	+	–	–	–
	Supervised	Context FHMM [29]	–	+	–	–	–	–	–
	Unsupervised	[5]	–	+	+	–	–	–	–
	Semi-supervised	HMM/DHMM [30]	+	+	+	+	–	–	+
	Unsupervised	Non-parametric HSMM [11]	+	–	+	+	–	–	–
	Unsupervised	IFDM [36]	+	+	+	+	–	–	–
	Unsupervised	NFHMM [10]	+	+	+	+	–	–	–
	Unsupervised	IUFHMM [37]	+	+	+	+	–	–	–

References

1. M. Baranski, J. Voss, Nonintrusive appliance load monitoring based on an optical sensor, in *2003 IEEE Bologna Power Tech Conference Proceedings*, vol. 4 (IEEE, Piscataway, 2003), p. 8
2. K.S. Barsim, R. Streubel, B. Yang, An approach for unsupervised non-intrusive load monitoring of residential appliances, in *Proceedings of the 2nd International Workshop on Non-Intrusive Load Monitoring* (2014)
3. M. Berges, E. Goldman, H.S. Matthews, L. Soibelman, K. Anderson, User-centered nonintrusive electricity load monitoring for residential buildings. J. Comput. Civ. Eng. **25**(6), 471–480 (2011)
4. Bilan électrique (2014). http://www.rte-france.com/sites/default/files/bilan_electrique_2014.pdf. Accessed 18 Aug 2016
5. R. Bonfigli, E. Principi, M. Fagiani, M. Severini, S. Squartini, F. Piazza, Non-intrusive load monitoring by using active and reactive power in additive factorial hidden Markov models. Appl. Energy **208**, 1590–1607 (2017)
6. C. Dinesh, S. Makonin, I.V. Bajic, Incorporating time-of-day usage patterns into non-intrusive load monitoring, in *2017 IEEE Global Conference on Signal and Information Processing (GlobalSIP)* (IEEE, Piscataway, 2017), pp. 1110–1114
7. D. Egarter, V.P. Bhuvana, W. Elmenreich, PALDi: Online load disaggregation via particle filtering. IEEE Trans. Instrum. Meas. **64**(2), 467–477 (2015)
8. G.W. Hart, Nonintrusive appliance load monitoring. Proc. IEEE **80**(12), 1870–1891 (1992)
9. S. Inagaki, T. Egami, T. Suzuki, H. Nakamura, K. Ito, Nonintrusive appliance load monitoring based on integer programming. Electr. Eng. Jpn. **174**(2), 18–25 (2011)
10. R. Jia, Y. Gao, C.J. Spanos, A fully unsupervised non-intrusive load monitoring framework, in *2015 IEEE International Conference on Smart Grid Communications (SmartGridComm)* (IEEE, Piscataway, 2015), pp. 872–878
11. M.J. Johnson, A.S. Willsky, Bayesian nonparametric hidden semi-Markov models. J. Mach. Learn. Res. **14**, 673–701 (2013)
12. J. Kelly, W. Knottenbelt, Neural NILM: deep neural networks applied to energy disaggregation, in *Proceedings of the 2nd ACM International Conference on Embedded Systems for Energy-Efficient Built Environments* (ACM, New York, 2015), pp. 55–64
13. H. Kim, M. Marwah, M. Arlitt, G. Lyon, J. Han, Unsupervised disaggregation of low frequency power measurements, in *Proceedings of the 2011 SIAM International Conference on Data Mining* (SIAM, 2011), pp. 747–758
14. J.Z. Kolter, T. Jaakkola, Approximate inference in additive factorial HMMs with application to energy disaggregation, in *Artificial Intelligence and Statistics* (2012), pp. 1472–1482
15. J.Z. Kolter, S. Batra, A.Y. Ng, Energy disaggregation via discriminative sparse coding, in *Advances in Neural Information Processing Systems* (2010), pp. 1153–1161
16. H. Lange, M. Bergés, The neural energy decoder: energy disaggregation by combining binary subcomponents, in *NILM2016 3rd International Workshop on Non-Intrusive Load Monitoring* (2016). nilmworkshop.org
17. G.-Y. Lin, S.-C. Lee, J.Y.-J. Hsu, W.-R. Jih, Applying power meters for appliance recognition on the electric panel, in *2010 5th IEEE Conference on Industrial Electronics and Applications* (IEEE, 2010), pp. 2254–2259
18. S. Makonin, Approaches to non-intrusive load monitoring (NILM) in the home (2012)
19. S. Makonin, Investigating the switch continuity principle assumed in non-intrusive load monitoring (NILM), in *2016 IEEE Canadian Conference on Electrical and Computer Engineering (CCECE)* (IEEE, Piscataway, 2016), pp. 1–4
20. S. Makonin, F. Popowich, Nonintrusive load monitoring (NILM) performance evaluation. Energy Effic. **8**(4), 809–814 (2015)

21. S. Makonin, I.V. Bajic, F. Popowich, Efficient sparse matrix processing for nonintrusive load monitoring (NILM), in *2nd International Workshop on Non-Intrusive Load Monitoring* (2014)
22. S. Makonin, F. Popowich, I.V. Bajić, B. Gill, L. Bartram, Exploiting HMM sparsity to perform online real-time nonintrusive load monitoring. IEEE Trans. Smart Grid **7**(6), 2575–2585 (2016)
23. A. Marchiori, D. Hakkarinen, Q. Han, L. Earle, Circuit-level load monitoring for household energy management. IEEE Pervasive Comput. **10**(1), 40–48 (2011)
24. L. Mauch, B. Yang, A new approach for supervised power disaggregation by using a deep recurrent LSTM network, in *2015 IEEE Global Conference on Signal and Information Processing (GlobalSIP)* (IEEE, Piscataway, 2015), pp. 63–67
25. C. Nalmpantis, D. Vrakas, Machine learning approaches for non-intrusive load monitoring: from qualitative to quantitative comparation. Artif. Intell. Rev. **52**(1), 217–243 (2018)
26. T.A. Nguyen, M. Aiello, Energy intelligent buildings based on user activity: a survey. Energy Build. **56**, 244–257 (2013)
27. L.K. Norford, S.B. Leeb, Non-intrusive electrical load monitoring in commercial buildings based on steady-state and transient load-detection algorithms. Energy Build. **24**(1), 51–64 (1996)
28. J. Page, D. Robinson, N. Morel, J.-L. Scartezzini, A generalised stochastic model for the simulation of occupant presence. Energy Build. **40**(2), 83–98 (2008)
29. F. Paradiso, F. Paganelli, D. Giuli, S. Capobianco, Context-based energy disaggregation in smart homes. Future Internet **8**(1), 4 (2016)
30. O. Parson, S. Ghosh, M.J. Weal, A. Rogers, Non-intrusive load monitoring using prior models of general appliance types, in *Proceedings of the Twenty-Sixth AAAI Conference on Artificial Intelligence* (2012)
31. O. Parson, S. Ghosh, M. Weal, A. Rogers, An unsupervised training method for non-intrusive appliance load monitoring. Artif. Intell. **217**, 1–19 (2014)
32. R. Raudjärv, L. Kuskova, Energy consumption in households. https://www.stat.ee/dokumendid/68628. Accessed 5 May 2019
33. A.G. Ruzzelli, C. Nicolas, A. Schoofs, G.M.P. O'Hare, Real-time recognition and profiling of appliances through a single electricity sensor, in *2010 7th Annual IEEE Communications Society Conference on Sensor Mesh and Ad Hoc Communications and Networks (SECON)* (IEEE, Piscataway, 2010), pp. 1–9
34. H. Salem, M. Sayed-Mouchaweh, A.B. Hassine, A review on machine learning and data mining techniques for residential energy smart management, in *2016 15th IEEE International Conference on Machine Learning and Applications (ICMLA)* (IEEE, Piscataway, 2016), pp. 1073–1076
35. The World Business Council for Sustainable Development, Transforming the market: energy efficiency in buildings, survey report. https://www.wbcsd.org/programs/cities-and-mobility/energy-efficiency-in-buildings/resources/transforming-the-market-energy-efficiency-in-buildings. Accessed 5 July 2019
36. I. Valera, F. Ruiz, L. Svensson, F. Perez-Cruz, Infinite factorial dynamical model, in *Advances in Neural Information Processing Systems* (2015), pp. 1666–1674
37. I. Valera, F.J.R. Ruiz, F. Perez-Cruz, Infinite factorial unbounded-state hidden Markov model. IEEE Trans. Pattern Anal. Mach. Intell. **38**(9), 1816–1828 (2016)
38. M. Weiss, A. Helfenstein, F. Mattern, T. Staake, Leveraging smart meter data to recognize home appliances, in *2012 IEEE International Conference on Pervasive Computing and Communications (PerCom)* (IEEE, Piscataway, 2012), pp. 190–197
39. M. Zeifman, Disaggregation of home energy display data using probabilistic approach. IEEE Trans. Consum. Electron. **58**(1), 23–31 (2012)
40. M. Zeifman, K. Roth, Nonintrusive appliance load monitoring: review and outlook. IEEE Trans. Consum. Electron. **57**(1), 76–84 (2011)

41. C. Zhang, M. Zhong, Z. Wang, N. Goddard, C. Sutton, Sequence-to-point learning with neural networks for non-intrusive load monitoring, in *Thirty-Second AAAI Conference on Artificial Intelligence* (2018)
42. B. Zhao, K. He, L. Stankovic, V. Stankovic, Improving event-based non-intrusive load monitoring using graph signal processing. IEEE Access **6**, 53944–53959 (2018)
43. A. Zoha, A. Gluhak, M.A. Imran, S. Rajasegarar, Non-intrusive load monitoring approaches for disaggregated energy sensing: a survey. Sensors **12**(12), 16838–16866 (2012)

Part II
Artificial Intelligence for Reliable Smart Power Systems

Chapter 6
Neural Networks and Statistical Decision Making for Fault Diagnosis in Energy Conversion Systems

Gerasimos Rigatos, Dimitrios Serpanos, Vasilios Siadimas, Pierluigi Siano, and Masoud Abbaszadeh

6.1 Introduction

Energy conversion systems that include DC-DC converters are often used for energy conversion purposes [1–4]. DC-DC converters find several applications in DC electric power generation systems such as photovoltaics. DC power transmission systems, in the control of motors as well as of several appliances, and in the actuation of robots and in the traction systems of electric vehicles [5–8]. DC-DC converters undergo failures and condition monitoring for them is important so as to prolong their life-time and to avoid critical conditions [9–12]. Condition monitoring of DC-DC converters is part of the wider fault diagnosis problem of the electricity grid [13–15]. There have been several attempts to solve the fault detection and isolation problem for DC-DC converters [16–20]. These are mainly grouped in two classes: (1) model-based approaches [21–24], and (2) model-free approaches [25–28]. In the case of (1) it is considered that a dynamic model of the DC-DC converters is available and this allows to develop software that emulates the fault-free functioning of the converters, for instance in the form of state-observers

G. Rigatos (✉)
Unit of Industrial Automation, Industrial Systems Institute, Rion Patras, Greece
e-mail: grigat@ieee.org

D. Serpanos · V. Siadimas
Department of Electrical and Computer Engineering, University of Patras, Rion Patras, Greece
e-mail: serpanos@isi.gr

P. Siano
Department of Management and Innovations Systems, University of Salerno, Fisciano, Italy
e-mail: psiano@unisa.it

M. Abbaszadeh
GE Global Research, General Electric Company, Niskayuna, NY, USA
e-mail: masouda@ualberta.ca

M. Sayed-Mouchaweh (ed.), *Artificial Intelligence Techniques for a Scalable Energy Transition*, https://doi.org/10.1007/978-3-030-42726-9_6

or filters [29–32]. In the case of (2) there is no available model of the DC-DC converter's dynamics; however, one can generate a nonparametric model in the form of a neural or neurofuzzy network through the processing of raw data of the converter's input and output measurements [33–36]. To perform fault diagnosis, in both cases (1) and (2) the sequence of the real outputs of the converter is compared against the output of the estimated outputs which are generated by the previously mentioned state-observer/filter or the neural network [37–41]. The fault threshold definition problem has been also the subject of intensive research [42–45]. In this area, the local statistical approach to fault diagnosis, allows for optimal selection of the fault threshold and for the early detection of incipient parametric changes [46–50].

In the present chapter, a model-free approach for fault diagnosis of an energy conversion system that consists of a DC-DC converter connected to a DC motor is developed. This method relies on neural modeling of the energy conversion system's dynamics and on the use of a statistical decision making procedure for detecting the existence of a failure [1]. To develop the dynamic model of this energy conversion system, data sets are generated consisting of input–output measurements accumulated from its functioning at different operating conditions. The neural model comprises (1) a hidden layer of basis functions having the form of Gauss–Hermite polynomials, (2) a weights output layer [41]. Weights' adaptation and learning take place in the form of a first-order gradient algorithm. The approximation error is minimized. The neural model that is obtained through this learning procedure is considered to represent the fault-free functioning of the DC-DC converter and of the DC motor system. Next, real output measurements from the energy conversion system are used for diagnosing the appearance of faults.

The real values of the system's outputs are subtracted from the estimated values which come from the neural model. Thus the residuals' sequence is generated [1, 37]. The residuals' data set undergoes statistical processing. It is shown that the sum of the squares of the residuals' vector, being multiplied by the inverse of the associated covariance matrix, stands for a stochastic variable which follows the χ^2 distribution. Next, with the use of the 96% or of the 98% confidence intervals of the χ^2 distribution, one can define fault thresholds which designate with significant accuracy and undebatable certainty the appearance of a fault in the DC-DC converter and DC motor energy conversion system. As long as the value of the aforementioned stochastic variable remains within the upper and the lower bound of the confidence interval it can be concluded that the functioning of the DC-DC converter and of the DC motor remains normal [1]. On the other side, whenever the previously noted bounds are exceeded it can be concluded that the functioning of this energy conversion system has been subject to fault. Moreover, by using different outputs from the energy conversion system, and by repeating the statistical test into subspaces of its state-space model, one can also achieve fault isolation which shows the specific part of the energy conversion system that has exhibited the failure (for instance failure in its mechanical part that is the DC motor, or failure in its electrical part that is the DC-DC converter).

The structure of the paper is as follows: in Sect. 6.2 the dynamic model of the energy conversion system that comprises the DC-DC converter and the DC motor is analyzed and the associated state-space description is given. In Sect. 6.3 modeling of the dynamics of the energy conversion system is performed with the use of Gauss–Hermite neural networks. In Sect. 6.4 a statistical decision making method which is based on the χ^2 distribution is proposed as a tool for performing fault diagnosis in the energy conversion system. In Sect. 6.5 the performance of this fault diagnosis scheme is tested through simulation experiments considering the functioning under different operating conditions of the energy conversion system that incorporates the DC-DC converter and the DC motor. Finally, in Sect. 6.6 concluding remarks are stated.

6.2 Dynamic Model of the Energy Conversion System

Energy conversion systems comprising DC motors controlled through DC-DC converters can be found in several applications, as for instance in photovoltaics-powered pumps or in desalination units (Fig. 6.1). Control is implemented through a pulse-width-modulation (PWM) approach [1]. The equivalent circuit of the system that is formed after connecting a DC motor to a DC-DC (buck) converter is depicted in Fig. 6.2.

Pulse width modulation (PWM) is applied for the converter's control. The amplitude of the output voltage V_o is determined by the duty cycle of the PWM. The on/off state of the switch Q sets voltage u to E or to 0 for specific time intervals within the sampling period. The ratio between the time interval in which $u = E$

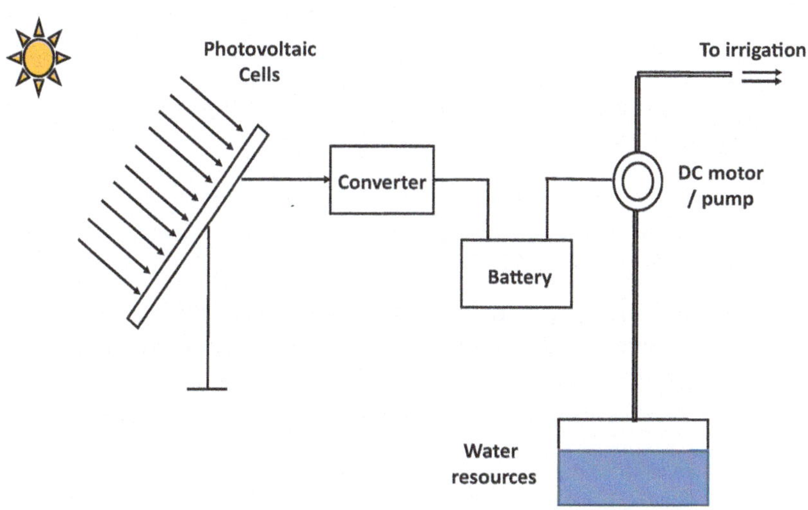

Fig. 6.1 Energy conversion system turning solar power into mechanical power

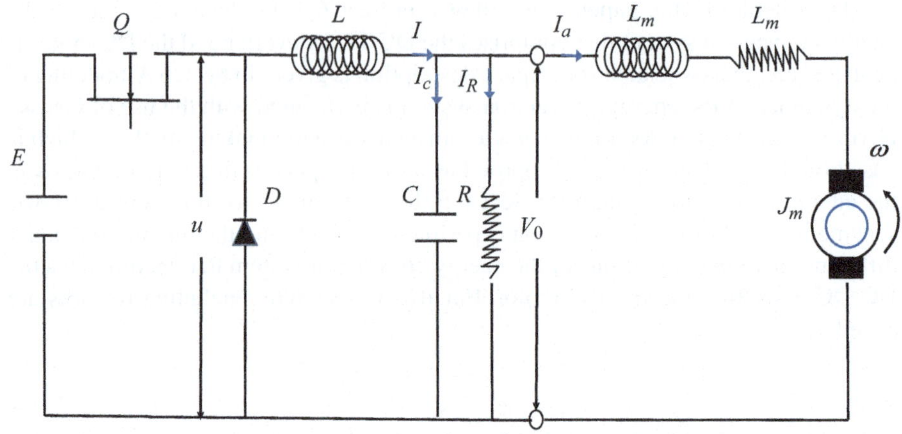

Fig. 6.2 Circuit of the DC-DC converter connected to a DC motor

and the sampling period defines the duty cycle. By varying the duty cycle one can control the voltage output V_o, as if a variable input voltage E was applied to the circuit.

The dynamics of the electrical part of the circuit comes from the application of Kirchhoff's laws. It holds

$$L\frac{dI}{dt} + V_c = u$$

$$L_m\frac{dI_a}{dt} + R_m I_a + K_e\omega = V_c$$

$$I = I_c + I_R + I_a \; or \; I = C\frac{dV_c}{dt} + \frac{V_c}{R} + I_a \tag{6.1}$$

The dynamics of the mechanical part of the circuit comes from the laws of rotational motion. It holds that

$$\dot{\theta} = \omega$$

$$J\dot{\omega} = -B\omega + K_a I_a + \tau_L \tag{6.2}$$

where τ_L is the load's torque. By defining the state variables $x_1 = \theta$, $x_2 = \omega$, $x_3 = I$, $x_4 = V_c$, and $x_5 = I_a$ one obtains the following state-space model:

$$\dot{x}_1 = x_2$$

$$\dot{x}_2 = -\frac{B}{J}x_2 + \frac{K_a}{J}x_5 + \frac{1}{J}\tau_d$$

$$\dot{x}_3 = -\frac{1}{L}x_4 + \frac{1}{L}u$$

$$\dot{x}_4 = \frac{1}{C}x_3 - \frac{1}{RC}x_4 + \frac{1}{C}x_5$$

$$\dot{x}_5 = \frac{1}{L_m}x_4 - \frac{R_m}{L_m}x_5 + \frac{K_e}{L_m}x_2 \tag{6.3}$$

Without loss of generality it is considered that the load's torque is $\tau_L = \text{mgl}\sin(x_1)$ (the motor is considered to be lifting a rod of length l having a mass m at its end), the previous state-space description is written in the following matrix form:

$$\begin{pmatrix} \dot{x}_1 \\ \dot{x}_2 \\ \dot{x}_3 \\ \dot{x}_4 \\ \dot{x}_5 \end{pmatrix} = \begin{pmatrix} x_2 \\ -\frac{B}{J}x_2 + \frac{K_a}{J}x_5 + \frac{1}{J}\text{mgl}\sin(x_1) \\ -\frac{1}{L}x_4 \\ \frac{1}{C}x_3 - \frac{1}{RC}x_4 + \frac{1}{C}x_5 \\ \frac{K_e}{L_m}x_2 + \frac{1}{L_m}x_4 - \frac{R_m}{L_m}x_5 \end{pmatrix} + \begin{pmatrix} 0 \\ 0 \\ \frac{1}{L} \\ 0 \\ 0 \end{pmatrix} u \tag{6.4}$$

6.3 Modeling of the Energy Conversion System with the Use of Neural Networks

6.3.1 Feed-Forward Neural Networks for Nonlinear Systems Modeling

The proposed fault diagnosis approach for energy conversion systems that exhibit nonlinear dynamics, can be implemented with the use of feed-forward neural networks. The idea of function approximation with the use of feed-forward neural networks (FNN) comes from generalized Fourier series. It is known that any function $\psi(x)$ in a L^2 space can be expanded, using generalized Fourier series in a given orthonormal basis, i.e. [1, 37]

$$\psi(x) = \sum_{k=1}^{\infty} c_k \psi_k(x), \ a \leq x \leq b \tag{6.5}$$

Truncation of the series yields in the sum

$$S_M(x) = \sum_{k=1}^{M} a_k \psi_k(x) \tag{6.6}$$

If the coefficients a_k are taken to be equal to the generalized Fourier coefficients, i.e. when $a_k = c_k = \int_a^b \psi(x)\psi_k(x)dx$, then Eq. (6.6) is a mean square optimal approximation of $\psi(x)$.

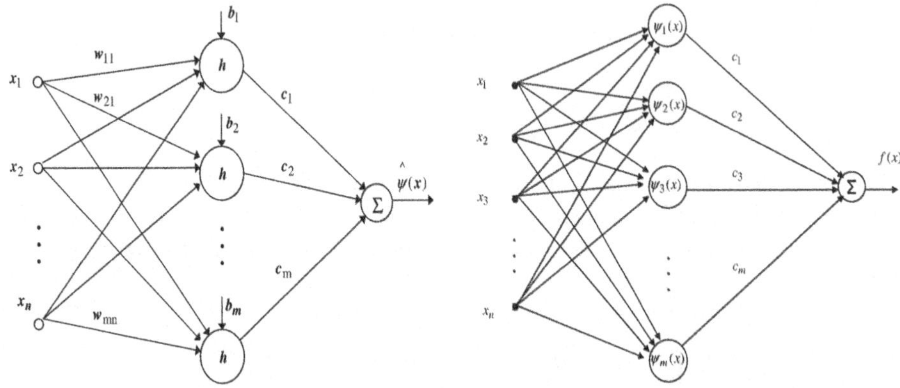

Fig. 6.3 (**a**) Feed-forward neural network (**b**) Neural network with Gauss–Hermite basis functions

Unlike generalized Fourier series, in FNN the basis functions are not necessarily orthogonal. The hidden units in a FNN usually have the same activation functions and are often selected as sigmoidal functions or Gaussians. A typical feed-forward neural network consists of n inputs x_i, $i = 1, 2, \ldots, n$, a hidden layer of m neurons with activation function $h : R \to R$ and a single output unit (see Fig. 6.3a). The FNN's output is given by

$$\psi(x) = \sum_{j=1}^{n} c_j h\left(\sum_{i=1}^{n} w_{ji} x_i + b_j\right) \tag{6.7}$$

The root mean square error in the approximation of function $\psi(x)$ by the FNN is given by

$$E_{RMS} = \sqrt{\frac{1}{N}\sum_{k=1}^{N}\left(\psi\left(x^k\right) - \hat{\psi}\left(x^k\right)\right)^2} \tag{6.8}$$

where $x^k = [x_1^k, x_2^k, \ldots, x_n^k]$ is the k-th input vector of the neural network. The activation function is usually a sigmoidal function $h(x) = \frac{1}{1+e^{-x}}$ while in the case of radial basis functions networks it is a Gaussian [41]. Several learning algorithms for neural networks have been studied. The objective of all these algorithms is to find numerical values for the network's weights so as to minimize the mean square error E_{RMS} of Eq. (6.8). The algorithms are usually based on first and second order gradient techniques. These algorithms belong to: (1) batch-mode learning, where to perform parameters update the outputs of a large training set are accumulated and the mean square error is calculated (back-propagation algorithm, Gauss–Newton method, Levenberg–Marquardt method, etc.), (2) pattern-mode learning, in which training examples are run in cycles and the parameters update is carried out each time a new datum appears (Extended Kalman Filter algorithm) [42].

Unlike conventional FNN with sigmoidal or Gaussian basis functions, Hermite polynomial-based FNN remain closer to Fourier series expansions by employing activation functions which satisfy the property of orthogonality [41]. Other basis functions with the property of orthogonality are Hermite, Legendre, Chebyshev, and Volterra polynomials [41].

6.3.2 Neural Networks Using Gauss–Hermite Activation Functions

6.3.2.1 The Gauss–Hermite Series Expansion

Next, as orthogonal basis functions of the feed-forward neural network Gauss–Hermite activation functions are considered [1, 37]:

$$X_k(x) = H_k(x)e^{\frac{-x^2}{2}}, \; k = 0, 1, 2, \cdots \tag{6.9}$$

where $H_k(x)$ are the Hermite orthogonal functions (Fig. 6.4). The Hermite functions $H_k(x)$ are also known to be the eigenstates of the quantum harmonic oscillator. The general relation for the Hermite polynomials is

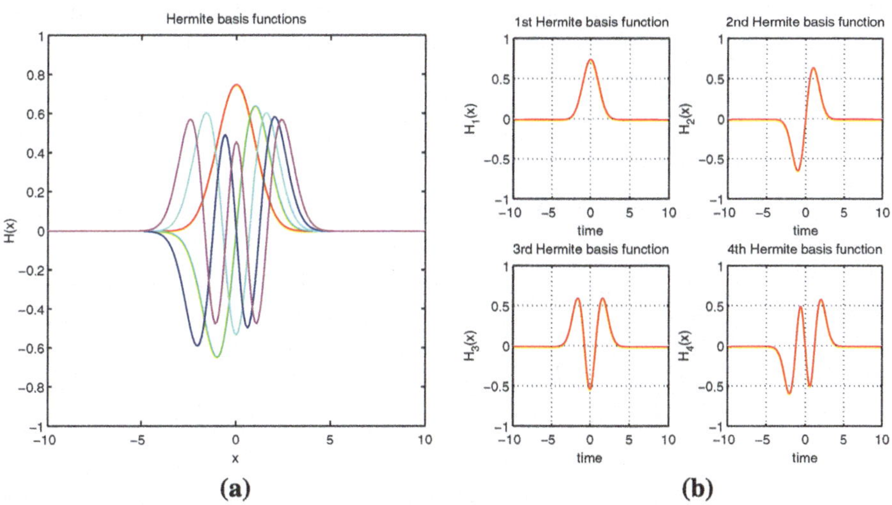

Fig. 6.4 (**a**) First five one-dimensional Hermite basis functions (**b**) Analytical representation of the 1D Hermite basis function

$$H_k(x) = (-1)^k e^{x^2} \frac{d^{(k)}}{dx^{(k)}} e^{-x^2} \tag{6.10}$$

According to Eq. (6.10) the first five Hermite polynomials are

$$H_0(x) = 1, \quad H_1(x) = 2x, \quad H_2(x) = 4x^2 - 2, \quad H_3(x) = 8x^3 - 12x,$$
$$H_4(x) = 16x^4 - 48x^2 + 12$$

It is known that Hermite polynomials are orthogonal, i.e. it holds

$$\int_{-\infty}^{+\infty} e^{-x^2} H_m(x) H_k(x) dx = \begin{cases} 2^k k! \sqrt{\pi} \text{ if } m = k \\ 0 \text{ if } m \neq k \end{cases} \tag{6.11}$$

Using now, Eq. (6.11), the following basis functions can be defined [41]:

$$\psi_k(x) = \left[2^k \pi^{\frac{1}{2}} k!\right]^{-\frac{1}{2}} H_k(x) e^{-\frac{x^2}{2}} \tag{6.12}$$

where $H_k(x)$ is the associated Hermite polynomial. From Eq. (6.11), the orthogonality of basis functions of Eq. (6.12) can be concluded, which means

$$\int_{-\infty}^{+\infty} \psi_m(x) \psi_k(x) dx = \begin{cases} 1 \; if \; m = k \\ 0 \; if \; m \neq k \end{cases} \tag{6.13}$$

Moreover, to achieve multi-resolution analysis Gauss–Hermite basis functions of Eq. (6.12) are multiplied with the scale coefficient α. Thus the following basis functions are derived [41]:

$$\beta_k(x, \alpha) = \alpha^{-\frac{1}{2}} \psi_k\left(\alpha^{-1} x\right) \tag{6.14}$$

which also satisfy the orthogonality condition

$$\int_{-\infty}^{+\infty} \beta_m(x, \alpha) \beta_k(x, \alpha) dx = \begin{cases} 1 \; if \; m = k \\ 0 \; if \; m \neq k \end{cases} \tag{6.15}$$

Any function $f(x)$, $x \in R$ can be written as a weighted sum of the above orthogonal basis functions, i.e.

$$f(x) = \sum_{k=0}^{\infty} c_k \beta_k(x, \alpha) \tag{6.16}$$

where coefficients c_k are calculated using the orthogonality condition

$$c_k = \int_{-\infty}^{+\infty} f(x)\beta_k(x,\alpha)dx \tag{6.17}$$

Assuming now that instead of infinite terms in the expansion of Eq. (6.16), M terms are maintained, then an approximation of $f(x)$ is achieved. The expansion of $f(x)$ using Eq. (6.16) is a Gauss–Hermite series. Equation (6.16) is a form of Fourier expansion for $f(x)$. Equation (6.16) can be considered as the Fourier transform of $f(x)$ subject only to a scale change. Indeed, the Fourier transform of $f(x)$ is given by

$$F(s) = \frac{1}{2\pi}\int_{-\infty}^{+\infty} f(x)e^{-jsx}dx \Rightarrow f(x) = \frac{1}{2\pi}\int_{-\infty}^{+\infty} F(s)e^{jsx}ds \tag{6.18}$$

The Fourier transform of the basis function $\psi_k(x)$ of Eq. (6.12) satisfies [41]

$$\Psi_k(s) = j^k\psi_k(s) \tag{6.19}$$

while for the basis functions $\beta_k(x,\alpha)$ using scale coefficient α it holds that

$$B_k(s,\alpha) = j^k\beta_k\left(s,\alpha^{-1}\right) \tag{6.20}$$

Therefore, it holds

$$f(x) = \sum_{k=0}^{\infty} c_k\beta_k(x,\alpha) \overset{F}{\Rightarrow} F(s) = \sum_{k=0}^{\infty} c_k j^n\beta_k\left(s,\alpha^{-1}\right) \tag{6.21}$$

which means that the Fourier transform of Eq. (6.16) is the same as the initial function, subject only to a change of scale. The structure of a feed-forward neural network with Hermite basis functions is depicted in Fig. 6.3b.

6.3.2.2 Neural Networks Using 2D Hermite Activation Functions

Two-dimensional Hermite polynomial-based neural networks can be constructed by taking products of the one-dimensional basis functions $B_k(x,\alpha)$. Thus, setting $x = [x_1, x_2]^T$ one can define the following basis functions [1, 37]:

$$B_k(x,\alpha) = \frac{1}{\alpha}B_{k_1}(x_1,\alpha)B_{k_2}(x_2,\alpha) \tag{6.22}$$

These two-dimensional basis functions are again orthonormal, i.e. it holds

$$\int d^2x B_n(x,\alpha)B_m(x,\alpha) = \delta_{n_1m_1}\delta_{n_2m_2} \tag{6.23}$$

The basis functions $B_k(x)$ are the eigenstates of the two-dimensional harmonic oscillator and form a complete basis for integrable functions of two variables. A two-dimensional function $f(x)$ can thus be written in the series expansion:

$$f(x) = \sum_{k_1,k_2}^{\infty} c_k B_k(x, \alpha) \tag{6.24}$$

The choice of an appropriate scale coefficient α and maximum order k_{max} is of practical interest. The coefficients c_k are given by

$$c_k = \int dx^2 f(x) B_k(x, \alpha) \tag{6.25}$$

Indicative basis functions $B_2(x, \alpha)$, $B_6(x, \alpha)$, $B_9(x, \alpha)$, $B_{11}(x, \alpha)$ and $B_{13}(x, \alpha)$, $B_{15}(x, \alpha)$ of a 2D feed-forward neural network with Hermite basis functions are depicted in Figs. 6.5, 6.6, and 6.7. Following, the same method N-dimensional Hermite polynomial-based neural networks ($N > 2$) can be constructed. The associated high-dimensional Gauss–Hermite activation functions preserve the properties of orthogonality and invariance to Fourier transform.

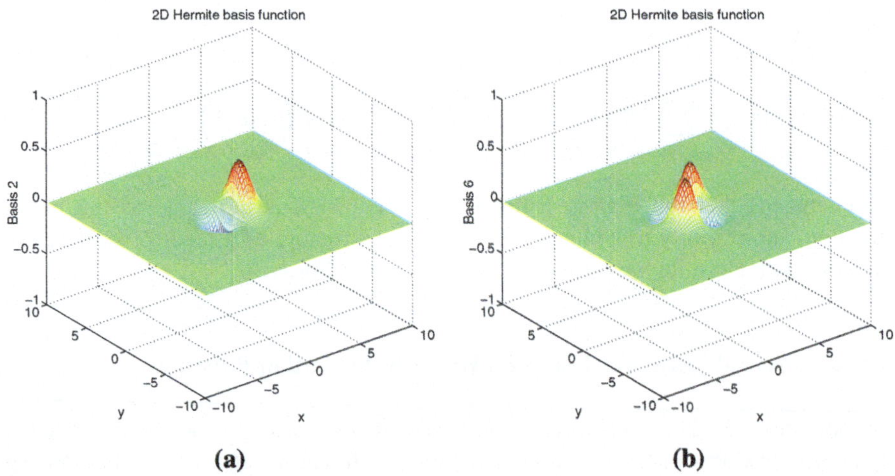

Fig. 6.5 2D Hermite polynomial activation functions: (**a**) basis function $B_2(x, \alpha)$ (**b**) basis function $B_6(x, \alpha)$

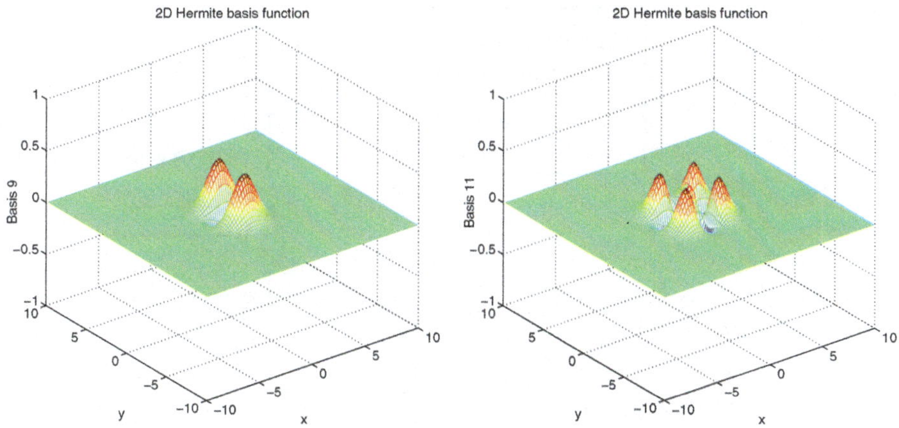

(a) **(b)**

Fig. 6.6 2D Hermite polynomial activation functions: (**a**) basis function $B_9(x, \alpha)$ (**b**) basis function $B_{11}(x, \alpha)$

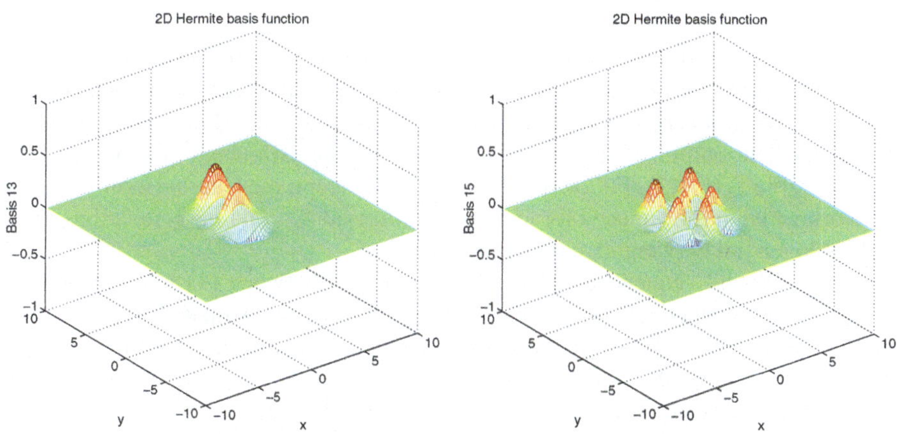

(a) **(b)**

Fig. 6.7 2D Hermite polynomial activation functions: (**a**) basis function $B_{13}(x, \alpha)$ (**b**) basis function $B_{15}(x, \alpha)$

6.4 Statistical Fault Diagnosis Using the Neural Network

6.4.1 Fault Detection

A Gauss–Hermite neural network has been used to learn the dynamics of the energy conversion system. For each functioning mode, the training set comprised $N = 2000$ vectors of input–output data. In each vector the input data were $x_1(k)$, $x_1(k -$

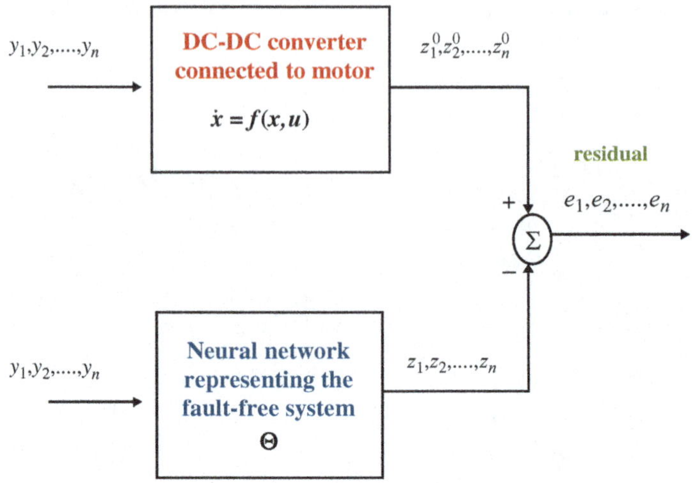

Fig. 6.8 Residuals' generation for the energy conversion system that comprises the DC-DC converter and DC motor, with the use of a neural network

1), $x_1(k-2)$ standing for the three most recent values of the first state variable x_1 of the energy conversion system. It was considered that each input variable can be expressed in a series expansion form using the first four Gauss–Hermite basis functions. The output of the neural network was the estimated value of the first state variable $\hat{x}_1(k+1)$. Following the previous concept, the Gauss–Hermite neural network which has been used for learning, comprised $4^3 = 64$ basis functions in its hidden layer and 64 weights in its output layer.

The residuals' sequence, that is, the differences between (1) the real outputs of the energy conversion power unit and (2) the outputs estimated by the neural network (Fig. 6.8) is a discrete error process e_k with dimension $m \times 1$ (here $m = N$ is the dimension of the output measurements vector). Actually, it is a zero-mean Gaussian white-noise process with covariance given by E_k [1].

A conclusion can be stated that the energy conversion system which comprises the DC-DC converter and the DC motor has not been subjected to a fault. To this end, the following *normalized error square* (NES) is defined [1]:

$$\epsilon_k = e_k^T E_k^{-1} e_k \tag{6.26}$$

The sum of this normalized residuals' square follows a χ^2 distribution, with a number of degrees of freedom that is equal to the dimension of the residuals' vector. The form of the χ^2 distribution for various degrees of freedom is shown in Fig. 6.9. An appropriate test for the normalized error sum is to numerically show that the following condition is met within a level of confidence (according to the properties of the χ^2 distribution):

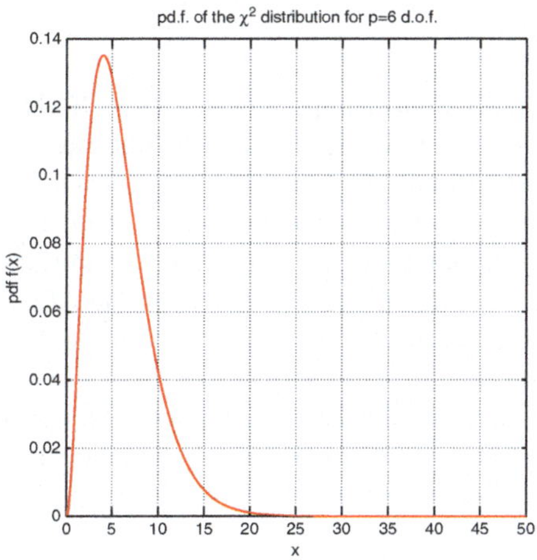

(a)

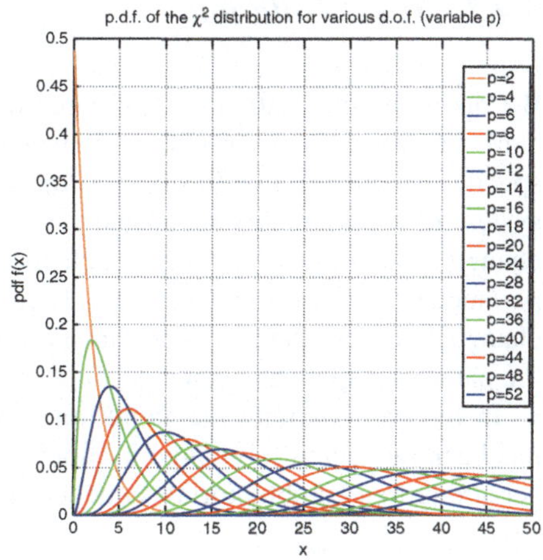

(b)

Fig. 6.9 (**a**) Probability density function of the χ^2 distribution for $p = 6$ degrees of freedom, (**b**) Probability density function of the distribution for several values of the degrees of freedom (variable p)

$$E\{\epsilon_k\} = m \tag{6.27}$$

This can be achieved using statistical hypothesis testing, which is associated with confidence intervals. A 95% confidence interval is frequently applied, which is specified using $100(1-a)$ with $a = 0.05$. Actually, a two-sided probability region is considered cutting-off two end tails of 2.5% each. For M runs the normalized error square that is obtained is given by

$$\bar{\epsilon}_k = \frac{1}{M}\sum_{i=1}^{M}\epsilon_k(i) = \frac{1}{M}\sum_{i=1}^{M}e_k^T(i)E_k^{-1}(i)e_k(i) \tag{6.28}$$

where ϵ_i stands for the i-th run at time t_k. Then $M\bar{\epsilon}_k$ will follow a χ^2 density with Mm degrees of freedom. This condition can be checked using a χ^2 test. The hypothesis holds, if the following condition is satisfied:

$$\bar{\epsilon}_k \in [\zeta_1, \zeta_2] \tag{6.29}$$

where ζ_1 and ζ_2 are derived from the tail probabilities of the χ^2 density. For example, for $m = 1$ (dimension of the measurements vector) and $M = 2000$ (total number of the output vector's samples) one has $\chi^2_{Mm}(0.025) = 1878$ and $\chi^2_{Mm}(0.975) = 2126$. Using that $M = 100$ one obtains $\zeta_1 = \chi^2_{Mm}(0.025)/M = 0.948$ and $\zeta_2 = \chi^2_{Mm}(0.975)/M = 1.052$.

6.4.2 Fault Isolation

By applying the statistical test into the individual components of the DC-DC converter and DC motor energy conversion unit, it is also possible to find out the specific component that has been subjected to a fault [1]. For an energy conversion system of n parameters suspected for change one has to carry out n χ^2 statistical change detection tests, where each test is applied to the subset that comprises parameters $i-1$, i and $i+1$, $i=1,2,\dots,n$. Actually, out of the n χ^2 statistical change detection tests, the one that exhibits the highest score are those that identify the parameter that has been subject to change.

In the case of multiple faults one can identify the subset of parameters that has been subjected to change by applying the χ^2 statistical change detection test according to a combinatorial sequence. This means that

$$\binom{n}{k} = \frac{n!}{k!(n-k)!} \tag{6.30}$$

tests have to take place, for all clusters in the energy conversion system, that finally comprise n, $n-1$, $n-2$, ..., 2, 1 parameters. Again the χ^2 tests that give the highest scores indicate the parameters which are most likely to have been subjected to change.

As a whole, the concept of the proposed fault detection and isolation method is simple. The sum of the squares of the residuals' vector, weighted by the inverse of the residuals' covariance matrix, stands for a stochastic variable which follows the χ^2 distribution. Actually, this is a multi-dimensional χ^2 distribution and the number of its degrees of freedom is equal to the dimension of the residuals' vector. Since, there is one measurable output of the energy conversion system, the residuals' vector is of dimension 1 and the number of degrees of freedom is also 1. Next, from the properties of the χ^2 distribution, the mean value of the aforementioned stochastic variable in the fault-free case should be also 1. However, due to having sensor measurements subject to noise, the value of the statistical test in the fault-free case will not be precisely equal to 1 but it may vary within a small range around this value. This range is determined by the confidence intervals of the χ^2 distribution. For a probability of 98% to get a value of the stochastic variable about 1, the associated confidence interval is given by the lower bound $L=0.934$ and by the upper bound $U=1.066$. Consequently, as long as the statistical test provides an indication that the aforementioned stochastic variable is in the interval $[L,U]$ the functioning of the power unit can be concluded to be free of faults. On the other side, when the bounds of the previously given confidence interval are exceeded it can be concluded that the power unit has been subject to a fault or cyber-attack. Finally, by performing the statistical test into subspaces of the DC-DC converter and DC motor energy conversion unit's state-space model, where each subspace is associated with different components, one can also achieve fault isolation. This signifies that the specific component that has caused the malfunctioning of the power unit can be identified.

6.5 Simulation Tests

The performance of the proposed fault diagnosis method for the energy conversion system that comprises the DC-DC converter and the DC motor has been tested through simulation experiments. Six different functioning modes have been considered. These are depicted in Figs. 6.10, 6.11, and 6.12. The induced changes and disturbances in the energy conversion system were incipient and as confirmed by the previously noted diagrams they can be hardly distinguished by human supervisors of the system. By applying the previously analyzed statistical test which relies on the properties of the χ^2 distribution, results have been obtained about the detection of failures in the energy conversion system. These results are depicted in Figs. 6.13, 6.14, 6.15, 6.16, 6.17, and 6.18. Actually, two types of fault detection tests are presented: (1) consecutive statistical tests carried out on the energy conversion system and (2) mean value of the statistical tests.

The number of monitored outputs in the energy conversion system was $n = 1$. Consequently, according to the previous analysis, in the fault-free case the mean value of the statistical test should be very close to the value 1. In particular considering 98% confidence intervals, the bounds of the normal functioning of the system were $U = 1.066$ (upper bound) and $L = 0.934$ (lower bound). As long as the mean value of the statistical test remains within the bounds of this confidence interval it can be concluded that the functioning of the energy conversion system remains normal. This is shown in the test cases 2, 4, and 5. On the other side,

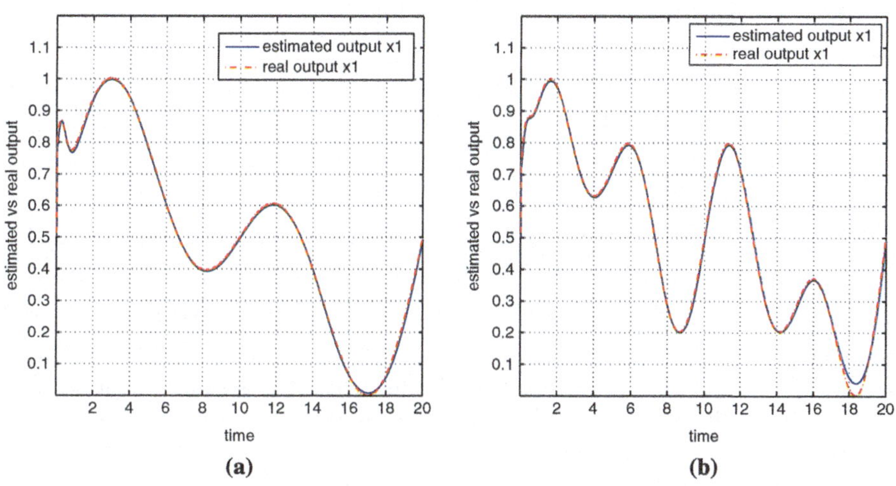

Fig. 6.10 (**a**) Test case 1 (tracking of setpoint 1 and fault at the output of the energy conversion system): Real value (blue) of the system's output x_1 vs estimated value $\hat{x}_1$ (red), (**b**) Test case 2 (tracking of setpoint 2 and no-fault): Real value (blue) of the system's output x_1 vs estimated value $\hat{x}_1$ (red)

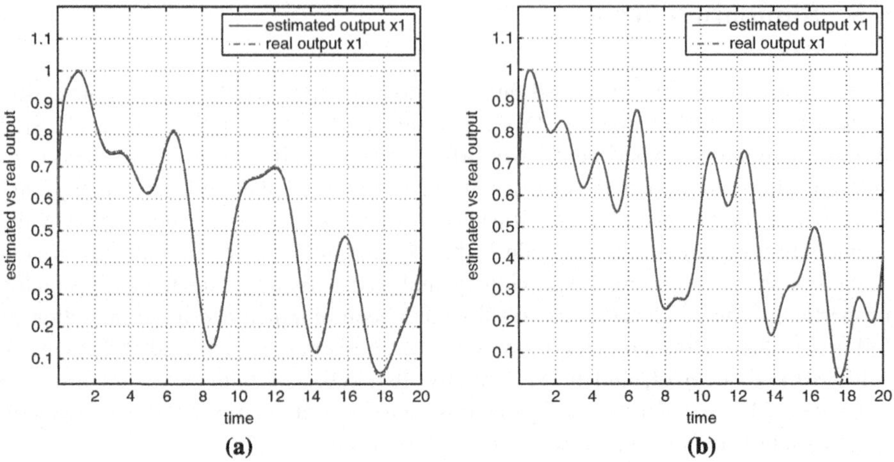

Fig. 6.11 (**a**) Test case 3 (tracking of setpoint 3 and fault at the output of the energy conversion system): Real value (blue) of the system's output x_1 vs estimated value $\hat{x}_1$ (red), (**b**) Test case 4 (tracking of setpoint 4 and no-fault): Real value (blue) of the system's output x_1 vs estimated value $\hat{x}_1$ (red)

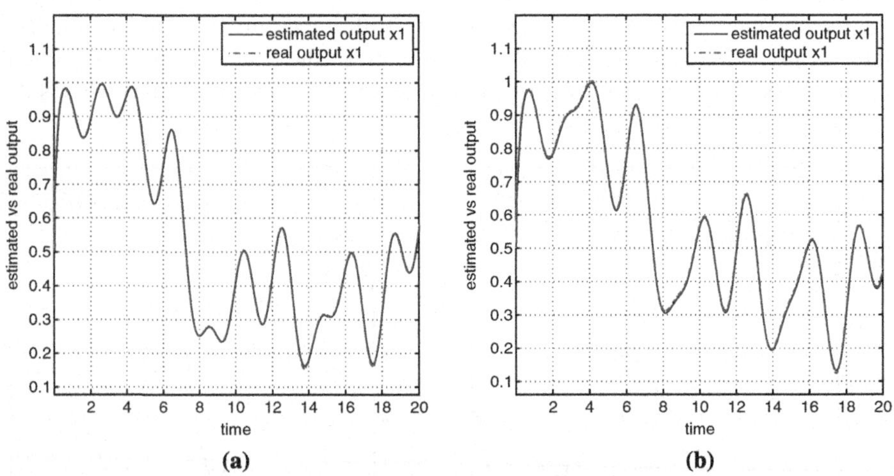

Fig. 6.12 (**a**) Test case 5 (tracking of setpoint 5 and no-fault): Real value (blue) of the system's output x_1 vs estimated value $\hat{x}_1$ (red), (**b**) Test case 6 (tracking of setpoint 6 and fault at the output of the energy conversion system): Real value (blue) of the system's output x_1 vs estimated value $\hat{x}_1$

whenever the value of the statistical test exceeds persistently the aforementioned bounds, then one can infer the appearance of a parametric change or of a strong external perturbation. This is shown in test cases 1, 3, and 6. Finally, it is noted that by performing the statistical test in subspaces of its state-space model one can also achieve fault isolation. For instance, it is possible to conclude the existence of a

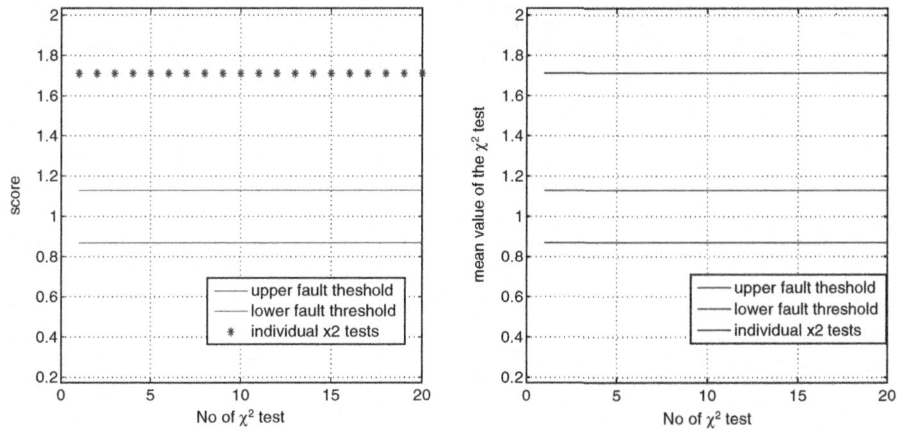

(a) (b)

Fig. 6.13 Test case 1: fault (additive disturbance) at the output of the energy conversion system, (**a**) individual χ^2 tests and related confidence intervals, (**b**) mean value of the χ^2 test and related confidence intervals

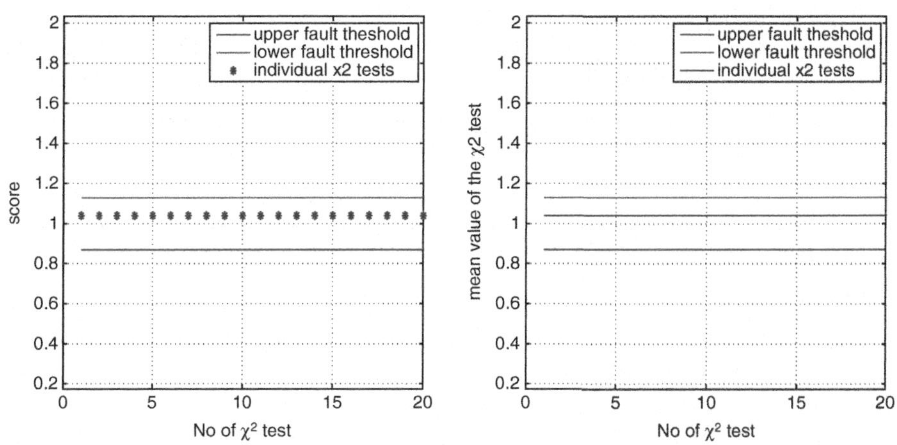

(a) (b)

Fig. 6.14 Test case 2: no-fault, (**a**) individual χ^2 tests and related confidence intervals, (**b**) mean value of the χ^2 test and related confidence intervals

failure in the mechanical part of the energy conversion unit (that is the DC-motor), or the existence of a failure in the electrical part of the energy conversion system (that is the DC-DC converter).

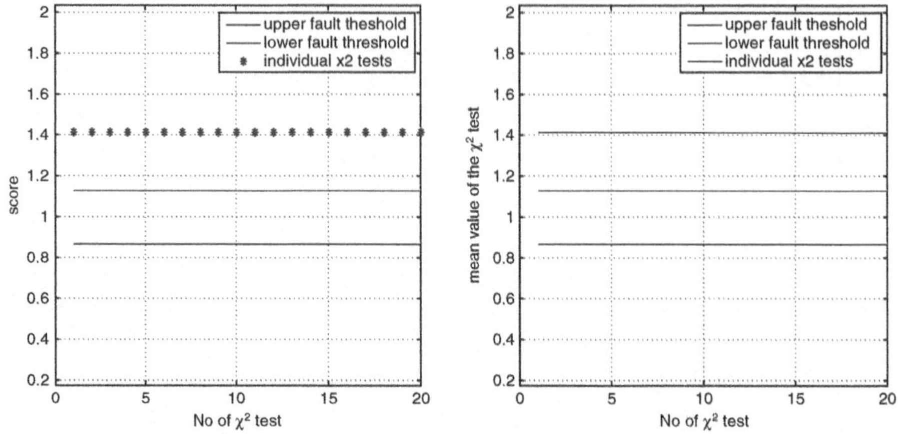

(a) (b)

Fig. 6.15 Test case 3: fault (additive disturbance) at the output of the energy conversion system, (**a**) individual χ^2 tests and related confidence intervals, (**b**) mean value of the χ^2 test and related confidence intervals

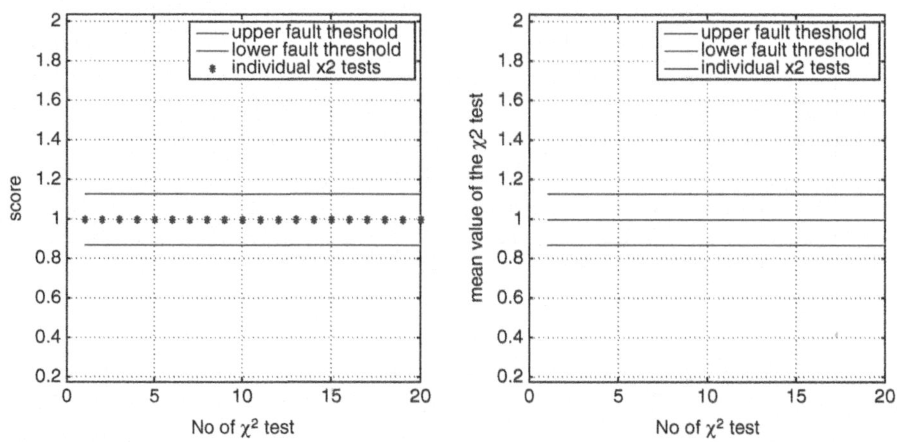

(a) (b)

Fig. 6.16 Test case 4: no-fault, (**a**) individual χ^2 tests and related confidence intervals, (**b**) mean value of the χ^2 test and related confidence intervals

6.6 Conclusions

The chapter has proposed neural modeling and statistical decision making approach for performing condition monitoring in energy conversion systems. As a case study the dynamic model of a solar power unit is considered, when this is connected to DC-DC converter and the latter is used to control precisely a DC motor. First, a neural model of the functioning of the energy conversion unit in the fault-free case is generated. The neural model comprises a hidden layer of Gauss–Hermite

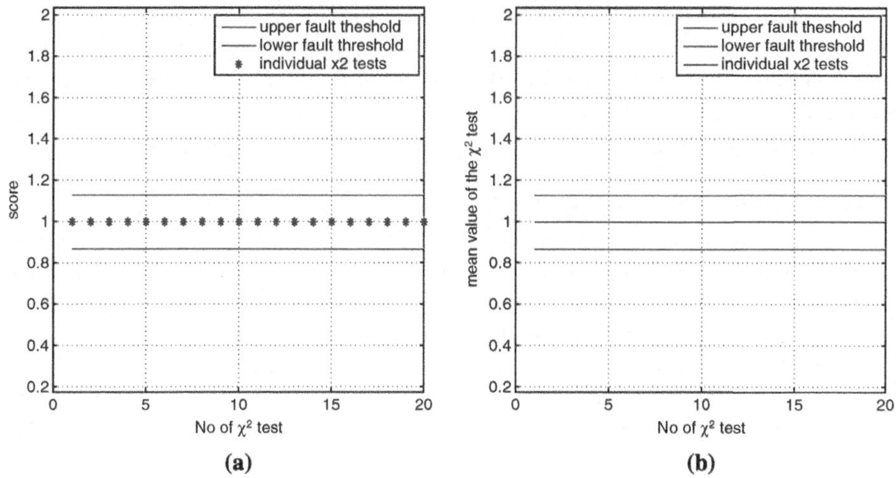

Fig. 6.17 Test case 5: no-fault, (**a**) individual χ^2 tests and related confidence intervals, (**b**) mean value of the χ^2 test and related confidence intervals

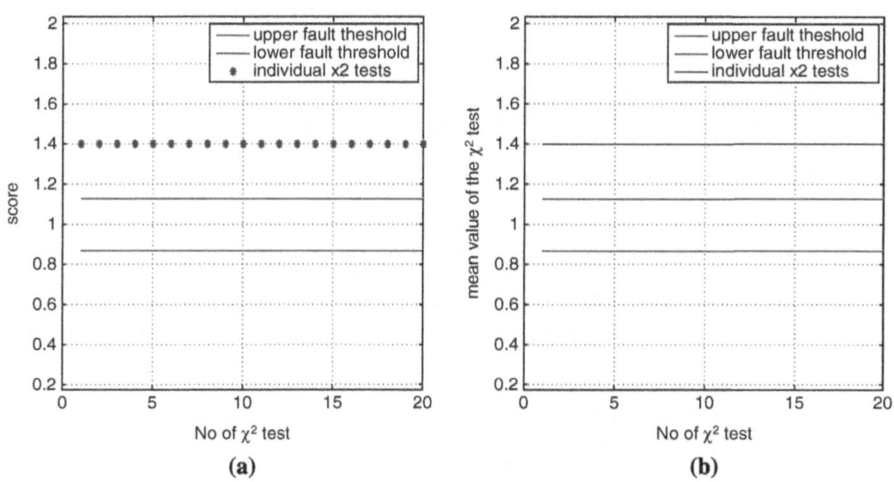

Fig. 6.18 Test case 6L fault (additive disturbance) at the output of the energy conversion system, (**a**) individual χ^2 tests and related confidence intervals, (**b**) mean value of the χ^2 test and related confidence intervals

polynomial activation functions and an output layer of linear weights. The neural network is trained from input–output data of the energy conversion system with the use of a first-order gradient algorithm. Next, by comparing the sequence of the real outputs of the monitored system against the estimated outputs of the neural network the residuals' sequence is generated. The statistical processing of the residuals allows for solving the fault diagnosis problem for the energy conversion unit.

It is shown that the sum of the residuals' vector, being multiplied by the inverse of the associated covariance matrix, stands for a stochastic variable (statistical test) which follows the χ^2 distribution. By exploiting the statistical properties of this distribution one can define a 96% or a 98% confidence interval which reveals in an almost certain, precise, and undebatable manner whether a fault (parametric change) has taken place in the energy conversion unit. As long as the previously noted stochastic variable falls within the upper and the lower bound of the confidence interval one can infer that the functioning of the energy conversion unit remains normal. On the other side, when these bounds are persistently exceeded one could infer that the DC-DC converter and DC-motor energy conversion unit has been subjected to failure. Finally, by performing the statistical test in subspaces of the state-space model of the energy conversion unit fault isolation can be also achieved.

References

1. G. Rigatos, *Intelligent Renewable Energy Systems: Modelling and Control* (Springer, Berlin, 2016)
2. H. Wang, X. Pei, Y. Wu, Y. Xiang, Y. Kang, Switch fault diagnosis method for series-parallel forward DC-DC converter system. IEEE Trans. Ind. Electron. **66**(6), 4684–4695 (2019)
3. X. Pei, S. Nie, Y. Kang, Switch short-circuit fault diagnosis and remedial strategy for full-bridge DC-DC converters. IEEE Trans. Power Electron. **30**(2), 996–1005 (2015)
4. S.Y. Kim, K. Nam, H.S. Song, H.G. Kim, Fault diagnosis of a ZVS DC-DC converter based on DC-link current pulse shapes. IEEE Trans. Ind. Electron. **55**(3), 1491–1495 (2008)
5. M. Dhimish, V. Holmes, B. Mehradi, M. Dales, Simultaneous fault detection algorithm for grid-connected photovoltaic plants. IET Renew. Power Gener. **11**(2), 1565–1575 (2018)
6. R. Platon, J. Martel, N. Wiidruff, T.Y. Chan, Online fault detection in PV systems. IEEE Trans. Sustainable Energy **6**(4), 1200–1207 (2015)
7. A. Triki-Lahiani, A.B.B. Abdelghani, I. Slama-Belkhadjia, Fault detection and monitoring system for photovoltaic installations: a review, in *Renewable and Sustainable Energy Reviews*, vol. 82 (Elsevier, Amsterdam, 2018), pp. 2680–2692
8. S. Lu, B.T. Phang, D. Zhang, A comprehensive review on DC arc faults and their diagnosis methods in photovoltaic systems, in *Renewable and Sustainable Energy Reviews*, vol. 89 (Elsevier, Amsterdam, 2018), pp 89–98
9. S. Nie, X. Pei, Y. Chen, Y. Kang, Fault diagnosis of PWM DC-DC converters based on magnetic component voltages equation. IEEE Trans. Power Electron. **29**(9), 4978–4989 (2014)
10. H Song, F. Wang, W. Tipton, A fault detection and protection scheme for three-level DC-DC converters based on monitoring flying capacitor voltage. IEEE Trans. Power Electron. **27**(2), 685–697 (2012)
11. A. Gnanasaravanan, M. Rajaram, Artificial neural network for monitoring the asymmetric half-bridge DC-DC converter. Electr. Power Energy Syst. **43**, 788–792 (2012). Elsevier
12. W. Xiaowei, S. Guobing, G. Jie, W. Xiangiang, W. Yanfang, M. Khesti, H. Zhigue, Z. Zhigang, High impedance fault detection method based on improved complete ensemble empirical mode decomposition for DC distribution network. Electr. Power Energy Syst. **107**, 538–556 (2019) Elsevier
13. V.M. Catterson, E.M. Davidson, S.D.J. McArthur, Embedded intelligence for electrical network operation and control. IEEE Intell. Syst. **26**(2), 38–45 (2011)
14. V.M. Catterson, E.M. Davidson, S.D.J. McArthur, Agents for active network management and condition monitoring in the smart grid, in *Proceeding of the First International Workshop on*

Agent Technologies for Energy Systems (ATES), Toronto, Canada (2010)
15. G. Rigatos, N. Zervos, D. Serpanos, V. Siadimas, P. Siano, M. Abbaszadeh, Fault diagnosis of gas-turbine power units with the Derivative-free nonlinear Kalman Filter. Electr. Power Syst. Res. **174**, 105810 (2019). Elsevier
16. X. Pei, S. Nie, Y. Chen, Y. Kang, Open-circuit fault diagnosis and fault tolerant strategies for full bridge DC-DC converters. IEEE Trans. Power Electron. **27**(5), 2550–2575 (2012)
17. G. Cirrincione, M. Cirrincione, D. Guilbert, A. Mohammadi, V. Ranodzo, Power switch open-circuit fault detection in an interleaved DC/DC buck converter for electrolyzer applications by using curvilinear component analysis, in *2018 21st International Conference on Electrical Machines and Systems, Jeju, Korea (IEEE ICEMS 2018)* (2018)
18. J. Hannanen, J. Hankanen, J.P. Strom, T. Karakanen, P. Silvenoinen, Capacitor aging detection in a DC-DC converter output stage. IEEE Trans. Ind. Appl. **52**(4), 3234–3243 (2016)
19. O. Sun, Y. Wang, Y. Jiang, A novel fault diagnosis approach for DC-DC converters based on CSA-DBN. IEEE Access **8**, 6273–8285 (2018)
20. K. Ambushaid, V. Pickert, B. Zahawi, New circuit topology for fault tolerant H-bridge DC-DC converter. IEEE Trans. Power Electron. **15**(6), 1509–1516 (2010)
21. T. Kamel, Y. Biletskiy, L. Chang, Fault diagnosis for industrial grid-connected converters in the power distribution system. IEEE Trans. Ind. Electron. **62**(19), 6496–6507 (2015)
22. K. Bi, Q. An, J. Duan, K. Gai, Fast diagnostic method of open circuit fault for modular multilevel DC-DC converter. IEEE Trans. Power Electron. **32**(5), 3292–3296 (2017)
23. S.M. Alavi, M. Saif, S. Fekriasi, Multiple fault detectiom and isolation in DC-DC converters, in *IEEE IECON 2014, 40th Annual Conference of IEEE on Industrial Electronics, Texas, Dallas* (2014)
24. Z. Jin, Z. Lin, C.M. VEng, X. Lin, Real-time response-based fault analysis and prognostic techniques of non-isolated DC-DC converters. IEEE Access **7**, 67996–68009 (2019)
25. M. Dhimish, V. Holmes, Fault detection algorithms for grid-connected photovoltaic plants. Sol. Energy **137**, 236–245 (2016). Elsevier
26. M. Dhimish, V. Holmes, M. Dales, Parallel fault-detection algorithm for grid-connected photovoltaic plants. Renew. Energy **113**, 94–111 (2017). Elsevier
27. M. Dhimish, V. Holmes, B. Mehradi, M. Dales, P. Mather, Photovoltaic fault detection algorithm based on theoretical curves modelling and fuzzy classification systems. Energy **140**, 276–290 (2017). Elsevier
28. A. Mohammadi, D. Guilbert, A. Gaillard, D. Bougain, D. Khabouri, A Djerdir, Fault diagnosis between PEM fuel cell and DC-DC converter using neural network for automotive applications, in *Proceeding of 39th IEEE International Conference on Industrial Electronics, Vienna, Austria (IEEE IECON 2015)* (2013)
29. M.M. Morato, D.J. Regner, Pr.C. Mendes, J.E. Noremy-Rice, C. Bordons, Fault analysis, detection and estimation for a microgrid via H_2/H_∞ LPV observers. Electr. Power Energy Syst. **105**, 823–825 (2019). Elsevier
30. F. Bento, A.J. Marques Cardoso, A comprehensive survey on fault diagnosis and fault tolerance of DC-DC converters. Chin. J. Electr. Eng. **4**(3), 1–12 (2018)
31. H. Reneaudineau, J.P. Martin, B. Nahid-Mabarakeh, S. Pierfederici, DC-DC converters dynamic modelling with state observer-based parameter estimation. IEEE Trans. Power Electron. **30**(6), 3356–3363 (2015)
32. Z. Cen, P. Stewart, Condition parameter estimation for photovoltaic buck converters based on adaptive model observers. IEEE Trans. Reliab. **66**(1), 148–162 (2017)
33. A. Luchietta, S. Manetti, M.C. Piccirilli, A. Reatti, F. Corti, M. Catelani, L. Ciani, M.K. Kaziemierczuk, MLMVNN for parameter fault detection in PWM DC-DC converters and its applications for back and boost DC-DC converters. IEEE Trans. Instrum. Meas. **68**(2), 439–449 (2019)
34. I.M. Karmacharya, R. Gokaraju, Fault location in ungrounded photovoltaic system using wavelets and ANN. IEEE Trans. Power Delivery **33**(2), 549–560 (2018)
35. Q. Yang, J. Li, S. LeBland, C. Wang, Artificial neural network-based fault detection and fault location in the DC microgrid. Energy Procedia **103**, 129–134 (2016). Elsevier

36. M. Dhimish, V. Holmes, B. Mehradi, M. Dales, Comparing Mamdani Sugeno fuzzy logic and RBF ANN network for PV fault detection. Renew. Energy **117**, 257–274 (2017). Elsevier
37. G.G. Rigatos, P. Siano, A. Piccolo, A neural network-based approach for early detection of cascading events in electric power systems. IET J. Gener. Transm. Distrib. **3**(7), 650–665 (2009)
38. R. Isermann, *Fault-Diagnosis Systems: An Introduction from Fault Detection to Fault Tolerance* (Springer, Berlin, 2006)
39. C.J. Harris, X. Hang, Q. Gon, *Adaptive Modelling, Estimation and Fusion from Data: A Neurofuzzy Approach* (Springer, Berlin, 2002)
40. L.X. Wang, *A Course in Fuzzy Systems and Control* (Prentice-Hall, Englewood, 1998)
41. G.G. Rigatos, S.G. Tzafestas, Neural structures using the eigenstates of the Quantum Harmonic Oscillator, in *Open Systems and Information Dynamics*, vol. 13(1) (Springer, Berlin, 2006)
42. G. Rigatos, Q. Zhang, Fuzzy model validation using the local statistical approach. Fuzzy Sets Syst. **60**(7), 882–904 (2009). Elsevier
43. G. Rigatos, P. Siano, A. Piccolo, Incipient fault detection for electric power transformers using neural modeling and the local statistical approach to fault diagnosis, in *IEEE SAS 2012, 2012 IEEE Sensors Applications Symposium, University of Brescia, Italy* (2012)
44. Q. Zhang, M. Basseville, A. Benveniste, Early warning of slight changes in systems special issue on statistical methods in signal processing and control. Automatica **30**(1), 95–113 (1994)
45. Q. Zhang, A. Benveniste, Wavelet networks. IEEE Trans. Neural Netw. **3**(6), 869-898 (1993)
46. M. Basseville, I. Nikiforov, *Detection of Abrupt Changes* (Prentice Hall, Englewood, 1993)
47. A. Benveniste, M. Basseville, G. Moustakides, The asymptotic local approach to change detection and model validation. IEEE Trans. Autom. Control **32**(7), 583–592 (1987)
48. M. Basseville, A. Benveniste, Q. Zhang, *Surveilliance d' installations industrielles: démarche générale et conception de l' algorithmique* (IRISA Publication Interne No 1010, 1996)
49. Q. Zhang, Fault detection and isolation with nonlinear black-box models, in *Proceeding of SYSID, Kitakyushu, Japan* (1997)
50. Q. Zhang, M. Basseville, A. Benveniste, Fault detection and isolation in nonlinear dynamic systems: a combined input-output and local approach. Automatica **34**(11), 1359–1373 (1998). Elsevier

Chapter 7
Support Vector Machine Classification of Current Data for Fault Diagnosis and Similarity-Based Approach for Failure Prognosis in Wind Turbine Systems

Samir Benmoussa, Mohand Arab Djeziri, and Roberto Sanchez

7.1 Introduction

Wind turbines (WTs) are complex systems, subject to a hostile environment that promotes accelerated aging of components, which increases maintenance cost and operational expenditure (OPEX). So, the development of methods of fault diagnosis and remaining useful life (RUL) estimation will make it possible to substitute preventive or corrective maintenance strategies by a conditional maintenance strategy to reduce the downtime for maintenance, as well as the number and the cost of interventions.

Although many physical models have been developed for the design control of WTs, fault diagnosis and prognosis methods based on physical models such as observer-based methods, parity space, or analytical redundancy are not widely used compared to data-driven approaches because they require system observability, a condition often unverified because of the cost of the required instrumentation and the difficulty of deploying some sensors on the system. Data-driven methods are the most used in this field, but the cost of maintenance and the high number of downtimes for maintenance show the limitations of these approaches, related to the facts that the databases available for learning are incomplete and do not cover all

S. Benmoussa (✉)
LASA, Badji-Mokhtar Annaba University, Annaba, Algeria
e-mail: samir.benmoussa@univ-annaba.dz

M. A. Djeziri
LIS, UMR 7020, Marseille, France
e-mail: mohand.djeziri@lis-lab.fr

R. Sanchez
Facultad de Ingenieria Electrica, Universidad Michoacana de San Nicolas de Hidalgo, Morelia, Mexico
e-mail: rtsanchez@dep.fie.umich.mx

M. Sayed-Mouchaweh (ed.), *Artificial Intelligence Techniques for a Scalable Energy Transition*, https://doi.org/10.1007/978-3-030-42726-9_7

the normal, degraded, and failure modes of the system. In addition, the correlations between the measured variables and the degradation phenomena are not formally demonstrated, but based on expert knowledge.

To address these issues, a hybrid method for fault diagnosis and failure prognosis is proposed in this work. The proposed method is implementable on all existing WTs as it does not require additional sensor placement, since it is based on the current measurement available on all WTs. To compensate the unavailability of data for learning, the deep physical knowledge of the phenomena of transformation of wind power into mechanical power and then into electrical power, the phenomena of power conservation and dissipation, as well as the hardware components that make up the wind turbine (blades, hub, main Bearing, main shaft, gearbox, brake, high speed shaft, and generator) and their interactions, are used in this work to build a physical model using the Bond Graph (BG) methodology. The causal and structural properties of the bond graph allow this model to be viewed as a directed graph, while respecting the cause-to-effect relationships. These properties are used to demonstrate the causal relation between the measured variable and the considered degradation, i.e., to demonstrate that the measured variable carries information about the degradation. The model is then used to generate a database that covers all the features needed for fault diagnosis and prognosis. Structured and unstructured uncertainties are taken into account by generating thresholds in the classification stage.

The fault diagnosis task is based on the classification of the attributes like the root mean square (RMS) and the absolute mean extracted from the current measurements of the WT generator using a multi-class support vector machine (MC-SVM) classifier. The result of the classifier is taken as the fault detection and location decision. In some cases, and due to an overlapping in the extracted attributes, the location of the fault is not correctly determined. To provide an accurate and a reliable decision on the fault class, a criterion based on the rate of appearance of a class on a moving window is used. The idea is to compare this rate with a fixed threshold a priori defined. Regarding the fault prognosis, the proposed method for RUL estimation is based on the similarity measurement between the reference attributes of the failure operation, identified offline, and the current attributes calculated continuously online at each sampling time. This geometric approach based on Euclidean distance calculation does not require prior knowledge about the profile of the degradation process.

7.2 Related Works

To assess the operation of WT systems, several methods have been developed. They can be divided into four main approaches: Reliability based approach, Similarity-based approach, Model based approach, and Data-driven based approach. Data-driven approaches, including vibration signals processing, statistical and artificial intelligence, are the most used. Since the WT system is an assembly of mechanical and electrical systems, subjected to strong vibrations whose features change depending on the conditions of use and the operating state of the system, the vibration-based methods for fault diagnosis are widely used [21]. These methods use velocity and acceleration measurements and signal processing techniques such as Fast Fourier Transform (FFT) [9, 24, 32] or time-frequency analysis [5, 16, 29, 31]. These methods require the installation of sensors and equipment in each part of the WT, hence the efficiency and cost effectiveness of these methods remain an open field for researchers. To overcome the issue related to the instrumentation cost, methods based on demodulation techniques of current generator signals have emerged [1, 10], as they need few sensors for implementation. These techniques are limited to mechanical faults that lead to stator current amplitude modulation such as air gap eccentricity and bearing wear.

When enough information is available on the degradation profile and its matching with the use conditions, stochastic models as the Wiener and Gamma processes are used [7, 18] for modeling fault indicators generated using techniques of multivariate analysis as the principal component analysis [8] or independent component analysis for non-Gaussian and nonlinear processes [34]. These methods are preferred over the vibration-based methods when the measured signals are non-stationary. However, an accurate and well identified model is necessary.

AI methods for fault diagnosis of WTs are based on attributes extraction from the historical data in different system operation and their classification in normal and faulty operation classes. Examples of used classifiers are: SVM [14], auto-adaptive dynamical clustering (AuDyC) [25], and artificial neural network (ANN) [33]. The used data are often provided by a supervisory control and data acquisition (SCADA) system information, which is standard equipment on WTs. The state of the system operation can be concluded through a thorough and rigorous analysis of these data, in association with an intelligent decision system. Chen et al.[4] use the SCADA information with an adaptive neuro-fuzzy inference system to detect a pitch fault. Verma [28] proposes a data-mining approach of SCADA operation data to monitor the WT components. The power curve and operation data obtained from SCADA are analyzed for fault diagnosis and prognosis in WT [27]. The most advantages of these methods are: the evolutionary aspect as learning can be improved throughout the operating life of the system whenever new real databases are available especially on systems like WTs, designed to operate for decades, the ability to use different types of data, the ability to integrate expert knowledge, and the availability of increasingly powerful data processing and calculation tools. However, in practice, the available data do not cover all the characteristics of the system in normal, degraded, and faulty operation.

In the light of the above, designing a fault diagnosis for WT system, one must take into consideration the complex and the nonlinear aspect of the system, its architecture measurements, and the available historical data in different operation.

Regarding the fault prognosis for WT which is defined as the estimation of the remaining useful life (RUL), and the estimation of the risk of subsequent development or existence of one or more faulty mode (Norm ISO 13381-1:2004); few works exist in the literature. Most are based on SCADA data exploration in association with AI based methods. In these prognostic approaches, two modules are generally distinguished, a diagnostic module used to extract health indices from raw data and a prognostic module where time series prediction methods associated with online update algorithms. The projection of the evolution of health indices until the predefined thresholds of total failure allows the RUL prediction.

In [13], a data-mining approach is applied on data provided by SCADA to build models predicting possible faults. Based on the wind data and the power output data provided by SCADA, the fault-related data is analyzed at three levels: the existence of a status or a fault is predicted in level 1, the fault category is determined in level 2, and the specific fault is predicted in level 3. After a pre-processing of a raw data, several data-mining algorithms are applied to extract the prediction models, including the neural network (NN), the Standard Classification and Regression Tree (CART), the Boosting Tree Algorithm (BTA), and SVM. To compare the prediction results, three metrics are used: Accuracy, which provides the percentage of correctly made predictions, sensitivity, expressed as the percentage of correctly predicted faults, and specificity, defined as the percentage of correctly predicted normal instances. Chen et al. [3] propose a priori knowledge-based adaptive neuro-fuzzy inference system (APK-ANFIS) applied to pitch fault SCADA data. Zhao et al.[35] use SVM method to detect abnormality in WT generator based on SCADA data, and an autoregressive integrated moving average (ARIMA) based statistical model to conduct fault prognosis task.

7.3 Modeling and Structural Analysis

7.3.1 WT Modeling

The case study considered in this work is a 750 kW WT. It consists of a three-bladed rotor, a hub, a main bearing, a main shaft, a gearbox with a braking system, a high speed shaft, and a generator connected to a three-phase network as illustrated by the word BG of Fig. 7.1 [8, 22].

In the considered model, the blade is divided into two sections of lengths l and masses M and $M2$ (see Fig. 7.2). It is considered as a system with a flexible structure which can be divided into several sections, and which represents a rotational movement and a translation movement generated by aerodynamic forces. This

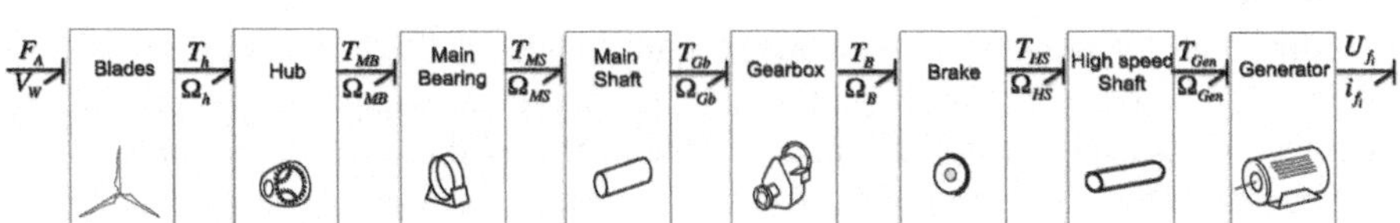

Fig. 7.1 Word BG of the WT system

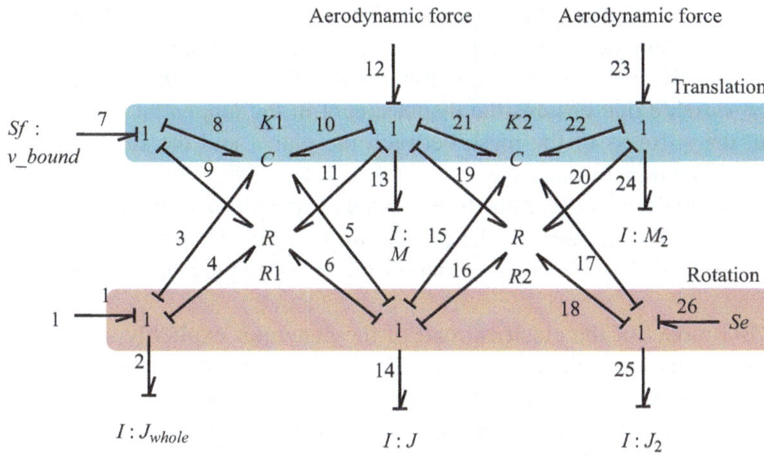

Fig. 7.2 BG model of the blade

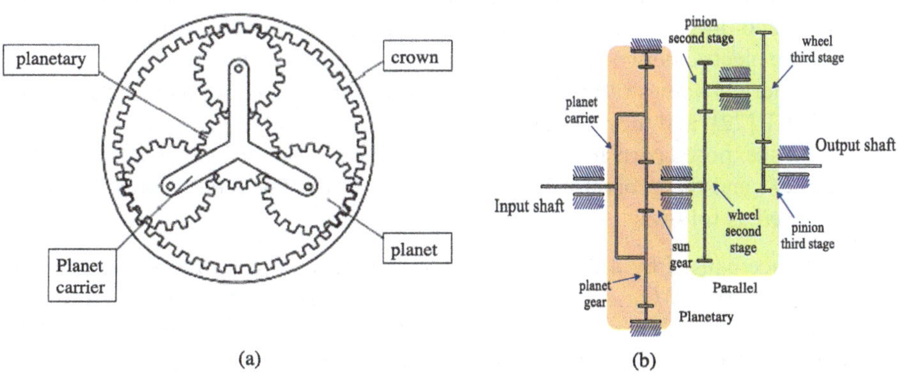

Fig. 7.3 Schematic of the gearbox. (**a**) cross section of WT gearbox (**b**) corresponding schema

dynamic can be modeled by a C-multiport element and a R-multiport element in order to take into the account the two movements, the blade rigidity (element C) and the structural damping (element R) of the blades. The aerodynamic load is caused by the flow (wind) that flows through the structure. The aerodynamic forces applied on the blades are calculated using the fundamental laws of aerodynamics, especially the blade element momentum theory (BEM). The latter is described in detail in [22].

The gearbox of the WT considered in this work consists of three stages: one planetary stage and two other parallel stages to increase the angular velocity (Fig. 7.3) with a complete gain of 60. The corresponding BG model (Fig. 7.4) shows the three planetary gears represented by their moment of inertia $Jpi_{(i=1,\ldots,3)}$. These planets are connected to the sun and the crown by meshes of stiffness $Kspi_{(i=1,\ldots,3)}$ and $Krpi_{(i=1,\ldots,3)}$, respectively. In the model, $Zj_{(j=p,s,r)}$ represents the number of teeth of each gear. The flux junctions 1_{1-3}, 1_4, 1_5, 1_6 represent the angular velocities of the planets, the support, the ring, and the sun, respectively. They are linked to one another by TF elements. For example, the planet 1 (see model of Fig. 7.4) is modeled by a junction 0_1, between the transformer TF_1, TF_2, and TF_3, which represents the force variable that ensures the displacement in the tangential direction. In this junction, the stiffness of the mesh between the planet and the sun is modeled by element C. The transformers $TF_1 : Zs$, $TF_1 : 1/Zs$, $TF_3 : Zp$ allow obtaining the rotation speed of the sun gear, the linear speed of revolution of the planet revolution around the sun gear as well as the planet autorotation. This structure is the same for junction 0_2 but considering the relationship between the planet, the ring, and the carrier.

The BG model of the electrical part of the generator explicitly shows the three phases of the stator and the rotor. Rs and Rr are the resistances of the stator and the rotor, the multiport IC represents the inductance (L_S, L_R) and the mutual inductance (L_M) between the stator and the rotor. The mechanical part is characterized by the inertia J_{ind}, the torque T (which is supplied by the gearbox stage, link 127)), and the electromagnetic torque Te (link 129). In order to allow a direct connection to the electrical network, a squirrel configuration of the rotor cage is used (connections 144, 146, and 148). Elements L and R, in series between the generator and the three-phase electric network (3 MTF elements), are used as load.

All numerical values of the BG elements are identified based on real data of the 750 kW WT [22], and they are given in Table 7.4.

The most exposed parts to defects in WT system are the rotor hub with the blades, the gearbox, and the generator [1, 6, 28]. In this work, these faults are considered and modeled as follows:

- Unbalance caused by a deformation of the blade, it is modeled by a modulated source of effort ($Mse : F_1$) whose excitation signal has the form $A_1 \sin(2\pi f_1)$ where A_1 is the unbalance amplitude and f_1 is the unbalance frequency equal to the hub rotation frequency.
- Unbalance in high speed shaft affects the shaft between the gearbox and the generator, it is modeled by a modulated source of effort ($Msf : F_2$) whose excitation signal has the form $A_2 \sin(2\pi f_2)$ where A_2 is the unbalance amplitude and f_2 is the unbalance frequency equal to the high speed shaft rotation frequency.
- Stator eccentricity in the generator modeled by a modulated source of effort ($Mse : F_3$) whose excitation signal has the form $A_3 \sin(2\pi f_3)$ where A_3 is the eccentricity amplitude and f_3 is the network frequency (50 hz)
- Electrical fault, which affects the stator resistive element $R : R_s$, modeled by a modulated source of effort ($Mse : F_4$). The considered electrical fault is an abrupt fault of an amplitude f_4 introduced at time t.

  To conduct the simulation, a RL load (596 kW, 447 kVAR) is connected to an induction machine of 1000 Hp, 460 V at 50 Hz. The following assumptions are considered:

  - Wind never exceeds its nominal values; this allows the wind turbine to operate without pitch control.

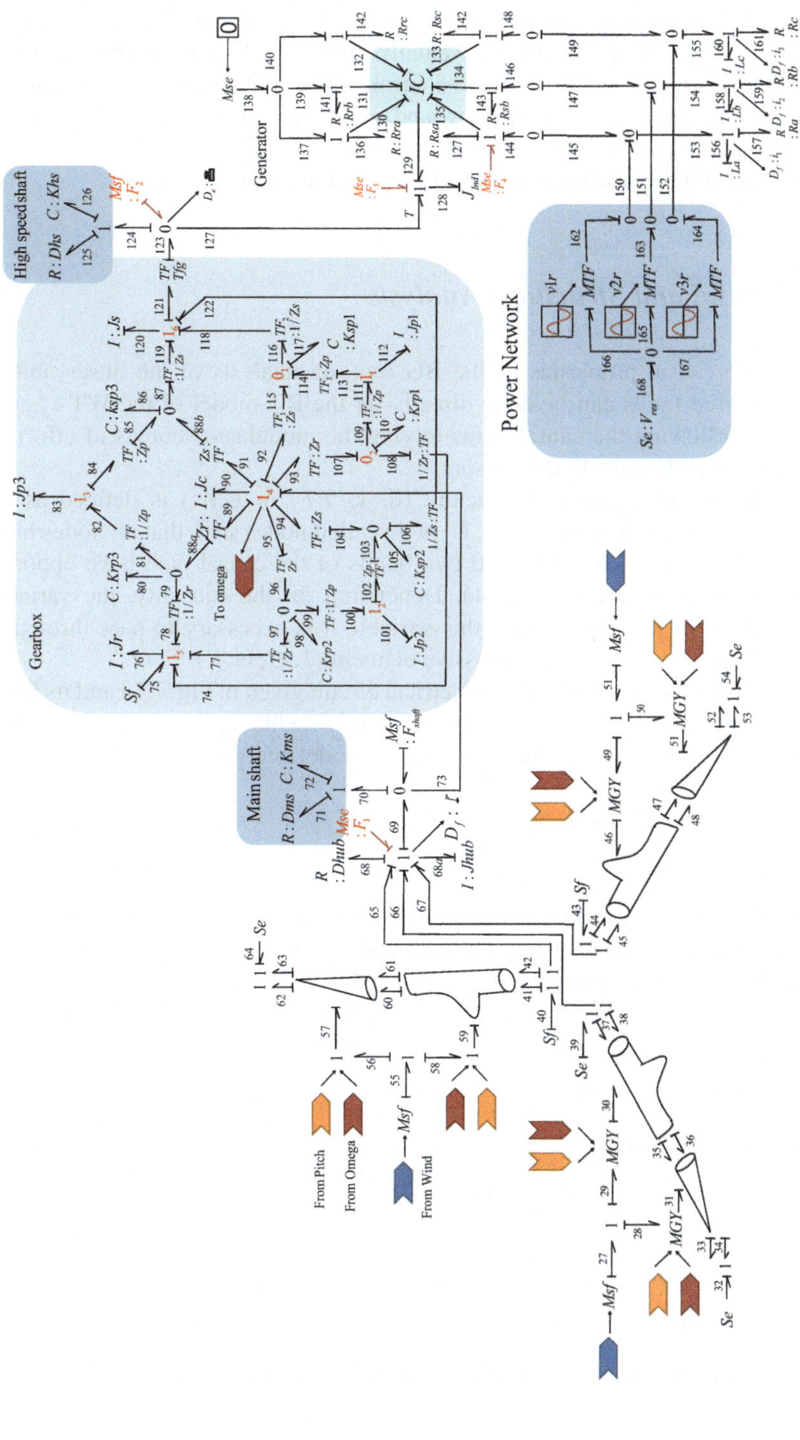

Fig. 7.4 BG model of the WT

- There are no power losses at the gearbox stage.
- A load is connected between generator and the power network, this allows to have an exchange of power. Actually, this consideration allows the wind turbine to work without speed control, since induction machine rotates depending on network frequency and pole number. It means that omega (angular velocity) is imposed by the generator.
- The sampling frequency is imposed by the simulation software.

7.3.2 Causal and Structural Analysis

Thanks to the causal properties of the BG tool, an analysis of the observability of the considered faults can be done directly on the BG model of the WT system (Fig. 7.4) by following the causal paths relying the modulated sources of effort or flow associated with faults to the sensors.

A causal path in a junction structure (0, 1, TF, or GY) is defined as an alternation of bonds and elements (R, C, or I) called nodes such that all nodes have a complete and correct causality, and two bonds of the causal path have opposite causal orientations in the same node. Depending on the causality, the variable crossed is effort or flow. To change this variable it is necessary to pass through a junction element GY, or through a passive element (I, C, or R).

Let us consider the example of an electrical circuit given in Fig. 7.5a and its bond graph model given in Fig. 7.5b. The causal path relying the input voltage $Se : E$ and the output voltage $De : U_C$ is illustrated on the model and given as follows:

$$\begin{aligned} Se : E \rightarrow e_1, e_2 \rightarrow I : L \rightarrow f_2, f_4, f_5 \rightarrow C : C \\ \rightarrow e_5, e_6 \rightarrow De : U_C \end{aligned} \tag{7.1}$$

The same reasoning is considered when the input of the system is a fault, the causal path makes it possible to visualize the fault path in the system up to the measured output, as well as the components of the system that will be affected by the presence

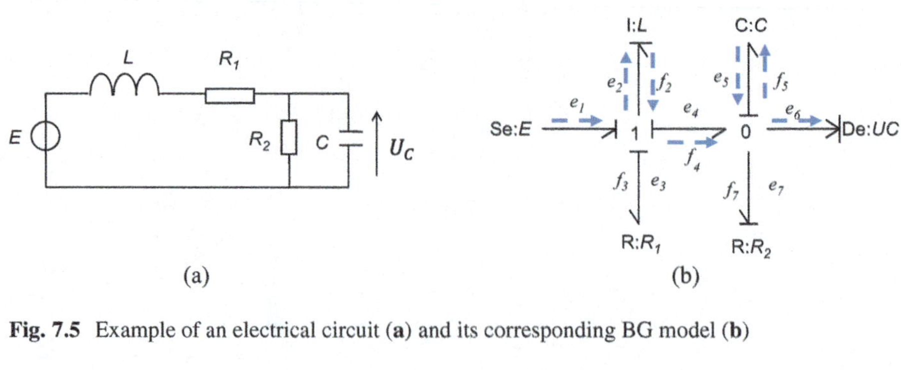

Fig. 7.5 Example of an electrical circuit (**a**) and its corresponding BG model (**b**)

of the fault. The presence of such paths permits to conclude on the detectability of these faults and the benefit of using only current sensors for fault diagnosis and prognosis since these sensors already exist on this system. Concerning the first fault, the following causal path is identified:

$$\begin{aligned}
&MSe: F1 \rightarrow e_{68a} \rightarrow I: J_{hub} \rightarrow f_{68a}, f_{69}, f_{70}, f_{72} \rightarrow C: kms\\
&\rightarrow e_{72}, e_{70}, e_{71}, e_{90} \rightarrow I: J_c \rightarrow f_{90}, f_{91}, f_{88b},, f_{86} \rightarrow C: ksp3\\
&\rightarrow e_{86}, e_{87}, e_{119}, e_{120} \rightarrow I: J_s \rightarrow f_{120}, f_{121}, f_{123}, f_{124}, f_{126} \rightarrow C: khs\\
&\rightarrow e_{126}, e_{124}, e_{127}, e_{128} \rightarrow I: J_{Ind1} \rightarrow f_{128}, f_{129} \rightarrow IC
\end{aligned} \tag{7.2}$$

From IC element, three causal paths are identified as follows:

$$\begin{aligned}
IC &\rightarrow f_{135}, f_{144}, f_{145}, f_{153} \rightarrow Df: i_1\\
IC &\rightarrow f_{134}, f_{146}, f_{147}, f_{154} \rightarrow Df: i_2\\
IC &\rightarrow f_{133}, f_{148}, f_{149}, f_{155} \rightarrow Df: i_3
\end{aligned} \tag{7.3}$$

The causal paths, identified above, from the unbalance fault to the three current sensors, prove that these sensors carry the unbalance fault. The same conclusion can be made for other considered faults (Eq. (7.4)).

$$\begin{aligned}
&Msf: F_2 \rightarrow f_{124}, f_{126} \rightarrow C: khs \rightarrow e_{126}, f_{124}, f_{127}, f_{128}\\
&\rightarrow J: Ind_1 \rightarrow e_{128}, e_{129} \rightarrow IC\\
&Msf: F_3 \rightarrow e_{128} \rightarrow I: J_{Ind1} \rightarrow e_{128}, e_{129} \rightarrow IC\\
&Msf: F_4 \rightarrow e_{135} \rightarrow IC
\end{aligned} \tag{7.4}$$

7.4 Proposed Method

The considered fault diagnosis and prognosis method is illustrated in Fig. 7.6 and consists of two stages, offline stage and online one:

- The offline stage is composed of two steps: step #1 simulation of healthy and faulty operations, where the bond graph model of the system in normal and faulty operations is used for the generation of data in different operation modes of the WT (normal, faulty, and failure). The faulty operation is considered as the minimal degraded mode and the failure one corresponds to the maximal degraded mode of the WT. In step #2, the attributes are extracted offline from the generated data and used for training a MC-SVM classifier (step #3). In this step, the SVM model of faulty and failure operations are identified and implemented in the online stage.
- The online stage, step #4, the fault diagnosis module assesses online the operation of the WT. Once a fault is detected and located, the failure prognosis module,

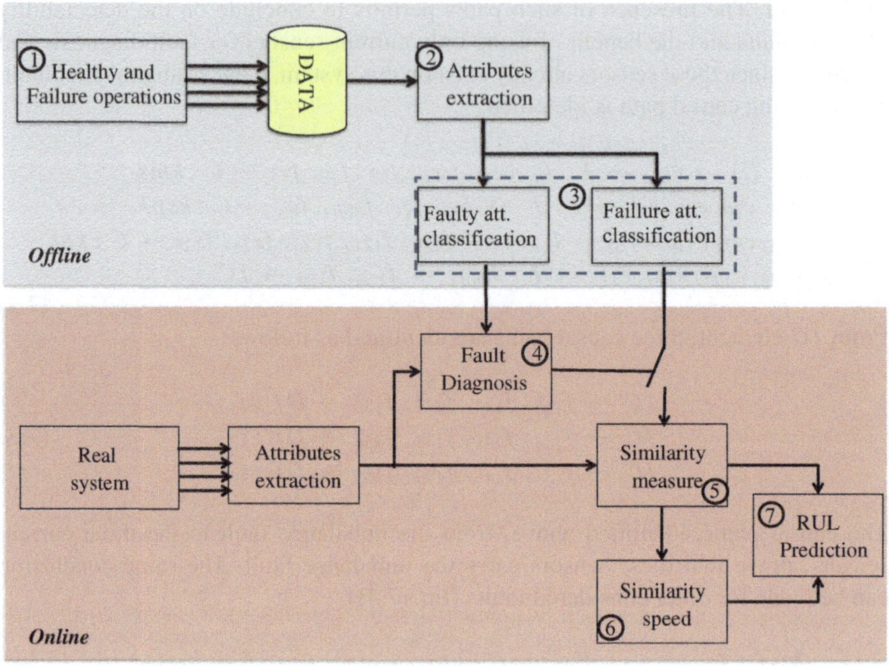

Fig. 7.6 The proposed fault diagnosis and prognosis diagram

summarized in three steps (step #5: similarity measure, step #6: similarity speed, step #7: RUL prediction), is triggered in order to estimate the RUL before observing the system failure.

7.4.1 Learning

7.4.1.1 Step #2: Attributes Extraction

The performance of the fault diagnosis based on data mining is related mainly to the available data on system operation, the choice of observed variables, and the attributes. Generally, diagnosis based on vibration analysis relies on frequency or time-frequency analysis of speed or acceleration measurements. In this work, the WT monitoring is developed based on current measurements of the three phases of the generator. This choice is motivated by the presence of such sensors on most WTs and no need to install additional sensors and equipment in each part of the system. Therefore the proposed method is interesting from the point of view of cost-effectiveness. In order to detect degradation from the observed variables (current), the use of temporal methods is more appropriate compared to the frequency one.

Table 7.1 Most used temporal attributes

Root mean square (RMS)	$\mathrm{RMS} = \sqrt{\frac{1}{N}\sum_{i=1}^{N} x_i^2}$
Standard deviation (std)	$\mathrm{std} = \sqrt{\frac{1}{N}\sum_{i=1}^{N} (x_i - \bar{x})^2}$
Kurtosis	$Vk = \frac{1/N \sum_{i=1}^{N} (x_i - \bar{x})^4}{\mathrm{RMS}^4}$
Skewness	$Sk = \frac{1/N \sum_{i=1}^{N} (x_i - \bar{x})^3}{\mathrm{std}^3}$
Peak value (Pv)	$Pv = \frac{1}{2}\left[\max(x_i) - \min(x_i)\right]$
Crest factor	$F_C = \frac{Pv}{\mathrm{RMS}}$
Form factor	$F_F = \frac{\mathrm{RMS}}{1/n \sum_{i=1}^{N} \lvert x_i \rvert}$
Impulse factor	$F_I = \frac{Pv}{1/n \sum_{i=1}^{N} \lvert x_i \rvert}$

The most used temporal attributes in the literature are: the root mean square (RMS), the standard deviation (std), the peak value (Pv), the Kurtosis, the skewness, the crest factor, the form factor, the impulse factor, Table 7.1 [17].

To identify degradation trends in temporal measurements for diagnosis and prognosis purposes, the RMS, the Pv, and the absolute mean ($\left|\frac{1}{N}\sum_{i=1}^{N} x_i\right|$) are the most suitable. The Kurtosis value is inversely proportional to the RMS, so it cannot describe the degradation. The skewness is a division of two quantities which represent the dispersion of the moving window around its mean value, so it does not describe the degradation trend. The crest factor, the form factor, and the impulse factor are obtained by division of two increasing quantities in the same magnitude, so they cannot represent the irreversible dynamics of degradation. In this work, only the absolute mean and the RMS value are used as attributes for classifying current signals in different WT operation (normal and faulty).

7.4.1.2 Steps #3: MC-SVM Classifier

In this work, the MC-SVM in its "one against one" variant [11] is used. For that, $n(n + 1)$binary SVM classifiers are built (n is the number of classes) and each one trains data from two classes. The classification results are obtained by a voting strategy: a pattern is classified to the class where the maximum number of votes is obtained. SVM is well suited for supervised and unsupervised learning, and its use in fault diagnosis purposes has shown its advantage compared to the conventional methods such as artificial neural networks or Bayesian network [30] since it is usually impossible to obtain data that cover all the faulty condition. Using SVM and due to ability in classification process, reliable and better results with small number of learning samples can be obtained. However, in the case of large dataset (a significant number of samples and classes), using standard SVM (by solving quadratic problem) is time and memory consuming and not suited in real time application. To overcome this issue, a decomposition algorithm proposed in [19] can be used.

The theory behind binary SVM (Algorithm 1) classification lies on separating data on learning data and testing sets. Let us consider that the learning data matrix (x) is composed of m attributes or variables representing monitoring indicators and a corresponding assigned label value $(y = Cl)$ $(l = 1, \ldots, n)$. SVM classifier aims to build a model which predicts the target class (y) of the testing data based only on the data attributes. This is performed by finding an optimal hyperplane (Fig. 7.7) optimizing a quadratic problem formalized in Eq. (7.5):

$$\begin{aligned} &\min J(a) = \frac{1}{2} \sum_{i=1}^{N} \sum_{j=1}^{N} a_i a_j g_i(x) g_j(x) k(x, x) - \sum_{i=1}^{N} a_i \\ &s.t: \sum_{i=1}^{N} a_i g_i(x) = 0,\ 0 \le a_i \le D \text{ for } i = 1, \ldots, N \end{aligned} \tag{7.5}$$

where $g(x) = 1$ if $x \in C1$ and $g(x) = -1$ if $x \in C2$, $a = [a_1, a_2, \ldots, a_N]^T$ are the Lagrange multipliers, D is the penalty parameter, and $k(x, x)$ is the Kernel function.

In online stage, the classification of a new sample data is performed by using the following decision function:

$$y = \begin{cases} 1, & \text{if } \operatorname{sign}\left(\sum_{i=1}^{S} a_i^S g_i^S k(x_i^S, x) + b \right) = 1 \\ -1, & \text{elsewhere} \end{cases} \tag{7.6}$$

where $b = \frac{1}{S} \sum_{j=1}^{S} \left(g_j^S - \sum_{i=1}^{S} a_i^S g_i^S k(x_i^S, x_j^S) \right)$.

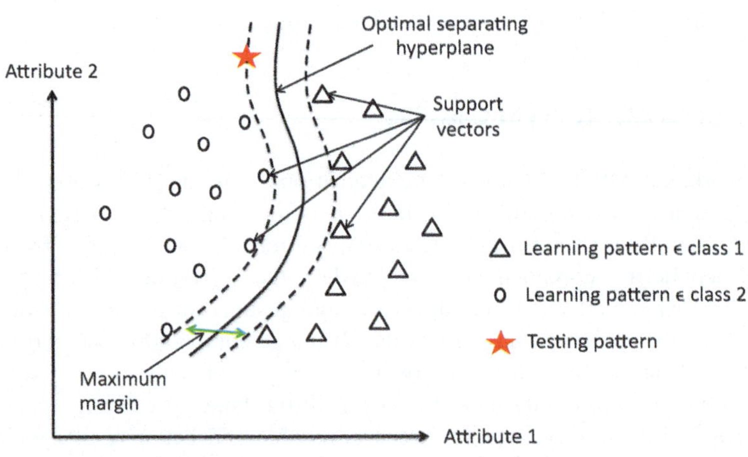

Fig. 7.7 Schematic dedicated to SVM classification

7.4.2 Step #4: Fault Detection and Location

The fault detection and location are the results of MC-SVM classification decision (Algorithm 2). Thus, a fault is detected if the classifier result passes, during operation, from the class of the normal operation $C1$ to the fault class $Ci/i = 2\ldots n$. The fault location takes as class the one identified by the SVM classifier. However, in some cases, the extracted attributes cannot be perfectly classified due to an overlapping in the attributes. In this case, an additional criterion should be designed to provide a precise and a reliable decision on the fault class. This criterion Fr_i is based on the rate of appearance of a class on a moving window, it is computed as follows:

$$Fr_i = \frac{\sum_{j=1}^{N} y == Ci}{N} \tag{7.7}$$

where y is the classifier result, and N is the number of sample in a window. The fault location is then determined if the occurrence of a class exceeds a previously fixed threshold th (Eq. (7.8)).

$$Fr_i \geq th \tag{7.8}$$

7.4.3 Fault Prognosis

The RUL prediction method (Algorithm 3) is based on the similarity measure $S(t)$ between the reference attributes x_r of the failure operation, identified offline, and the current attributes x_t calculated continuously online at each sampling time t. The main advantage of this method is the no need of the degradation model or trend.

7.4.3.1 Step #5: Similarity Measure

The considered metric for similarity measure is based on the Euclidean distance between x_r and x_t as follows:

$$S(t) = 1 - \sqrt{\sum_{i=1}^{m} (x_{ri} - x_{ti})^2} \tag{7.9}$$

where m is the number of considered attributes.

To get similarity score between [0, 1], data has to be normalized. The attributes are similar when the similarity score is equal to one. Else, they are different. The choice of the Euclidean distance is motivated by the fact that it is a derivable function [15], which is a peremptory property for the similarity speed calculation.

7.4.3.2 Step #6: Similarity Speed

The similarity speed v, expressed by Eq. (7.10), is obtained by the numerical differentiation of the similarity measure $S(t)$. It characterizes how the actual attributes move to the failure attributes and at which speed.

$$v(t) = \frac{S(t + \eta T_s) - S(t)}{\eta \mathrm{T}_s} \tag{7.10}$$

where T_s is the sampling time and η is the prediction window.

7.4.3.3 Step #7: RUL Estimation

To predict the RUL before observing a failure, both the similarity measure S and the similarity speed v are considered. The RUL is expressed as the quotient between the two variables as follows:

$$rul(t) = \left| \frac{S(t)}{v(t)} \right| \tag{7.11}$$

7.4.4 *Performance Evaluation*

In order to evaluate the accuracy and the performance of the proposed method, the prognosis horizon (PH) and the $\alpha - \lambda$ metrics, proposed by Saxena et al. [23], are used.

- The PH determines if the prediction performance meets the desired specifications. It ranges within $[0, \infty]$ and it is calculated using the following formula:

$$PH(i) = EOP - i \tag{7.12}$$

  It represents the difference between the actual time index (i) and the end of prediction time index (EOP). The latter is obtained when the prediction cross the failure threshold.
- The $\alpha - \lambda$ performance is calculated by using the following formula:

$$(1 - \alpha)\, rul^{*}(t) \leq rul(t_{\lambda}) \leq (1 + \alpha)\, rul^{*}(t) \tag{7.13}$$

where α is the accuracy bound and $t_\lambda = t_p + \lambda \left(EOL - t_p\right)$ such that t_p is time prediction and EoL represents End of Life time.

The $\alpha - \lambda$ performance determines if the accuracy prediction is within $\alpha * 100$ of the actual RUL at specific time instance t_λ.

7.5 Results and Discussion

The BG model of the WT is implemented under $20 - sim$ software [26] to generate a database describing the normal and faulty operations. The simulation is carried out using real wind data. These data are introduced in the simulator with a "data from file" block, which takes the numerical values from a table each step time. The simulations have been conducted with the backward differentiation formula method of $20 - sim$, and using an absolute integral error of $1e{-}4$ and a relative integral error of $1e{-}5$. From the generated database, samples representing different WT operations are used as the training dataset from which the RMS and $absolute\ mean$ are extracted. For MC-SVM procedures, the toolbox provided by Canu et al. [2] is used. The used kernel is the Gaussian one (Eq. (7.14)), the penalty parameter D, and the kernel parameter σ have been chosen by using the cross validation.

$$k\left(x_n, x_m\right) = \exp\left(-\frac{\|x_n - x_m\|^2}{\sigma}\right) \tag{7.14}$$

7.5.1 Fault Diagnosis Results

The projection in the attributes space of the extracted attributes in normal and faulty operations is given in Fig. 7.8, where the fault #1 corresponds to unbalance fault caused by a blade deformation, the fault #2 corresponds to the unbalance affecting high speed shaft. Fault #3 and #4 are the stator eccentricity in the generator and the electrical fault, which affects the stator resistive element $R : R_s$, respectively.

The classification results of the training data using the trained SVM model are given in Table 7.2, where the global classification accuracy rate is 83.57%. It could be observed that the classification accuracy rate of the samples in normal operation and faulty operation subject to fault #2 and #4 are high. In other words, under normal operation, the risk of a false alarm due to wrong classification is extremely low.

However, when the WT system is subject to fault #1 (resp, #3), the classification results swing between normal operation (C1), fault #3 (C4), and fault #1 (C2) as illustrated in Fig. 7.9. In that case, the classification accuracy is 51.39% for fault #1 and 66.49% for fault #3. This wrong classification is due to an overlapping in attributes extracted from current signals. This can be explained by the fact that an unbalance fault caused by a blade deformation and a stator eccentricity have the same effect on current signals of the generator. In this case, no decision can be

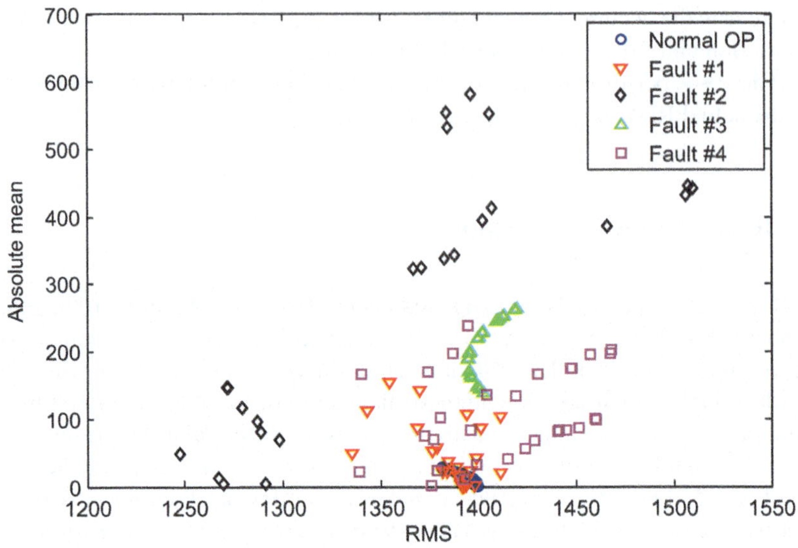

Fig. 7.8 Data projection in the attributes space

Table 7.2 Prediction class matrix

WT subjects to	Predicted class				
–	C1	C2	C3	C4	C5
Normal (C1)	100%	0	0	0	0
Fault #1 (C2)	31.26%	51.39%	0	17.33%	0
Fault #2 (C3)	0	0	100%	0	0
Fault #3 (C4)	2.92%	30.49%	0.08%	66.49%	0
Fault #4 (C5)	0	0	0	0	100%

made regarding the fault location. By using the additional criterion and threshold, the classification rate is improved and the fault is located. Figure 7.10 shows the prediction class of SVM classifier for WT system subject to fault #1 and #3. As shown in the figure, the classification results are improved to reach a classification accuracy rate of 64.06% for fault #1 and 85.71% for fault #3, which are acceptable values to conclude on fault location.

To highlight the performance of SVM-based fault diagnosis regarding the earlier detection of a fault, the following scenario is considered:

- Normal operation of WT from $t = 0$ h to $t = 40$ h,
- WT is subject to a fault #4 from $t = 40$ h to $t = 150$ h.

Figure 7.11 shows the classification result of the test scenario. The fault is detected and located in time. The wrong classification in the transition from normal operation to fault #4 operation is due to changes in the characteristics of the current measurement signals.

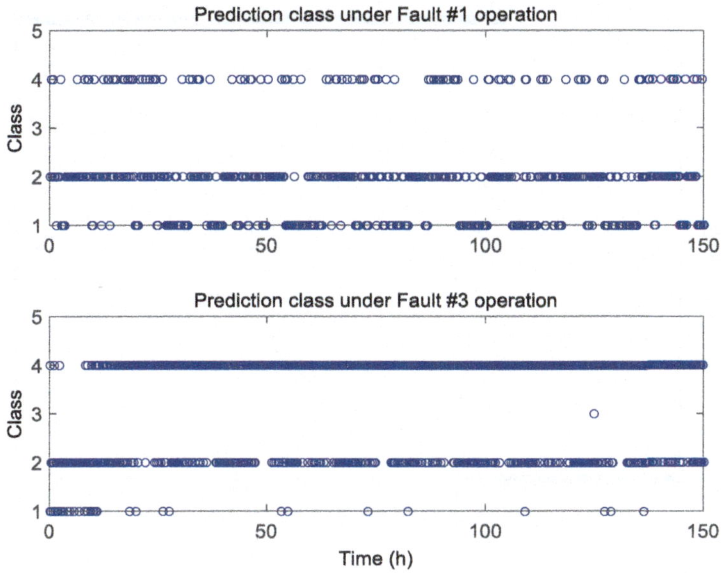

Fig. 7.9 Prediction results under fault #1 (class 2) / fault #3 (class 4) operation

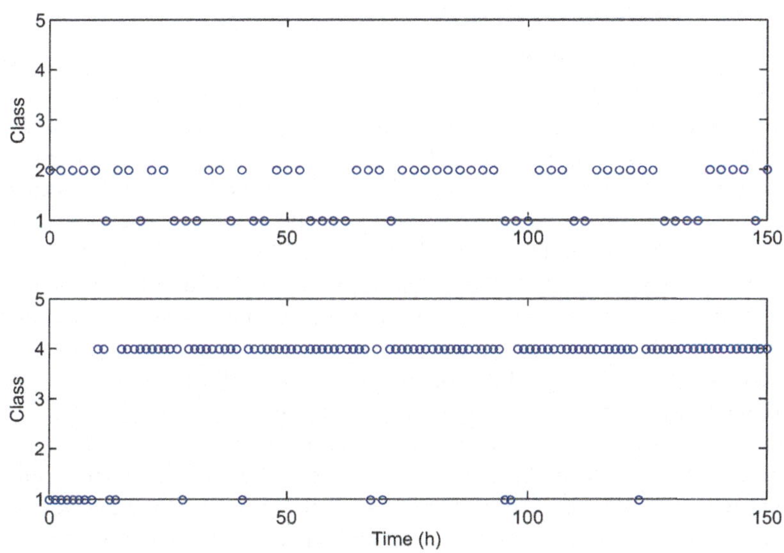

Fig. 7.10 Prediction class considering additional criteria under fault #1 (*C*2)/Fault #3 (*C*4) operation

Since a fault is detected and located, the fault prognosis is triggered to estimate the RUL before observing a failure in WT system. In the next section, the performance of the proposed method is analyzed.

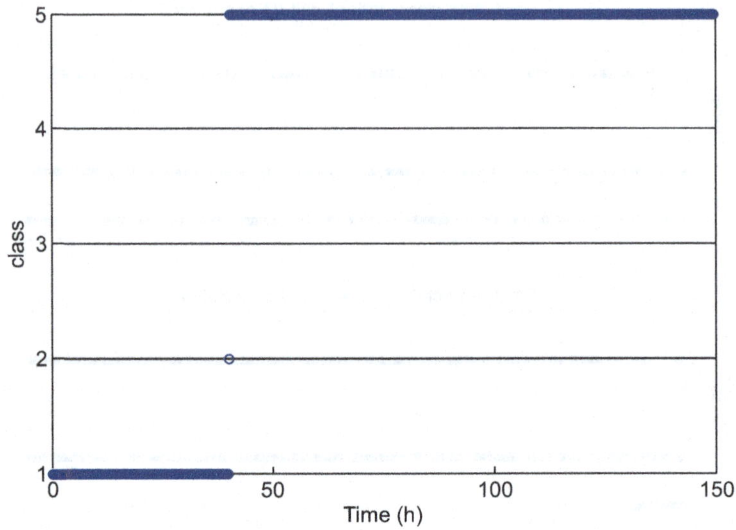

Fig. 7.11 Fault diagnosis result subject to fault #4

7.5.2 Fault Prognosis Results

The BG model of the WT is used for online introduction of a progressive degradation stator resistance (fault #4) until the total failure, with no consideration about the degradation profile. Once the fault is detected and located (Fig. 7.11), the similarity measure and speed similarity are calculated in order to estimate the RUL. Figures 7.12 and 7.13 show, respectively, the similarity measure score and similarity speed function of time (h). As illustrated by the similarity measure plot, the similarity score between data in actual operation and data in failure operation is about 0% at the beginning of the prediction (detection time), and then reaches 100% at the end of prediction. Figure 7.12 shows also that the similarity measure is linear in the considered case. Regarding the similarity speed, illustrated by Fig. 7.13, the evolution over time of the similarity speed is not constant.

The evolution of the RUL estimation over the time is given in Fig. 7.14, it shows that after a convergence time $t = 28\,\text{h}$, the estimated RUL joins the real one with small error. The performance of the proposed method is evaluated using the prognosis horizon and $\alpha - \lambda$ performance whose scores are given in Figs. 7.15 and 7.16, respectively. The PH is about 88 h, which is widely sufficient for condition-based maintenance scheduling. It can also be observed that the estimated RUL is inside the interval $(1 + \alpha)\,\text{RUL}*$ where $\alpha = 0.2$ and RUL$*$ is the reference RUL.

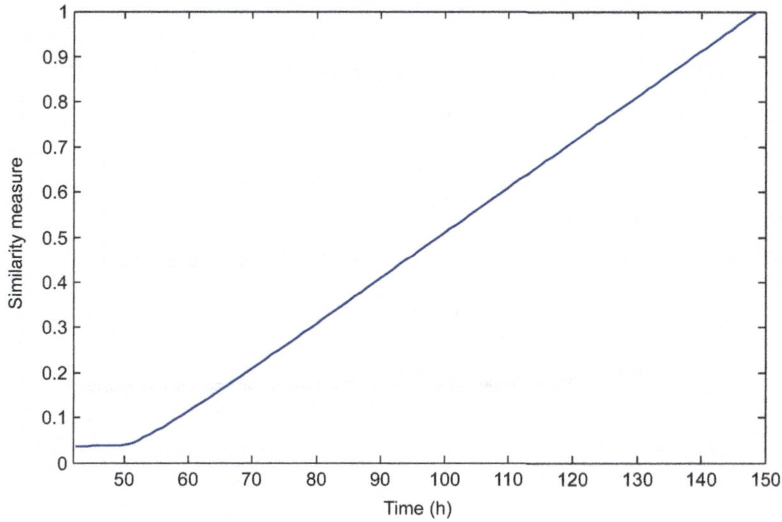

Fig. 7.12 Similarity measure

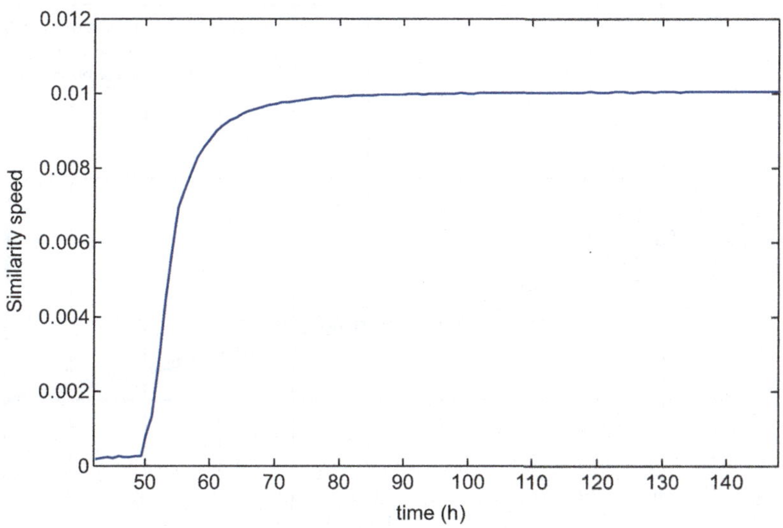

Fig. 7.13 Similarity speed

7.5.3 Discussion

To highlight the contribution of the proposed approach, examples of the most used approaches in the literature [12, 20] compared the approach proposed in this work, in terms of analysis tools, measured variables, and prior knowledge necessary for

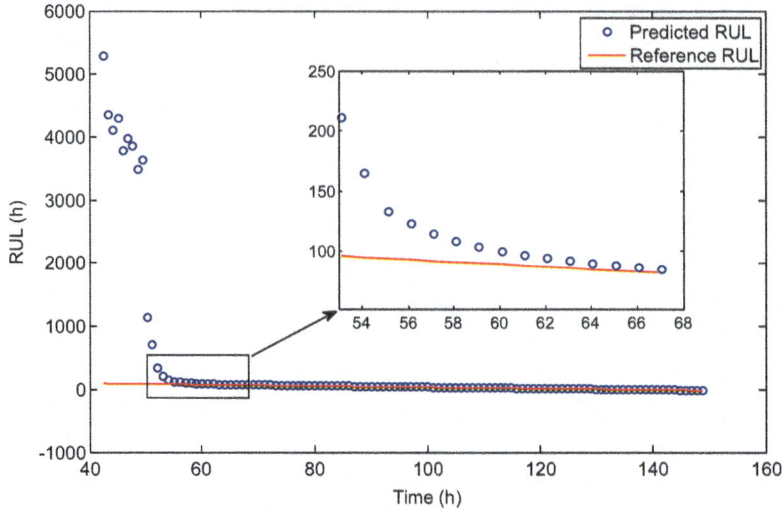

Fig. 7.14 RUL prediction

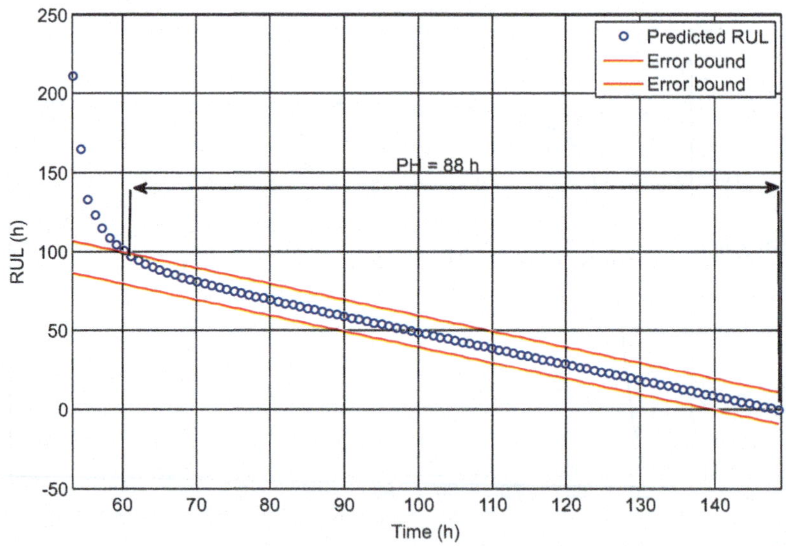

Fig. 7.15 RUL evaluation using prognosis horizon (10%)

implementation are summarized in Table 7.3. These articles are representative of the existing prognosis approaches for RUL estimation (vibratory analysis and trend analysis). This table illustrates the tools used in each case as well as the knowledge necessary for the implementation of each method. From this table, the following points are highlighted:

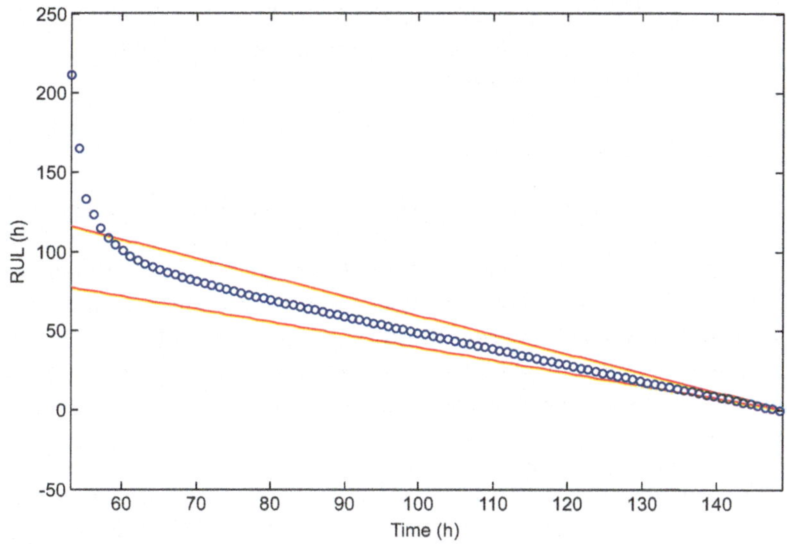

Fig. 7.16 RUL evaluation using $\alpha - \lambda$ ($\alpha = 0.2$)

Table 7.3 Comparison study

	Proposed method	Hu et al. [12]	Saidi et al. [20]
Power WT used	750 kW	1.5 MW	2.2 MW
A priori knowledge	Physical knowledge and data in normal operation	Data in normal and degraded operation	Data in normal and degraded operation
Used measurement	Current	Temperature	Vibration
HI modeling	Attributes extraction	Wiener process model	Attributes extraction
Trend modeling for RUL pred.	Geometric model-based	Inverse Gaussian distribution	Kalman smoother
Prognosis perf. analysis	PH and $\alpha - \lambda$	Relative error	Confidence bounds

- The method proposed by Hu et al. [12] uses the temperature measurement and the one proposed by Saidi et al. [20] uses the vibration measurement, but the cause-effect relationships between these measurements and the faults are not formally demonstrated. In the proposed method, these relations are formally demonstrated.
- The method [12] and method [20] use a Wiener process and a Kalman smoother, respectively. These two methods are easier to implement but require a priori knowledge of the degradation profile for the identification of the parameters as well as prior knowledge of the distribution of the data, they therefore require the availability of data describing partially or totally the profile of the degradation. The geometric approach proposed in this work does not require prior knowledge

of the profile of the degradation or distribution of the data. In addition, the use of the MC-SVM makes it possible to take into account the significant number of possible faults on the wind turbines.

The proposed method and the obtained results can be discussed according to the following points:

- The first point is the use of the physical model that is subject to uncertainties: in this work, the parameters of the system in normal operation are identified using real data for a 750 kW WT system. SVM tool allows taking into account unstructured uncertainties by the calculation of the accuracy rate and by the using of the additional criterion based on the idea of a threshold to improve the classification rate. In addition, the performance of the fault prognosis can be adjusted according to the data quality by increasing or decreasing the confidence interval (α).
- The second point is the use of the derivative: the speed similarity is calculated by differentiating the similarity measure. So, in the presence of the noise, a pre-processing (filtering) step of the data can be added to avoid the noise amplification.
- The last point is the transient operating modes: the fault diagnosis results shown in this work do not consider the transient operation as it is time-limited, it concerns only the starting and the stopping modes of the WT operation. Only data in permanent operating modes are considered.

7.6 Conclusion and Prospects

In this chapter, an approach of wind turbines fault diagnosis and prognosis, based on a physical modeling, a multi-class support vector machine classifier, and a similarity approach, is presented. The proposed method uses the current measurements available on all existing wind turbines, so its implementation does not require additional sensor placement. The lack of prior knowledge about the cause-to-effect relationships between the degradation phenomena and the measured variable is overcome in this work by the using of a bond graph model whose causal and structural properties are used to formally demonstrate that current measurements carry information on the considered degradation. The physical model is then used to supplement the existing database by generating data representing different operating modes of the system. The fault diagnosis task is performed by a multi-class support vector machine classification of attributes extracted from current measurement of wind turbine generator. The wrong attributes classification is avoided by comparing the fault class rate with a predefined threshold. The remaining useful life estimation is launched once degradation beginning is detected and located, and it is calculated without needing for prior knowledge on the degradation profile, by using the similarity measure and similarity speed between attributes of the current operation calculated online and attributes of the failure operation. The performance of the proposed fault diagnosis and prognosis method is demonstrated using evaluation metrics of fault prognosis performance showing the effectiveness of this method.

Appendix

Table 7.4 Wind turbine data

Section 7.1	Section 7.2
Blade structure data	
$E = 1.7e10$	$E = 1.7e10$
$l = 11.7\,\text{m}$	$l = 11.7\,\text{m}$
$M = 1208\,\text{kg}$	$M_2 = 487\,\text{kg}$
$J = 3.3\,\text{kg m}^2$	$J2 = 2.33\,\text{kg m}^2$
$\mu = 0.01$	$\mu = 0.01$
$J_{\text{whole}} = 1000\,\text{kg m}^2$	
Blade aerodynamic conversion	
$c_1 = 1.9$	$c_2 = 1$
$b_{t1} = 11$	$b_{t2} = 1.7$
$q_{\text{air}} = 1.225$	$q_{\text{air}} = 1.225$
$r_1 = 5.85\,\text{m}$	$r_2 = 17.5\,\text{m}$
$l_1 = 11.7\,\text{m}$	$l_2 = 11.7\,\text{m}$
Hub and main shaft	
$J_{hub} = 5000\,\text{kg m}^2$, $D_{hub} = 1000\,\text{N/m}$	
$K_{ms} = 3.67e7\,\text{N/m}$, $D_{ms} = 200\,\text{N/m}$	
Gearbox	
$J_s = 3.2\,\text{kg m}^2$, $J_r = 144.2\,\text{kg m}^2$	
$J_c = 59.1\,\text{kg m}^2$, $J_p = 3.2\,\text{kg m}^2$	
$Z_p = 39$, $Z_r = 99$, $Z_s = 21$, $K_{sp} = 16.9e9\,\text{N/m}$	
$K_{rp} = 19.2e9\,\text{N/m}$, $T_{fg} = 10.5$	
High speed shaft	
$Khs = 10e7\,\text{N/m}$, $Dhs = 1e - 3\,\text{N/m}$	
Generator	
$V_n = 460\,\text{V}$, $S = 746\,\text{kW}$, $f = 50\,\text{Hz}$,	$R_s = 4.92e{-}4\,\Omega$
$R_r = 2.7e{-}4\,\Omega$, $p = 2$, $L_{ls} = 4.66e{-}5\,\text{H}$, $L_m = 1.99e{-}3\,\text{H}$	$J_{ind} = 18.7\,\text{kg m}^2$

Algorithm 1 Binary SVM

- **Offline training**

1. Initialize D and σ,
2. Collect $x_1, x_2, \ldots, x_N$ distributed in the 1st and 2nd classes,
3. Solve the quadratic problem:

$$\begin{aligned} \min J(a) &= \frac{1}{2}\sum_{i=1}^{N}\sum_{j=1}^{N} a_i a_j g_i(x) g_j(x) k(x, x) - \sum_{i=1}^{N} a_i \\ s.t:& \sum_{i=1}^{N} a_i g_i(x) = 0,\ 0 \le a_i \le D\ for\ i = 1, \ldots, N \end{aligned} \tag{7.15}$$

where $g(x) = 1$ if $x \in C1$ and $g(x) = -1$ if $x \in C2$, $a = [a_1, a_2, \ldots, a_N]^T$ are the Lagrange multipliers, and $k(x, x)$ is the Gaussian Kernel given by:

$$k(x_n, x_m) = \exp\left(-\frac{\|x_n - x_m\|^2}{\sigma}\right) \tag{7.16}$$

4. Save support vectors $x_1^s, x_2^s, \ldots, x_S^s$ and corresponding g_n and a_n denoted by $\{g_n^s\}$ and $\{a_n^s\}$ for which $a_n > 0$, where S is the number of support vectors.

- **Online performing**
- collect new sample x

$$y = \begin{cases} 1 & , if sign\left(\sum_{i=1}^{S} a_i^S g_i^S k(x_i^S, x) + b\right) = 1 \\ -1 & , elsewhere \end{cases} \tag{7.17}$$

where $b = \frac{1}{S}\sum_{j=1}^{S}\left(g_j^S - \sum_{i=1}^{S} a_i^S g_i^S k(x_i^S, x_j^S)\right)$. S is the number of support vectors.

Algorithm 2 Fault diagnosis and decision

1. collect new sample x
2. divide x into N windows
3. y_j = perform MC-SVM for N new samples

for $i = 1$ to n **do**

- Compute Fr_i according to Eq. (7.7)

  if $Fr_i > Th_i$ **then**
    fault i is detected
  else
    No fault
  end if

end for

Algorithm 3 RUL estimation

1. Initialize Ts and η,
2. collect x_t
3. perform $MC - SVM$
4. a. Compute the similarity measurement S using Eq. (7.9),
   b. Compute the similarity speed υ using Eq. (7.10),
   c. Compute the RUL using Eq. (7.11),

References

1. Y. Amirat, V. Choqueuse, M. Benbouzid, Wind turbines condition monitoring and fault diagnosis using generator current amplitude demodulation, in *IEEE International Energy Conference and Exhibition*, pp. 310–315 (2010)
2. S. Canu, Y. Grandvalet, V. Guigue, A. Rakotomamonjy, SVM and kernel methods matlab toolbox, in *Perception Systemes et Information* (INSA de Rouen, Rouen, 2005)
3. B. Chen, P. Matthews, P. Tavner, Wind turbine pitch faults prognosis using a-priori knowledge-based ANFIS. Expert Syst. Appl. **40**(17), 6863–6876 (2013)
4. B. Chen, P. Matthews, P. Tavner, Automated on-line fault prognosis for wind turbine pitch systems using supervisory control and data acquisition. IET Renew. Power Gener. **9**(5), 503–513 (2015)
5. J. Chen, J. Pan, Z. Li, Y. Zi, X. Chen, Generator bearing fault diagnosis for wind turbine via empirical wavelet transform using measured vibration signals. Renew. Energy **89**, 80–92 (2016)
6. Z. Daneshi-Far, G.A. Capolino, H. Henao, Review of failures and condition monitoring in wind turbine generators, in *XIX International Conference on Electrical Machines* (2010)
7. M. Djeziri, L. Nguyen, S. Benmoussa, N. Msirdi, Fault prognosis based on physical and stochastic models, in *European Control Conference* (2016)
8. M. Djeziri, S. Benmoussa, R. Sanchez, Hybrid method for remaining useful life prediction in wind turbine systems. Renew. Energy **116**(B), 173–187 (2018). https://doi.org/10.1016/j.renene.2017.05.020
9. Z. Fenga, S. Qina, M. Liangb, Time-frequency analysis based on Vold-Kalman filter and higher order energy separation for fault diagnosis of wind turbine planetary gearbox under nonstationary conditions. Renew. Energy **85**, 45–56 (2016)
10. X. Gong, W. Qiao, Current-based online bearing fault diagnosis for direct-drive wind turbines via spectrum analysis and impulse detection, in *IEEE Power Electronics and Machines in Wind Applications*, pp. 1–6 (2012). https://doi.org/10.1109/PEMWA.2012.6316398
11. C. Hsu, C. Lin, A comparison of methods for multiclass support vector machines. IEEE Trans. Neural Netw. **13**(2), 415–425 (2002)
12. Y. Hu, H. Li, P. Shi, Z. Chai, K. Wang, X.X.Z. Chen, A prediction method for the real-time remaining useful life of wind turbine bearings based on the wiener process. Renew. Energy **127**, 452–460 (2018)
13. A. Kusiak, W. Li, The prediction and diagnosis of wind turbine faults. Renew. Energy **36**, 16–23 (2011)
14. N. Laouti, N. Sheibat-Othman, S. Othman, Support vector machines for fault detection in wind turbines, in *18th IFAC World Congress* (2011)
15. M.J. Lesot, Similarity, typicality and fuzzy prototypes for numerical data, in *In 6th European Congress on Systems Science, Workshop Similarity and Resemblance*, vol. 94, p. 95 (2005)
16. D.F. Lin, P.H. Chen, Fault recognition of wind turbine using EMD analysis and FFT classification, in *Third International Conference on Digital Manufacturing Automation*, pp. 414–417 (2012). https://doi.org/10.1109/ICDMA.2012.99

17. L. Nguyen, Approche statistique pour le pronostic de defaillance : Application a l'industrie du semi-conducteur. Ph.D. Thesis, Aix-Marseille University (2016)
18. L. Nguyen, M. Djeziri, B. Ananou, M. Ouladsine, J. Pinaton, Fault prognosis for batch production based on percentile measure and gamma process: application to semiconductor manufacturing. J. Process Control **48**, 72–80 (2016)
19. Edgar E. Osuna and Robert Freund and Federico Girosi, *Support vector machines: Training and applications*. A.I. MEMO 1602, MIT A. I. LAB (1997).
20. L. Saidi, J.B. Alia, M. Benbouzid, E. Bechhoferd, An integrated wind turbine failures prognostic approach implementing Kalman smoother with confidence bounds. Appl. Acoust. **138**(1), 199–208 (2018)
21. A. Salem, A. Abu-Siada, S. Islam, Condition monitoring techniques of the wind turbines gearbox and rotor. Int. J. Electr. Energy **2**(1), 53–56 (2014)
22. R. Sanchez, A. Medina, Wind turbine model simulation: a bond graph approach. Simul. Model. Pract. Theory **41**, 28–45 (2014)
23. A. Saxena, J. Celaya, B. Saha, S. Saha, K. Goebel, On applying the prognostic performance metrics, in *Annual Conference of the Prognostics and Health Management Society* (2009)
24. P.B. Sonawane, Fault diagnosis of windmill by FFT analyzer. Int. J. Innov. Eng. Technol. **4**(4), 47–54 (2014)
25. H. Toubakh, M. Sayed-Mouchaweh, Advanced data mining approach for wind turbines fault prediction, in *Proceedings of Second European Conference of the Prognostics and Health Management Society* (2014)
26. Twentesim: Users Manual of Twentesim (20sim). Comtollab Products Inc., AE Enschede (1996)
27. O. Uluyol, G. Parthasarathy, W. Foslien, K. Kim, Power curve analytic for wind turbine performance monitoring and prognostics, in *Annual Conference of the Prognostics and Health Management Society* (2011)
28. A. Verma, Performance monitoring of wind turbines: a datamining approach. Ph.D. Thesis, University of Iowa (2012)
29. X. Wang, V. Makis, M. Yang, A wavelet approach to fault diagnosis of a gearbox under varying load conditions. J. Sound Vib. **329**, 1570–1585 (2010)
30. A. Widodo, B. Yang, Support vector machine in machine condition monitoring and fault diagnosis. Mech. Syst. Signal Process. **21**, 2560–2574 (2007)
31. Q. Yang, D. An, EMD and wavelet transform based fault diagnosis for wind turbine gear box. Adv. Mech. Eng. **5**, 212836 (2013)
32. J.X. Yuan, L. Cai, Variable amplitude Fourier series with its application in gearbox diagnosis. Part II: experiment and application. Mech. Syst. Signal Process. **19**(5), 1067–1081 (2005)
33. Z. Zhang, K. Wang, Wind turbine fault detection based on SCADA data analysis using ANN. Adv. Manuf. **2**(1), 70–78 (2014)
34. Y. Zhang, J. An, H. Zhang, Monitoring of time-varying processes using kernel independent component analysis. Chem. Eng. Sci. **88**, 23–32 (2013)
35. Y. Zhao, D. Li, A. Dong, J. Lin, D. Kang, L. Shang, Fault prognosis of wind turbine generator using SCADA data, in *2016 North American Power Symposium (NAPS)*, pp. 1–6 (2016). https://doi.org/10.1109/NAPS.2016.7747914

Chapter 8
Review on Health Indices Extraction and Trend Modeling for Remaining Useful Life Estimation

Mohand Arab Djeziri, Samir Benmoussa, and Enrico Zio

8.1 Introduction

Because of the growing demands of equipment availability, performance, and maintenance, the scientific community has been developing methods for forecasting failures, and for the estimation of the Remaining Useful Life (RUL) for scheduling Condition-Based Maintenance (CBM) and Predictive Maintenance (PM). The National Aeronautics and Space Administration (NASA) was among the first to work on prognosis, because in the aerospace field, prognosis of failure can avoid catastrophes. Performance evaluation is a key element in fault prognosis and several methods have been proposed, based on different evaluation criteria. The work presented in [63, 100, 101] goes in the direction of a standardization of these criteria and proposes performance metrics applicable to different methods of fault prognosis. These metrics allow, on the one hand, establishing design requirements by quantifying acceptable performance limits and on the other hand, comparing different methods. In [101], a structured synthesis of the used metrics for the evaluation of the performance of fault prognosis methods that adapt to different application domains is presented, including Prognosis Horizon (PH),

M. A. Djeziri (✉)
LIS, UMR 7020, Marseille, France
e-mail: mohand.djeziri@lis-lab.fr

S. Benmoussa
LASA, Badji-Mokhtar Annaba University, Annaba, Algeria
e-mail: samir.benmoussa@univ-annaba.dz

E. Zio
Energy Department, Politecnico di Milano, Milan, Italy

Foundation Électricité de France (EDF), Centrale Supélec, Université Paris Saclay, Chatenay-Malabry, France
e-mail: enrico.zio@polimi.it; enrico.zio@ecp.fr

M. Sayed-Mouchaweh (ed.), *Artificial Intelligence Techniques for a Scalable Energy Transition*, https://doi.org/10.1007/978-3-030-42726-9_8

Alpha-Lambda Performance, Relative Accuracy (RA), and Convergence Rate. Fault prognosis methods are compared with respect to these metrics.

From the methodological viewpoint, several classifications of fault prognosis methods are proposed, like the pyramidal classification proposed in [19], which provides a classification into three approaches: expert approaches, physical model-based approaches, and data-driven approaches. The originality of this classification is related to the fact that these approaches are positioned in a pyramidal organization chart according to the scope of application, cost, and complexity of each approach. The evolution of hardware and software resources for data acquisition, storage, and processing has favored the widening of the application scope and the accuracy of the data-driven methods of fault prognosis. In addition, hybrid approaches have emerged to benefit from the combination of these approaches [35, 36].

This paper focuses on fault prognosis with a horizontal approach, which offers the advantage of relating fault diagnosis and prognosis. Unlike other existing reviews [45, 57, 103, 108], which only focus on RUL prediction, this paper offers a review of approaches that deal with the problem including fault diagnosis and allowing RUL estimation also when the only available data relate to normal operation. Since the health indices (HIs) generated by the methods initially used for fault diagnosis are not all usable for failure prognosis, evaluation methods of the properties that a HI must satisfy to be usable for RUL estimation, namely the Monotonicity, Trendability, and Prognosability [10, 24], are presented and then used to evaluate the usability for failure prognosis of HIs generated by fault diagnosis methods. The methods are presented in their basic version to facilitate the understanding of the ideas and the formal analysis, followed by indications and references to their extensions for particular practical cases.

As illustrated in Fig. 8.1, the structure of the horizontal approaches has two main parts: HI generation based on condition monitoring and HI trend modeling for RUL estimation. Correspondingly, this paper is organized as follows: the studied framework and definitions are presented in Sect. 8.2. Section 8.3 is devoted to formal description of methods for the definition of HIs and an analysis of their use for the estimation of the RUL. Then, the techniques for the estimation of the RUL by trend modeling are presented in Sect. 8.4, with an analysis of their complexity and performance. The purpose is to provide an overview of existing techniques and guidance for choosing approaches according to the field of application and available

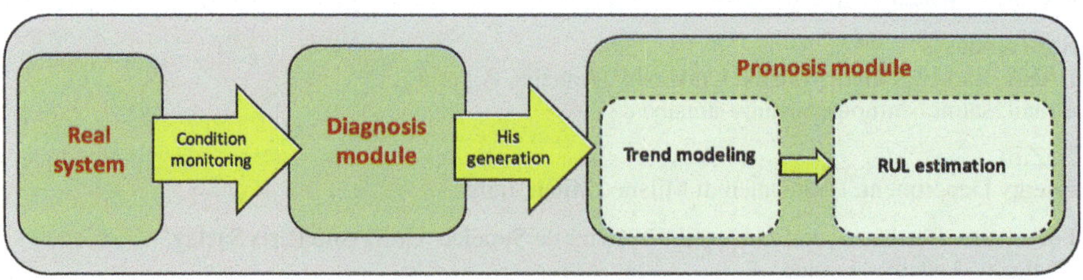

Fig. 8.1 Structure of the horizontal approach of fault prognosis

knowledge (physical knowledge, expert knowledge, data-driven). In the horizontal approach, the methods used for generating the HIs can be completely different from the method used to model trends for RUL estimation. For this reason, in this work, we have opted for a separate classification of the two parts: a classification of the methods used for the generation of HIs and a classification of the HIs trend modeling techniques for RUL estimation. In both parts, metrics are proposed for performance evaluation.

8.2 Study Framework

By definition, a fault is an unauthorized and unexpected deviation from the normal condition, whereas a degradation refers to the deterioration of performance in an irreversible manner. Degradation becomes failure when performance falls below a critical threshold defined in the functional specification of the equipment: the system is no longer able to perform the required function. According to the international standard (ISO 13381-1:2004), fault prognosis is defined as the estimation of the Remaining Useful Life (RUL) or the End of Life (EoL), and the estimation of the risk of subsequent development or existence of one or more faulty modes. However, in the literature, the definition of the fault prognosis concept is adapted to the context, the objectives, and the field of application, among these interpretations:

- Wang et al. [114]: In the industrial and manufacturing areas, prognosis is interpreted to answer the question: what is the RUL of a machine or a component once an impending failure condition is detected and identified.
- Mathur et al. [78]: Prognosis is an assessment of the future health.
- Lebold et al. [64]: Prognostics is the ability to perform a reliable and sufficiently accurate prediction of the RUL of equipment in service. The primary function of prognostics is the projection into the future of the current health state of equipment, taking into account the estimate of future usage profiles.
- Byington et al. [19]: Prognostics is the ability to predict the future condition of a machinery based on the current diagnostic state of the machinery and its available operating and failure history data.
- Jardine et al. [57]: Prognostics deals with fault prediction before it occurs. Fault prediction is a task to determine whether a fault is impending and estimate how soon and how likely a fault will occur.
- Muller et al. [87]: Prognostics is the ability to predict the future state of an item from its present, its past, its degradation laws, and the maintenance actions to be investigated.

In recent publications, the notion of prognosis is increasingly associated with the estimation of the RUL:

- Tobon et al. [109]: Fault prognostics can be defined as the prediction of when a failure might take place.

- Gucik-Derigny [50]: The prognosis consists in predicting the evolution of the future state of health of a system and estimating the remaining lifetime of a system before one or more failures appear on the system.
- Singleton et al. [104]: Effective diagnostic and prognostic tools are essential for timely fault detection and Remaining Useful Life prediction.
- Sun et al. [107]: Prognostics usually focuses on the prediction of the failure time or the Remaining Useful Life of a system or component in service by analysis of data collected from sensors.
- Lee et al. [66]: Prognostics can be interpreted as the process of health assessment and prediction, which includes detecting incipient failures and predicting RUL.
- Lim et al. [69]: Prognostics is the analysis of the symptoms to predict future conditions and Remaining Useful Life.

It can be noticed that the references cited above define prognosis as the prediction of the RUL based on an analysis of the monitoring condition data and the current state of the system.

8.2.1 Formal Definitions of the RUL

The RUL is sometimes also called Remaining Service Life, Residual Life, or Remnant Life [57], and refers to the time left before observing a failure given the current machine age and condition, and the past and future operation profile. In [103], the RUL of an asset or system is defined as the time-span from the current time to the end of the useful life. In [105], the RUL at any time t is defined as the remaining lifetime of a unit given that it is running at time t and given all the available information related to the unit at time t. Two main mathematical definitions of the RUL can be found in the literature, depending on the method used for estimating this quantity and depending on the available information: a definition of the RUL as a function of the condition monitoring (CM) and a definition of the RUL as a function of the reliability function (RF).

8.2.1.1 Definition of the RUL as a Function of CM

The definitions of the RUL given above are in agreement with the formal definition given in Jardine et al. [57], where the RUL is defined as a function of the CM of the system $(Z(t))$, which gathers all the prior knowledge on the past operating state of the system as well as the co-variables that describe its current operating state, and is expressed as follows:

$$\mathrm{RUL}(t|Z_t) = T - t\,|T > t,\ Z(t) \tag{8.1}$$

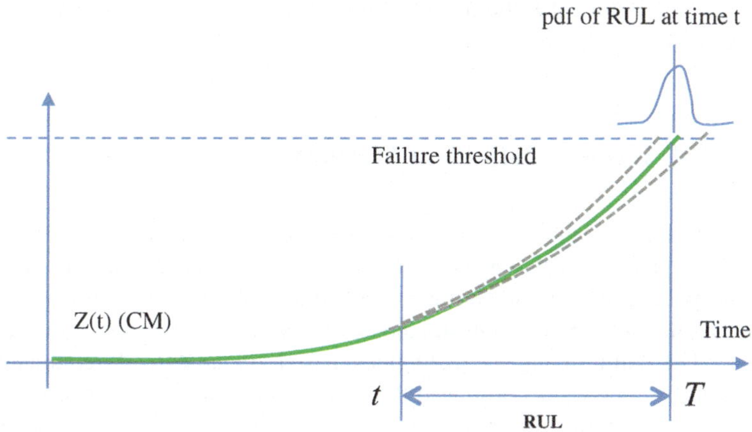

Fig. 8.2 Illustration of RUL

with:

- T: random variable of time to failure
- t: current age
- Z(t): past condition profile up to the current time.

This expression is illustrated in Fig. 8.2. According to the knowledge in $Z(t)$, the RUL can be calculated as a deterministic, statistic (as an expectation), or probabilistic variable (as a probability density function). The dashed gray envelope defines the margin of uncertainty about future operating conditions and the system environment.

8.2.1.2 Definition of the RUL as a Function of RF

In [8, 9], it is stated that information from condition monitoring can be included in reliability analysis by considering the hazard rate function as a probabilistic function. There are several methods for calculating the conditional and unconditional reliability functions (RFs) and for computing the Remaining Useful Life (RUL) as a function of the current conditions. In classical reliability, the RF is calculated mainly for two cases: as an unconditional RF, assuming that the item has not yet been put into operation ($P(T > t)$), and as a conditional RF, assuming that the item has not yet failed up to sometime x ($P(T > t|T > x)$).

Assuming that the system is operating at time t, the RUL is expressed in [112] as a time v for which the probability that the state of the system Z at time $t + v$, noted $P[Z(t + v) \geq L|Z(t)]$ approximates the probability of failure q assumed to be known. The RUL expression is given as follows:

$$\text{RUL}(t, q) = sup\{v : P[Z(t + v) \geq L|Z(t)] \leq q\} \quad (8.2)$$

where L is the failure threshold and $P[Z(t+v) \geq L|Z(t)]$ is defined as the reliability of the system.

8.3 Health Indices Definition Methods

Existing methods of fault diagnosis have been reviewed in recent years, such as [45, 46, 108]. The most recent is proposed by D. Gao et al. [45], in which the authors propose a first classification of fault diagnosis methods into two groups: hardware redundancy-based fault diagnosis and analytical redundancy-based fault diagnosis. The fault diagnosis techniques based on analytical redundancy are structured in five classes according to the mathematical tools and prior knowledge used: model-based fault diagnosis, signal-based fault diagnosis, knowledge-based fault diagnosis, hybrid fault diagnosis, and active fault diagnosis. In this work, the methods for defining health indices are gathered in two classes: physical model-based, data analysis and signal processing. The focus is on methods that can be used for the generation of health indices sensitive to progressive degradation and whose trend can be modeled for RUL estimation. The presented methods of HI generation are summarized in Table 8.1.

8.3.1 Physical Model-Based Methods

These methods are based on a physical representation of the process. They require a good understanding of the behavior of the system but does not require the availability of data on the operation of the system in degraded modes. Physical models are usually described by partial differential equations [3, 67] or state representation equations [72, 98]. Once the physical model is available, the behavior of the current process is compared with that of the model in normal operation to detect the start of degradation. After degradation has been detected, trend models are used to predict the evolution of degradation over time until reaching a failure threshold, usually predefined. The RUL corresponds to the time taken by the degradation to reach this failure threshold.

8.3.1.1 Analytical Redundancy Relations

The analytical redundancy relations (ARRs) are obtained from an over-constrained system by eliminating the unknown variables [18], assuming that all functions are differentiable with respect to their arguments. For a nonlinear deterministic system (Eq. (8.3)) where $x \in \Re^n$ is the state vector, $u \in \Re^{m_u}$ is the control vector, $d \in \Re^{m_d}$ is the disturbance vector, $\theta \in \Re^{m_f}$ is a fault vector, and $y \in \Re^p$ is the measurement vector:

Table 8.1 Analysis table of HI generation methods

	Approach	Physical knowledge on the system	Physical knowledge on the degradation process	Data describing the system features	Data describing the degradation	Complexity of implementation	Needed instrumentation	Consider changes in operating modes
Model-based	ARRs	Necessary	Necessary	Data in normal operation are necessary for parameter identification	Modeling and parameter estimation steps are complex	Modeling and parameter estimation steps are complex	Applicable on over-constrained systems	Yes
	Parity space	Necessary	Necessary	Data in normal operation are necessary for parameter identification	Modeling and parameter estimation steps are complex	Modeling and parameter estimation steps are complex	Applicable on over-constrained systems	Yes
	Observers theory	Necessary	Necessary	Data in normal operation are necessary for parameter identification	Modeling and parameter estimation steps are complex	Modeling and parameter estimation steps are complex	Applicable on observable systems	Yes
	Parameter estimation	Necessary	Necessary	Data in normal operation are necessary for parameter identification	Modeling and parameter estimation steps are complex	Modeling and parameter estimation steps are complex	Applicable on observable systems	Yes
	Algebraic methods	Necessary	Necessary	Data in normal operation are necessary for parameter identification	Modeling and parameter estimation steps are complex	Modeling and parameter estimation steps are complex	Applicable on observable systems	Yes

(continued)

Table 8.1 (continued)

Approach			Physical knowledge on the system	Physical knowledge on the degradation process	Data describing the system features	Data describing the degradation	Complexity of implementation	Needed instrumentation	Consider changes in operating modes
Data-based	Time	PCA SI EMD	Not necessary	Not necessary	Data describing all the system features is necessary	Necessary to check sensitivity to degradation	Easy	No instrumentation constraints	No need a learning of all operating modes
	Time-Fre-quency	FT DWT-HT EMD	Not necessary	Not necessary	Data describing all the system features is necessary	Necessary to define decomposition level	Easy	No instrumentation constraints	No need a learning of all operating modes
	Data-driven ARR	AR-models NN SVR	Not necessary	Not necessary	Data describing all the system features is necessary	Necessary for clustering	Learning step is complex	No instrumentation constraints	Yes by learning of all operating modes

$$\begin{aligned} \dot{x} &= h(x, u, d, \theta) \\ y &= g(x, u, d, \theta) \end{aligned} \tag{8.3}$$

the derivative of order q of the output y gives rise to the following set of $(q + 1)p$ constraints:

$$y^{(q)} = \gamma^q \left(x, \bar{u}^{(q)}, \bar{d}^{(q)}, \bar{\theta}^{(q)}\right) \tag{8.4}$$

where $\bar{u}^{(q)} \in \Re^{(q+1)m_u}$, $\bar{d}^{(q)} \in \Re^{(q+1)m_d}$, and $\bar{\theta}^{(q)} \in \Re^{(q+1)m_\theta}$.

Under the condition that $(q + 1)p > n + (q + 1)m_d$ and the Jacobian $\left[\frac{\partial\gamma^{(q)}}{\partial x} \; \frac{\partial\gamma^{(q)}}{\partial d}\right]$ is of rank $n + (q + 1)m_d$ [22], both the state x and the unknown input d can be eliminated, leading to the set of ARRs.

$$HI = r\left(\bar{y}^{(q)}, \bar{u}^{(q)}, \bar{\theta}^{(q)}\right) = 0 \tag{8.5}$$

In normal operation, Eq. (8.5) is true, whereas it is not in presence of a fault. Equation (8.5) shows that the set of residuals r is a function of the set of parameters θ identified on the system in normal operation, and corresponding to well-identified hardware components or physical phenomena. The appearance of a progressive degradation in the system manifests a progressive deviation of the parameter affected by the degradation from its nominal value identified during normal operation. Thus, the residuals that are a function of this parameter progressively deviate from zero, enabling the detection of the start of degradation. The use of the failure signature matrix makes it possible to check the isolability of the degradation: even if the isolation of the degraded component or physical phenomenon is not always possible, the subsystem that degrades in a complex system is often possible. The identification of the component or subsystem at the origin of the degradation is a relevant knowledge for practical purposes, exploitable in the modeling of the degradation trend for RUL estimation. In addition, ARRs can be generated automatically using a bipartite graph [18] or a bond graph model [60], and are easy to implement once the parameters of the state model have been identified.

8.3.1.2 Parity Space

This method is applicable to linear state models and consists in eliminating the internal variables of the system by projection onto an input-output representation space, called parity space [48]. It is generally applied in a discrete time space, taking measurements over a time interval called observation window. Information redundancy is, thus, created without resorting to successive derivations of the measurements. Consider the following example of a system described by the following state model (linear or linearized around an operating point):

$$\begin{cases} \dot{x}(t) = Ax(t) + Bu(t) \\ x(t) = Cx(t) \end{cases} \tag{8.6}$$

First, the observability matrix O_{obs} of the system is computed by Eq. (8.7), using the individual observability matrix of each sensor [73]:

$$\begin{array}{l} O_{obs} = \left(C_1 \; C_2 \cdots C_p \right) \\ \text{with } C_i = \left(c_i \; c_i A \cdots c_i A^{n-1} \right) \end{array} \tag{8.7}$$

The observability matrix is, then, used to calculate the left null observability matrix noted W, which is not unique. In practice, it is not possible to calculate a matrix W that is perfectly orthogonal to the matrix O_{obs}, adding thus a further uncertainty to the structured and unstructured uncertainties of the model and giving rise to non-zero residuals.

After computing the matrix W, the observability matrix is reformulated in terms of inputs, outputs, and their derivatives. The derived observability given for the ith output is

$$O_i = \begin{bmatrix} y(t) \\ \frac{d}{dt} y(t) \\ \vdots \\ \frac{d^n}{dt^n} y(t) \end{bmatrix} - \begin{bmatrix} 0 & 0 & \cdots & 0 \\ c_i B & 0 & \cdots & 0 \\ \vdots & \vdots & \ddots & \vdots \\ c_i A^{n-1} B & c_i AB & \cdots & c_i B \end{bmatrix} \begin{bmatrix} u(t) \\ \frac{d}{dt} u(t) \\ \vdots \\ \frac{d^{n-1}}{dt^{n-1}} y(t) \end{bmatrix} \tag{8.8}$$

The residuals are calculated using Eq. (8.9), multiplying the global derived observability noted O_D, computed for all outputs with the matrix W:

$$\begin{array}{l} HI = R = W O_D \\ \text{with } O_D = \left[o_1 \; o_2 \cdots o_p \right]^T \end{array} \tag{8.9}$$

After analyzing the equations, especially the observability matrix O_{obs} and the global derived observability matrix O_D, it is noted that the C_i and the O_i are functions of the state matrix A, whose parameters represent the physical elements of the system (physical components or physical phenomena). Thus, the occurrence of a system degradation will cause a variation of the parameters of the matrix A and, consequently, a deviation of the residuals from their values in normal operation. The residuals are, therefore, sensitive to the degradation of the system. However, the causal relationship between the residuals R and the variations of the parameters of the matrix A is not explicit: it is drawn in the process of projection in the parity space. For this reason, the parity space is used only for the detection of sensor faults, with an extension to the actuators, under the strong assumption that there are no system faults.

8.3.1.3 Observer Methods

Observers theory is widely used in the literature for the estimation of observable but unmeasured states [43, 82]. It has been used in fault diagnosis for the generation of health indices through the development of the unknown inputs observers. The stability and convergence analysis, the gains calculation, and assumptions on matrix rank and model inversion have been the subject of several research works [82, 84, 86] and will not be detailed in this work. Only the relevance of the use of the health indices generated using an observer in the context of fault prognosis is analyzed.

Two kinds of models are most used for the synthesis of observers in the framework of fault diagnosis and prognosis. The first model, given in Eq. (8.10) below, allows the simultaneous description of the state of the system and the degradation. These models are interconnected and are of multiple time scales, in order to highlight the difference in the evolution between the fast dynamics of the system behavior and the slow evolution of the degradation:

$$\begin{aligned}\dot{x} &= f(x, \lambda(\theta), u) \\ \dot{\theta} &= \epsilon g(x, \theta) \\ y &= Cx + Du + v\end{aligned} \tag{8.10}$$

where $x \in \mathbb{R}^n$ is the set of state variables associated with the fast dynamics of the system; $\theta \in \mathbb{R}^m$ is the set of slow-dynamic variables related to the degradation of the system; $u \in \mathbb{R}^l$ is the input vector. The parameter vector $\lambda \in \mathbb{R}^q$ is a function of θ. The rate constant $0 < \epsilon \ll 1$ defines the time scale separation between fast dynamics and slow drift. $y \in \mathbb{R}^p$ is the output vector and v is the measurements noise.

The general pattern of the observer-based fault prognosis begins with the joint estimation of the state and the unknown input, with precision, and in a finite time. Then, the estimation of the RUL is carried out by a time projection of the evolution of the slow and fast dynamics until the total failure. The finite-time convergence of an observer is a less common notion in the literature than asymptotic convergence; yet it is necessary in the context of fault prognosis. New conditions of stability and convergence in finite time have been developed in Lyapunov theory, and presented in [16, 17, 85] and [86]. Methods for the synthesis of observers with finite-time convergence have been proposed in [38, 56] and [82] for linear systems, and in [79–81] and [84] for nonlinear systems. In the case of observers with unknown input and finite-time convergence, synthesis work has been developed in [97] for the linear case, and in [43] for sliding mode observers.

Although the model of Eq. (8.10) is closest to the reality of the degradation phenomenon and its interaction with the state of the system, it is strongly nonlinear and, moreover, the interaction between the state and the degradation is difficult to formalize (to model). Thus, the most used model for the generation of health indices is the second type of model given in Eq. (8.11) below [71], where the nonlinear system considered consists of a linear part exploited for the synthesis of the observer's gain, and a nonlinear part satisfying some more or less restrictive assumptions:

$$\begin{cases} \dot{x} = l(x, f, u) \\ y = h(x, u) + W(u) f \end{cases} \tag{8.11}$$

$x \in R^n$ is the state vector. $u \in U$ is the known input vector. $y \in R^p$ is the output vector. $f \in R^m$ is the vector of unknown inputs whose number is equal to or less than the number of measurements (mp). W is the transfer matrix of the degradation to the output and it is a function of the conditions of use (of the input u). Once the state x is estimated, it is possible to derive an estimate of the unknown input as follows:

$$HI = \hat{f} = W_1^+(u)\left(y_1 - h_1(\hat{x}, u)\right) \tag{8.12}$$

where W_1^+ is the pseudo-inverse of W.

Thus, finite-time convergence is a necessary condition for the use of the unknown input observer for fault prognosis. Equation (8.10) also shows that the prior knowledge of the effect of degradation on the system is necessary, as it makes it possible to calculate the matrix W, which must be invertible. As the name implies, degradation is considered to be an unknown input, implying that any change in the dynamics of the system is considered as a degradation. Since the degradation is introduced in an additive way into the model of the system, it can be assumed that:

- The health index aggregates all the faults that may occur in the system.
- The prior knowledge necessary for the calculation of the matrix W can be used to identify the nature of the degradation.

8.3.1.4 Algebraic Methods

In the algebraic framework, the HIs are expressed as an algebraic equation of the system's variables (u and y) and their derivatives. In fact, in Fliess's theory [40], and differently from Kalman's theory, a nonlinear system is defined by a differential field extension L/K finitely generated, where L is the system field which contains the system variables and the algebraic equations that links the variables; K is the ground field that contains the coefficients of the system.

The input u of a system L/K is a set $u = \{u_1, \ldots, u_m\}$ of L such that the extension $L/K < u >$ is differentially algebraic, which means that any element $\omega \in L$ satisfies an algebraic differential equation over $K < u >$ of the form $P(\omega, u_1, \ldots, u_m, \ldots, \dot{u}_1, \ldots, \dot{u}_m)$. The output of a system L/K is a set $y = \{y_1, \ldots, y_p\}$ of L. A transcendence basis $x = \{x_1, \ldots, x_n\}$ of the differential field extension $L/K < u >$ is the state of the system L/K. Any component of the derivative $\dot{x} = \{\dot{x}_1, \ldots, \dot{x}_n\}$ and y are $K < u >$ algebraic on x, which leads to the following generalized state variables representation:

$$\begin{cases} F_1\left(\dot{x}_1, x, \ldots, u, \dot{u}, \ldots, u^{(\alpha_1)}\right) = 0 \\ \vdots \\ F_n\left(\dot{x}_n, x, \ldots, u, \dot{u}, \ldots, u^{(\alpha_n)}\right) = 0 \\ H_1\left(y_1, x, \ldots, u, \dot{u}, \ldots, u^{(\beta_1)}\right) = 0 \\ \vdots \\ H_p\left(y_p, x, \ldots, u, \dot{u}, \ldots, u^{(\beta_p)}\right) = 0 \end{cases} \tag{8.13}$$

where F_i,H_j are polynomials over k and $\alpha_i, \beta_j \in N$; x is called a generalized state.

In the presence of faults (f), the nonlinear system is denoted as an algebraic differential field extension $k(u : y, f)/k(u, f)$ [27]. If the fault f is a differential algebraic equation with coefficients over the field $K(u, y)$, then it is said to be diagnosable. In other words, the fault variable is written in polynomial form, as function of the input variables, the output variables and their respective derivatives as follows:

$$f = h\left(u, \dot{u}, \ddot{u}, \ldots, u^{(m)}, y, \dot{y}, \ddot{y}, \ldots, y(n)\right) \tag{8.14}$$

The residuals $(\hat{r})$ given by Eq. (8.15) below are used to evaluate the obtained fault indicator (Eq. (8.14)):

$$\hat{r} = s^{-n}\frac{d^n}{ds^n}\hat{F} \tag{8.15}$$

where the sign $\hat{()}$ means that the variable is written in the Laplace domain.

The following steps are used to obtain the residuals [11, 41]:

- Put the fault indicator in Laplace domain,
- Differentiate the result n times with respect to s in order to eliminate the initial conditions, which may be unknown,
- Multiply by s^{-n} and return back to time domain.

These residuals are performed by using the integrals of the measured signals. In the case of noisy signals, these integrals produce a filtering effect. The derivative with respect to s of order n $(\frac{d^n}{ds^n})$ in Laplace domain results in a multiplication by $(-1)^n t^n$ in the time domain, and the multiplication by s^{-n} in Laplace domain corresponds to an integration of order n in the time domain.

The fault diagnosis based on the algebraic approach is mainly applied to linear systems, and some nonlinear systems for actuator and sensor fault diagnosis. In [15], the algebraic approach in association with a bond graph tool was extended to component fault diagnosis, under the assumption that the system inputs and outputs are fault free. The residuals of Eq. (8.15) reflect only the cumulative sum of the fault indicator from the degradation start until failure time, which means that the residuals do not reflect the degradation dynamics but only its effect. It should also be pointed out that this method does not need prior knowledge on the degradation nature and it has not been used yet for fault prognosis. Finally, this method can only handle additive faults and any change in the system dynamics can be considered as degradation.

8.3.1.5 Parameter Estimation Method

The fault diagnosis based on the parameter estimation consists to parametric identification of the system model using the system inputs (u) and outputs (y), and monitoring the estimated system parameters. For a nonlinear system described by the following state-space model:

$$\begin{cases} \dot{x} = h(x, u, \theta) + d \\ y = g(x) \end{cases} \tag{8.16}$$

where x is the state vector, h is the state function, g is the output function, and d represents the system disturbances that are assumed to be a bounded signal, and under the assumption that the system model parameters vary depending on the occurrence of a fault on the physical system. In the normal operation, θ takes the nominal values of the physical parameters; however, in faulty operation, the value of θ varies as a function of fault severity on the physical system. The model of Eq. (8.16) is, then, used for an on-line nonlinear parameter estimation problem, for which unknown fault parameters are estimated using system inputs and outputs, and appropriate approaches such as neural networks [115], fuzzy models [13], and Takagi–Sugeno (TS) models [88], and for linear systems, least-squares (LS) approaches are used. The estimation error (Eq. (8.17)) between the reference model parameters estimated in normal operation and the parameters estimated under faulty conditions is taken as HIs for diagnosis purposes [21].

$$HI = \hat{\theta} - \theta_n \tag{8.17}$$

The fault diagnosis via parameter estimation can handle only additive faults on parameters with slow rate dynamic. The main limit of this method is the difficulty of concluding on fault isolability conditions, since the parameters being estimated are model parameters and they do not represent the system physical parameters. This problem has been partially solved by studying the influence of each physical parameter on the model parameters [55]. As the gradual variations in degradation cause progressive changes in system parameters from their nominal values, which leads to a gradual deviation of HIs from zero, the HI trend can be analyzed to construct prediction models for RUL estimation.

8.3.1.6 Practical Constraints

Data Availability At the beginning of operation, data describing the system degradation process and expert knowledge are not available. In this case, the HIs generated from the physical knowledge are the most suitable. In the majority of the practical cases, the faulty operation is defined by thresholds that the parameters of the system must not exceed. These thresholds are used in the literature to estimate the failure thresholds for HIs whose parameters have a clear physical meaning [14, 31]. The estimation of the RUL is then performed using trend modeling methods that do not require prior knowledge of the dynamics of degradation presented in the second part of this chapter.

System Instrumentation The observability is a necessary condition for the implementation of the methods presented above, for the implementation of the AR method the system must be, in addition, over-constrained. This property is easily verifiable on a state model or a bond graph model. The identification of an optimal placement of sensors to obtain an observable system (or over-constrained for the application of the AR method) is also possible. But in practice, it is not always possible to place all the necessary sensors for the observability of a system, for reasons of cost, lack of space on the system, the non-availability of the sensor, and the consequences of placing a given sensor on the system. Among the practical cases of systems on which the authors of this paper have encountered difficulties of instrumentation for the application of the methods presented above: electric motors [14], where the torque sensor is rarely available, which makes it impossible to generate HIs whose electrical part and mechanical part are decoupled. Indeed, in the example of the HIs generated for the mechanical transmission system, the torque $\Gamma(t)$ is not measured, but, rather, it is estimated using the model of the interaction between the electrical part and the mechanical part of the brushless motor, given as follows:

$$\begin{aligned} e(t) &= k_e\dot{\theta}(t) \\ \Gamma(t) &= k_t i(t) \end{aligned} \tag{8.18}$$

where k_e is the electromotive force (EMF) and k_t is the motor torque constant. The use of the current variable to calculate the torque creates a matching of the HIs generated from the electrical part of the motor with the HI generated from the mechanical part. Consequently, it is not possible to locate the degraded subsystem. On thermal engines, in particular marine diesel engines [61], there are many sensors, but the system is not observable. The available sensors are mostly effort sensors (temperature, pressure), while flow sensors (volume flow, mass flow, heat flow, entropy flow) are not available. This is due to the unavailability of some sensors (such as the entropy flow sensors) and the consequence that the sensor placement may have on the system (for example, a mass-flow sensor must be inserted in the pipe, which may promote the appearance of fluid leaks). The same instrumentation constraints are encountered on the electricity production and management systems, and industrial equipment [75, 122].

8.3.2 Data-Driven and Signal Processing Methods

Among the data-driven methods of HIs generation for fault diagnosis, several are used also for the generation of HIs for failure prognosis [124]. Methods of multivariate analysis, such as principal component analysis (PCA) and its variants (IsoMap, PCA-Kernel, ...), are widely used [2, 68, 90, 96] due to the fact that they allow, in addition to generating the HI, extracting a degradation profile from raw data, assuming that this information is initially contained in the raw data. The time and frequency attributes of the measurement signals are also widely used, especially when the instrumentation of the system is poor, and is limited to just one or two sensors. These techniques allow a separation of the features contained in the signal highlighting dynamics which are not perceptible on the raw signal. The features presenting progressive trends that are not related to the normal operation of the system are often related to the process of degradation of the system, and can be exploited for the prognosis of failure [49, 51, 94]. Signal processing methods the most used for generating HIs are: statistical indices [74], empirical modes decomposition (EMD) [54], low pass filters [92], fast Fourier transform (FFT) [77], and wavelet decomposition [76]. In the area of failure prognosis, Ref. [74] uses statistical indices to extract the characteristics susceptible to failure and robust to noise from vibration data pump oil sands; Ref. [44] also uses these statistical indices on the raw data measured on bearings; Ref. [119] applies the EMD to bearings vibration signals to identify and diagnose faults. For the diagnosis of bearing faults from acceleration signals, Ref. [4] applies filtering with several levels of bandwidth to improve the signal-to-noise ratio. For the application of the wavelet transformation, Ref. [70] applies it to the voltage data of the rolling elements of a gearbox to characterize symptoms of early fatigue and cracking.

The principle of analytical redundancy can also be applied by creating redundancy through data-driven models, such as neural networks, support vector machine (SVM), and auto-regressive models. This technique is applied in [31] for the prognosis of failures of the embedded electronic systems, where a NARX neural network is used for the estimation of the consumed power and an ARMAX model is used for the estimation of the temperature. These two estimates are then compared to the measured values to generate health indices for fault diagnosis and failure prognosis. A multilayer perceptron (MLP) neural network is used in [47] for health condition monitoring of a wind turbine gearbox, and a recurrent neural network (RNN)is proposed in [7] for or early fault detection of gearbox bearings. In [102] an adaptive network-based fuzzy inference system (ANFIS) is implemented for wind turbine condition monitoring using normal behavior models.

8.3.2.1 Practical Constraints

Data Availability The data-driven methods presented above are all based on the assumption that data containing the degradation process is available, so they are complementary to the physical model-based methods that only require data from the normal operation of the system, used for parameter identification, and a physical knowledge of the system. These two types of approaches are complementary, covering thus a wide field of application.

Properties of the Generated HIs Unlike HIs generated using physical model-based methods, the properties analysis (Monotonicity, Trendability, and Prognosability) of HIs generated by data-driven methods has been the subject of several research works [10, 24, 25, 52]. Monotonicity is related to the irreversibility assumption of degradation phenomena. Trendability is related to the degradation profile, i.e. it is related to the fact that the HI value is representative of the degradation value at any moment of the evolution of the degradation. Prognosability is related to the amplitude of the HI corresponding to the total failure; this property is respected when the threshold of HI corresponding to the total failure is constant. The metrics presented below are the ones proposed in [24] and [10], as their score is easily interpretable (between 0 and 1), where 1 indicates the most satisfactory and 0 the less satisfactory level of the specific HI property:

$$\text{Monotonicity} = \frac{1}{N}\left|\sum_{i=1}^{N} M_i\right| \tag{8.19}$$

where M_i is the monotonicity of a single run-to-failure trajectory given by:

$$M_i = \frac{n_i^+}{n_i - 1} - \frac{n_i^-}{n_i - 1}, i = 1, \ldots, N \tag{8.20}$$

n_i is the total number of observations in the ith run-to-failure trajectory and n_i^+ (n_i^-) the number of observations characterized by a positive (negative) first derivative.

$$\text{Trendability} = \min\left(\left|corrcoef_{ij}\right|\right), i, j = 1, \ldots, N \tag{8.21}$$

$corrcoef_{ij}$ is the linear correlation coefficient between the ith and the jth run-to-failure trajectories. The computation of the correlation coefficient between two vectors requires that they are formed by the same number of patterns.

$$\text{Prognosability} = exp\left(\frac{-std\left(HI_{fail}\right)}{mean\left|HI_{start} - HI_{fail}\right|}\right) \tag{8.22}$$

where HI_{start} and HI_{fail} are the HI values at the beginning and end of the run-to-failure trajectories, respectively; $std\left(HI_{fail}\right)$ is standard deviation of the HI values at the end of the trajectories. $mean\left|HI_{start}\right|$ and $mean\left|HI_{fail}\right|$ are the average variation of the HI values between the beginning and the end of the trajectories, respectively.

Recent researches are directed towards the development of methods allowing the extraction of a set of features optimizing the scores of the three properties, as in [10], where the HIs identification is formulated as the problem of selecting the best combination of features to be used, and a multi-objective optimization that considers as objectives the metrics of Monotonicity, Trendability, and Prognosability. The proposed method is based on a binary differential evolution (BDE) algorithm for the multi-objective optimization.

The HI generation methods presented in this paper have been applied by the authors of this review paper on real cases. The details of the application of each method can be found in [14] and [35] for AR method, [33] for parity space method, [15] for algebraic methods, [36, 90] for PCA method, [89] for EMD method, [34, 89] for WD method, and [31, 32] for HI generated using machine learning methods. Table 8.1 summarizes the constraints of use, the advantages, and limitations of the methods presented above.

8.4 HI Trend Modeling for RUL Estimation

As illustrated in Fig. 8.3, the modeling approaches of HI trends for the estimation of RUL can be decomposed into three main families: physical approaches, data-driven approaches, and expert methods. Another classification is proposed in [2], where the RUL estimation methods are classified into: reliability based, similarity based, model based, and data-driven based approaches. The most used physical model form is the differential one, whose order and parameters are identified according to the physical knowledge and data available on the degradation process. The updating of the parameters makes it possible to compensate the modeling uncertainties and the adaptation of the model to changes in the degradation rate. Data-driven approaches are the most used and can be decomposed into five families: statistical models, stochastic models, deterministic models, probabilistic models, and machine learning model. The third family of trend modeling comprises those approaches that formalize the knowledge of industry experts through the tools of fuzzy logic and Bayes probabilities. Only data-driven methods able to include expert knowledge into the prediction models are presented in this paper.

8.4.1 Data-Driven Models

Four kinds of models are presented in this section:

- Stochastic models, especially continuous and discrete Markov processes.
- Probabilistic models, based on Bayes probability theory.

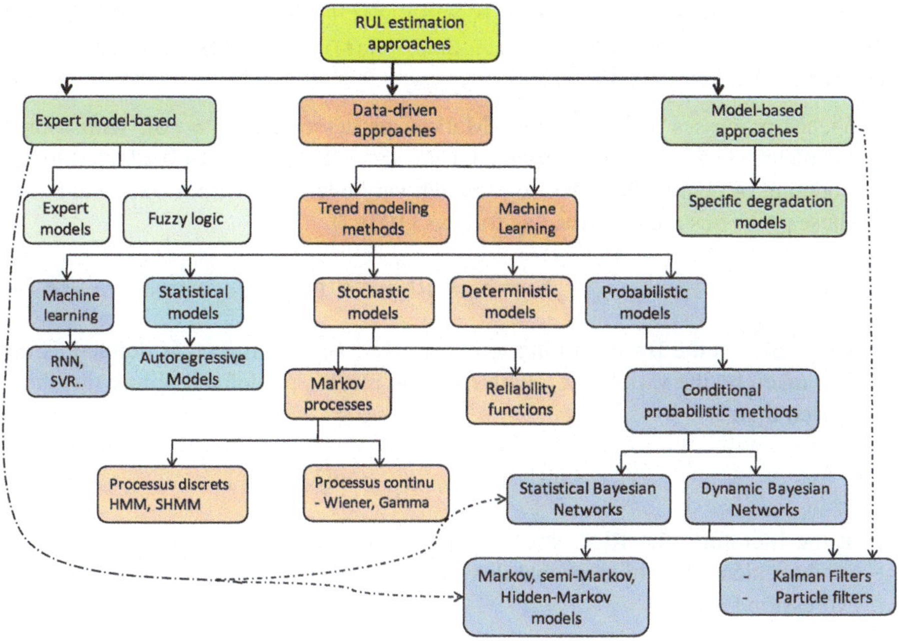

Fig. 8.3 Classification of RUL estimation approaches

- Statistical models, with a focus on the auto-regressive (AR) models and the auto-regressive moving average (ARMA) model, which is representative of these methods.
- Deterministic models, which are geometric models allowing to estimate the RUL as a deterministic variable.

8.4.1.1 Stochastic Models

Markov processes are the most used stochastic models for fault prognosis. These models describe processes without memory, where the probability of the future state X_n depends only on the current state X_{n-1} as shown in the following equation:

$$\mathrm{P}(X_n = x_n | X_{n-1} = x_{n-1}, \ldots, X_0 = x_0) = \mathrm{P}(X_n = x_n | X_{n-1} = x_{n-1}) \qquad (8.23)$$

$x_1, \ldots, x_n$ are linked to the different states of the system. These processes can be divided into two categories: continuous Markov processes and discrete Markov processes.

Continuous Markov Processes

The most common continuous Markov processes used in the literature for prognosis are Wiener and Gamma processes. The hypothesis of independent increments leads these processes to Markov properties because: $X(t+\Delta t) - X(t)$ is independent of $X(t)$ and $X(t+\Delta t) = X(t) + (X(t+\Delta t) - X(t))$; the process $\{X(t), t \geq 0\}$ is, therefore, a Markov process [91].

- Wiener processes are continuous Markov processes $[X_t, t > 0]$, with a drift parameter μ and a variance parameter σ^2, $\sigma > 0$. They are well adapted to the modeling of degradation processes which vary over time with a Gaussian noise. These processes are described as follows [103]:

$$X_t = x_0 + \mu t + \sigma B(t) \tag{8.24}$$

  where $B(t)$ is the Brownian motion. The RUL H_{t_i} at a time t_i is defined as the time taken by the variable X_t, with $t > t_i$ to reach a predefined failure threshold w such that:

$$H_{t_i} = \inf\{\Delta_{t_i} : X_{t_i+\Delta_{t_i}} \geq w | X_{t_i} < w\} \tag{8.25}$$

  In the literature, the RUL is often given with a confidence interval, obtained by the calculation of a probability density function given by the following expression [26]:

$$f_{H_{t_i}}(h_{t_i}) = \frac{w - X_{t_i}}{\sqrt{2\pi t_{t_i}^3 \sigma^2}} exp\left(-\frac{(w - X_{t_i} - \mu t_{t_i})^2}{2 t_{t_i} \sigma^2}\right) \tag{8.26}$$

  Many work apply this process and its variants [95, 105, 110, 111, 113], and particularly [116] which proposes a Wiener process with an updated drift parameter μ_t.
- Gamma process is the most appropriate for modeling a monotonic and gradual deterioration [91]. Reference [1] proposed to use it as a deterioration model randomly introduced over time [6, 23, 42, 62]. Mathematically, a random quantity X follows a Gamma distribution with a shape parameter $\upsilon > 0$ and a scale parameter $u > 0$ if its probability density function (PDF) is given as follows:

$$Ga(x|\upsilon, u) = \frac{u^{\upsilon}}{\Gamma(\upsilon)} x^{\upsilon-1} exp(-ux) I_{(0,\infty)}(x) \tag{8.27}$$

  where $I_{(0,\infty)}(x) = 1$ for $x \in (0, \infty)$ and $I_{(0,\infty)}(x) = 0$ for $x \notin (0, \infty)$, $\Gamma(\upsilon) = \int_{z=0}^{\infty} z^{\upsilon-1} e^{-z} dz$ is the Gamma function for $\upsilon > 0$.

  Given a non-decreasing function $\upsilon(t)$, Gamma process with the form function $\Upsilon(t) > 0$ and the scale parameter $u > 0$ is a continuous stochastic process with the following characteristics:

  1. $X(0) = 0$ with a probability of 1
  2. $X(\tau) - X(t) \sim Ga\big(\upsilon(\tau) - \upsilon(t), u\big)$ for all $\tau > t \geq 0$
  3. $X(t)$ has independent increments.

Let $X(t)$ be the deterioration at time $t, t \geq 0$: the PDF of $X(t)$ is as follows:

$$f_{X(t)}(x) = Ga(x|\upsilon(t), u) \tag{8.28}$$

HIs expectation and variance are as follows:

$$E(X(t)) = \frac{\upsilon(t)}{u}, \quad VAR(X(t)) = \frac{\upsilon(t)}{u^2} \tag{8.29}$$

A system is said to be faulty when its degradation reaches a predefined threshold S. From Eq. (8.28), the distribution of the failure time at time t is written as follows:

$$\begin{aligned} F(t) &= Pr\{T_S \leq t\} = Pr\{X(t) \geq S\} \\ &= \int_{x=S}^{\infty} f_{X(t)}(x)dx = \frac{\Gamma(\upsilon(t), Su)}{\Gamma(\upsilon(t))} \end{aligned} \tag{8.30}$$

where $\Gamma(a, x) = \int_{z=x}^{\infty} z^{a-1}e^{-z}dz$ is the incomplete gamma function with $x \geq 0$ et $a > 0$. The PDF of the failure time at time t is, thus:

$$f(t) = \frac{\partial}{\partial t}\left[\frac{\Gamma(\upsilon(t), Su)}{\Gamma(\upsilon(t))}\right] \tag{8.31}$$

The mean failure time and the average RUL are given in the following equations:

$$\mathcal{T}_t = \int_{t=0}^{+\infty} tf(t)dt \tag{8.32}$$

$$RUL_t = \mathcal{T}_t - t \tag{8.33}$$

The two Markov processes presented above are widely used to model degradation, covering the majority of degradation profiles: linear and nonlinear, noisy and monotonous. However, these processes require the calculation of a HI $X(t)$, which estimates the current level of degradation of the system and which can be calculated using one of the methods presented in Sect. 8.3. The main limitation is related to the central property of Markov models, called memoryless assumption, which is a relatively strong assumption and thus may lead to strong approximation for real applications. To overcome this issue, a reliability model can be developed to consider the changes in the operating modes of the systems [89]. This model is based on two assumptions: (1) the future value of the HI is a function of the current state of the system, given by the present value of the HI, time, operating modes assumed known, and external noises supposed to follow a Gaussian law; (2) the HI is non-negative and monotonous.

Given these two hypotheses, the dynamics of the HI can be described as follows:

$$\Delta X_t = \frac{\beta t^{\beta-1}}{\eta^\beta} exp(\gamma Z_t + \varepsilon) \tag{8.34}$$

where $\beta > 0$ is the shape parameter of the model, $\eta > 0$ is its scale parameter, $\gamma = [\gamma_1, \ldots, \gamma_m] \in \mathbb{R}^m$ is a vector of m elements, describing the influence of changes in operating modes $Z_t = [Z_{1,t}, \ldots, Z_{m,t}]$ on the degradation. The uncertainties of the model are represented by the variable ε assumed to follow a normal distribution $N(0, Q)$. The first term $\beta t^{\beta-1}/\eta^\beta$ depends on time and means that ΔX_t depends on the system aging.

The HI evolution X_t is defined as the accumulation of all segments ΔX_t:

$$X_t = \sum_{\tau=0}^{t} \Delta X_\tau \tag{8.35}$$

Based on the linearity of mathematical expectation, the value of the mathematical expectation of $X(t)$ is calculated as follows:

$$\begin{aligned} E[X_t] &= E\left[\sum_{\tau=0}^{t} \Delta X_\tau\right] \\ &= \sum_{\tau=0}^{t} E[\Delta X_\tau] \\ &= \sum_{\tau=0}^{t} E\left[\frac{\beta \tau^{\beta-1}}{\eta^\beta} exp(\varepsilon)\right] \\ &= \sum_{\tau=0}^{t} \frac{\beta \tau^{\beta-1}}{\eta^\beta} E[exp(\varepsilon)] \end{aligned} \tag{8.36}$$

$\epsilon \sim N(0, Q)$ being a normal distribution variable, $exp(\epsilon)$ is a log-normal distribution variable with mean value $exp(Q/2)$:

$$\begin{aligned} E[X_t] &= \sum_{\tau=0}^{t} \frac{\beta \tau^{\beta-1}}{\eta^\beta} exp(Q/2) \\ &= exp(Q/2) \sum_{\tau=0}^{t} \frac{\beta \tau^{\beta-1}}{\eta^\beta} \end{aligned} \tag{8.37}$$

Assuming that k is he RUL and $L > X_t$ the predefined failure threshold, the RUL can be estimated as follows:

$$
\begin{aligned}
P(k|X_t < L) = &\ P(X_{t+k} < L|X_t < L) \\
&= P\left(X_t + \sum_{i=t+1}^{t+k} \Delta X_i < L\right) \\
&= P\left(\sum_{i=t+1}^{t+k} \Delta X_i < L - X_t\right) \\
= &\ F_{\sum_{i=t+1}^{t+k} \Delta X_i}(L - X_t)
\end{aligned}
\tag{8.38}
$$

$F_{\sum_{i=t+1}^{t+k} \Delta X_i}(L - X_t)$ is the distribution function *(fr)* of the sum $S_k = \sum_{i=t+1}^{t+k} \Delta X_i$ to the value $L - X_t$.

Discrete Markov Processes

These methods are based on the principle of Markov chains for modeling processes that evolve through a finite number of states [37, 59]. By definition, it is assumed that the probability associated with each state, the probability associated with the transition from one state to another, and the probability of future failure can be estimated. The main property of Markov models is the assumption that the future state depends only on the current state, called conditionally independent or memoryless assumption. The most commonly used models for fault prognosis are the Hidden Markov models ($HMMs$), characterized by two parameters: (1) number of states of the system, (2) number of observations by state, and three probability distributions: (1) probability distribution of transitions between states, (2) probability distribution of observations, and (3) an initial probability distribution of states [12, 37, 83]. The HMM presents an appropriate mathematical model to describe the failure mechanisms of systems, which evolve in several degraded health states over the time prior to failure, as it can estimate the unobservable health states using observable sensor signals. The word "hidden" is related to the fact that the states are hidden from direct observations, so they only manifest themselves via a probabilistic behavior. HMM can exactly capture the characteristics of each state of the failure process, which is the basis of HMM prognosis [37]. These methods allow, thus, to model several operating conditions of the system and failure scenarios. However, their implementation requires a large amount of data and knowledge for learning, and the calculation intensity, which is proportional to the number of states, can become important for systems with several operating states. The three basic issues in HMMs implementation are: (1) Evaluation/Classification that represents what is the probability to get the model given an observation sequence, (2) Decoding/Recognition that represents what sequence of hidden states is the most optimal or is most probably the one that generates the given sequence of observations, and (3) Learning/Training that represents how to adjust the model parameters.

In the fault prognosis area, a widely used algorithm is the backward–forward algorithm, where the RUL at the time n can be defined as:

$$
X_n = \inf\left\{x_n : Y_{n+x_n} = N/Y_n \neq N\right\} \tag{8.39}
$$

where Y_n is the nth observation. The calculation of RUL using Markov chains usually involves the use of the phase-type distribution. As a result, the distribution and the expectation of the RUL are given as:

$$\Pr(\mathrm{X_n} = \mathrm{k}) = \alpha_n \tilde{P}^{\mathrm{k}-1}(I - \tilde{P})e \tag{8.40}$$

$$\mathrm{E}(\mathrm{X_n}) = \alpha_n (I - \tilde{P})^{-1} e \tag{8.41}$$

where

$$\mathrm{P} = \begin{pmatrix} \tilde{P} & P_0 \\ 0 & 1 \end{pmatrix} \text{ with } P_0 = \left(1 - \tilde{P}\right) e \tag{8.42}$$

HMM is suitable for nonlinear systems. It can estimate the data distribution of normal operation with nonlinear and multimodal characteristics, assuming that predictable fault patterns are not available. It is applicable to non-stationary systems. It has been widely applied in real applications. The main reason is that the plant operation condition can be divided into several meaningful states, such as "Good," "OK," "Minor Defects Only," "Maintenance Required," "Unserviceable," so that the state definition is closer to what is used in industry. It can be used for fault and degradation diagnosis on non-stationary signals and dynamical systems. It is appropriate for multi-failure modes [37, 59, 66].

The main limitation is related to the property of Markov models, i.e. the memoryless assumption. The health state visit time is assumed to follow an exponential distribution, which could be inappropriate for some cases. The transition probability among the system states in Markov models is often determined by empirical knowledge or by a large number of samples, which is not always available. A large amount of data is needed for accurate modeling [37, 59, 66].

The hidden semi-Markov model (HSMM) is an improved HMM, which overcomes the inherent limitation of assuming exponential distributions. Unlike the HMM, the HSMM does not follow the unrealistic Markov chain assumption and therefore provides more powerful modeling and analysis capability for real problems. In addition, the HSMM allows modeling the time duration of the hidden states and therefore is well suited for fault prognosis. A practical example is given in [28], where an approach for RUL estimation from heterogeneous fleet data under variable operating conditions is proposed in three steps:

- Identification of the degradation states of an homogeneous discrete-time finite-state semi-Markov model using unsupervised ensemble clustering approach.
- The maximum likelihood estimation (MLE) method and the Fisher information matrix (FIM) are used for parameter identification of the discrete Weibull distributions describing the transitions among the states and their uncertainties.
- The direct Monte Carlo (MC) simulation based on the degradation model is used to estimate the RUL of fleet equipment.

The proposed approach is applied to two case studies: heterogeneous fleets of aluminum electrolytic capacitors and turbofan engines. Another solution proposed in [53] to overcome the lack of knowledge on the condition monitoring is the online updating of the parameters of the degradation model formulated as a first-order Markov process. The originality of this work consists of the combination of Particle Filtering (PF) technique with a Kernel Smoothing (KS) one, for simultaneously estimating the degradation state and the unknown parameters in the degradation model, while significantly overcoming the problem of particle impoverishment.

8.4.1.2 Conditional Probabilistic Models

These models are based on Bayes theorem, which describes relationship between the conditional and marginal probabilities of two stochastic events A and B as follows:

$$P(A|B) = \frac{P(B|A)P(A)}{P(B)} \tag{8.43}$$

These methods describe the current state as a conditional probability function and, then, apply Bayes theorem to update the probability assessment of future behavior. The most used modeling tool is the Bayesian network, which is a probabilistic graphical model representing random variables in the form of an acyclic oriented graph. In the field of aeronautics, [39] uses the network with variables such as aircraft weight, landing speed, and brake operation to predict brake failure. In other research work, Bayesian networks are associated with the Kalman filter [22, 69] or particle filter [20, 93, 106] for failure prognosis.

8.4.1.3 Statistical Models

The ARMA, the ARIMA (*Auto-Regressive Integrated Moving Average*), and the ARMAX (*Auto-Regressive Moving Average eXogenous inputs*) models, initially used for time series prediction, have been used to estimate the RUL by considering the future value of the degradation as a linear function of system inputs, past observations, and random noise. To show how these methods are used for prognostics, let us take the example of the ARMA model. A time series $\{x_t|t = 1, 2, \ldots\}$ is generated by an ARMA model (p, q) as follows:

$$x_t = \sum_{i=1}^{p} \phi_i x_{t-i} + \sum_{j=0}^{q} \theta_j \epsilon_{t-j}, \quad (\theta_0 = 1) \tag{8.44}$$

where x_t is a series at the instant t, p and q are non-zero integers, p is the order of the auto-regressive part, q is the order of the moving average part, $\{\epsilon_t\}$ indicates the noise series, $\{\phi_i, i = 1, \ldots, p\}$ et $\{\theta_j, j = 1, \ldots, q\}$ are the parameters to be estimated.

To use this model for fault prognosis, the variable x_t is considered as the HI which represents the system condition state and the failure threshold D is supposed known. The RUL at instant t of the system is calculated by the following equation:

$$rul_t = \min\{\Delta t : x_{t+\Delta t} \geq D | x_t < D\} \tag{8.45}$$

Yan et al. [118] have used ARMA model for fault prognosis. An ARMA model is incorporated in a software for data fusion and prognosis [65]. An extension of an ARMA model by usingbootstrap forecasting is proposed in [117].

The use of these models is simple for prognosis. However, they assume that the future state of the system is a linear function of the system inputs, past observations, and noise, which is not often the case in reality. Moreover, their results are sensitive to the initial conditions, thus leading to an accumulation of systematic errors in the prediction.

8.4.1.4 Deterministic Models

This approach is supervised by the calculation of the Euclidean distance (d) between the actual status of the system, given by the actual HIs values and the faulty HIs identified offline. The degradation speed (v), which indicates how the degradation moves from the normal operation to the faulty one, is used to compute the RUL as follows:

$$rul\,(t) = \left|\frac{d\,(t)}{v\,(t)}\right| \tag{8.46}$$

To compute the distance $d(t)$ between the n HIs in real time operation and the barycenter of the identified faulty operating cluster ($C_f(c_1, c_2, \ldots, c_n)$), the following Euclid metric is considered:

$$d\,(t) = \sqrt{(r_1(t) - c_1)^2 + (r_2(t) - c_2)^2 + \cdots + (r_n(t) - c_n)^2} \tag{8.47}$$

where $r_1(t)$, $r_2(t)$,...$r_n(t)$ is a set of HIs defining the monitoring space. This set can be generated using one of the HI generation methods presented in the previous section.

The numerical differentiation of the distance variable d is taken to compute the degradation speed v:

$$v\,(t) = \frac{d\,(t + \Delta T) - d\,(t)}{\Delta T} \tag{8.48}$$

where ΔT is the sampling time for speed degradation computation. whose value is chosen so that the noise is not amplified.

The main interest of this method is the fact that no knowledge is required on the tendency or the pattern of degradation. It is accurate when the faulty operation is clearly identified and successfully applied to a wind turbine system [36] for RUL estimation.

Other deterministic models are used in the literature, especially when the degradation profile is known. The models are identified by using the fitting methods applied on the available profile of the degradation. Linear, exponential, and polynomial models are the most used [120].

8.4.1.5 Learning Techniques

Learning techniques are widely used in the literature for trend modeling in the field of failure prognosis. These regression models, such as neural networks and support vector regression (SVR), are scalable and able to accommodate nonlinear dynamics, but require a large amount of data for learning. For unsupervised learning cases, an example of using SVR for RUL prediction is proposed in [32] for failure prognosis of embedded systems. The prediction is realized using a SVR at a step of the evolution of the health index. The SVR expression is given as follows:

$$\hat{HI}(k+1) = \sum_{[i=(m-1)\tau+1]}^{N-1} \alpha_i^* K(HI(i), HI(k)) + b^* \tag{8.49}$$

where α_i^* are Lagrange multipliers and τ the delay. In this work, the standard SVR toolbox is used without making any special changes to the prediction of temporal overlays. The free parameters, C, ϵ, the size of the kernel (Gaussian), and the dipping dimension m are selected from a comprehensive search in the parameter space to optimize the performance of the prediction on the validation set. The available N observations are therefore shared between two sets of training and validation of respective sizes Ne and Nv. Values for which the prediction error at a step on the validation set is minimal are retained for the final prediction. Once the parameters are fixed, the prediction is made using all the N observations available. The predictions at several steps, i.e. for the values ($k \geq N+1$), are realized by the ratio of the prediction at one step, using the estimated vectors $\hat{HI}(k)$ at the previous iterations and not the observations themselves.

In addition to machine learning techniques, deep learning techniques like long short-term memory (LSTM), which can remember information for long periods of time, are used for trend modeling and RUL prediction. An application case is proposed in [121], where a long short-term memory recurrent neural network is used for RUL prediction of lithium-ion batteries.

8.4.2 Physical Models

RUL estimation based on a physical model consists of considering that degradation follows a parametric trend, which can take one of the following ordinary differential equations (ODE):

$$\begin{aligned} \dot{F} &= \beta_1 F \\ \dot{F} &= \beta_2 F^2 \\ \ddot{F} &= \beta_3 \dot{F} + F + \beta_4 \\ \ddot{F} &= \beta_5 \dot{F}^2 + \beta_6 F + \beta_7 \end{aligned} \tag{8.50}$$

where F is the fault component value describing the degradation and β_i $(i = 1 \ldots 7)$ represent the degradation model coefficients which are identified on-line by using the least square method [30] or particle filter [29, 58]. For example, the RUL associated with the degradation model of the form $\dot{F} = \beta_1 F$ is given by:

$$\text{RUL} = \frac{\ln\left(\frac{1-th}{1-HI_i}\right)}{\beta} - NT_s \tag{8.51}$$

where N represents the sample data, T_s is the sampling time, and th is the failure threshold.

Other trend modeling approaches for RUL estimation can be found in the literature, such as [99] where the RUL estimation is treated as an uncertainty propagation problem, [103] where the review is focused on statistical data-driven approaches, relying only on available past observed data and statistical models, [5] which provides practical options for prognostics so that beginners can select appropriate methods for their fields of application.

8.4.3 Practical Constraints

The purpose of the RUL estimate is to give to the maintenance experts' two pieces of information: the first is that the system will undergo the occurrence of a total failure, the second is to give a sufficient time horizon for the maintenance experts in order to plan a maintenance strategy. A metric proposed in [101], called Prognosis Horizon (PH), is used to evaluate this time horizon in a confidence interval which can be defined by the user. Prognosis Horizon (PH) ranges within $[0, \infty[$ and is calculated as follows:

$$PH(i) = EoP - CT \tag{8.52}$$

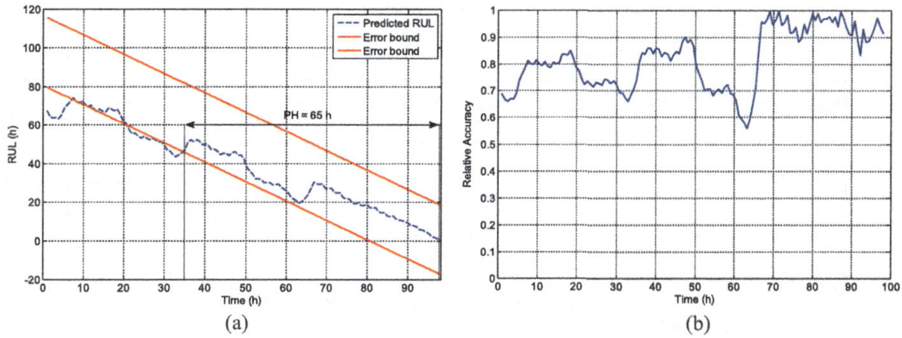

Fig. 8.4 Illustration of the PH RA metrics calculated for a wind turbine system in presence of an unbalance fault caused by a progressive deformation of the blade

It represents the difference between the Current Time index (CT) and the End of Prediction time index (EoP), obtained when the prediction crosses the failure threshold. A practical example of the PH calculation is illustrated in Fig. 8.4a which represents the PH calculated for a wind turbine system in presence of an unbalance fault caused by a progressive deformation of the blade. In this practical example, the HIs are generated using the PCA method and the trend modeling for RUL estimation is performed using a deterministic model, based on Euclidean distance [36]. Figure 8.4a shows that the obtained PH, by considering a confidence interval of 18%, is equal to 65 h. The maintenance expert can, then, make a decision, whether or not this HP is sufficient to plan a maintenance strategy in good conditions. If it judges that this PH is insufficient it is possible to increase it but by increasing the confidence interval.

To give the user an easily interpretable measurement tool of the confidence that can be given to the PH metric, another metric is proposed in [101] where the accuracy is quantified according to the real RUL. This metric is called relative accuracy (RA) and expressed as follows:

$$RA\,(t) = 1 - \frac{|RUL^*\,(t) - \mathrm{RUL}\,(t)|}{RUL^*\,(t)} \tag{8.53}$$

RUL^* is the real RUL. The range score of the RA metric is between [0, 1], and the best score is close to 1. A practical example of the results of this metric applied to the estimated RUL before the total degradation of the wind turbine system is given in Fig. 8.4b. It shows that the RA is greater than 0.7 in average over the PH, but has a great variability. All these measures will enable maintenance experts to assess the risks and make the right decisions for the maintenance of the systems.

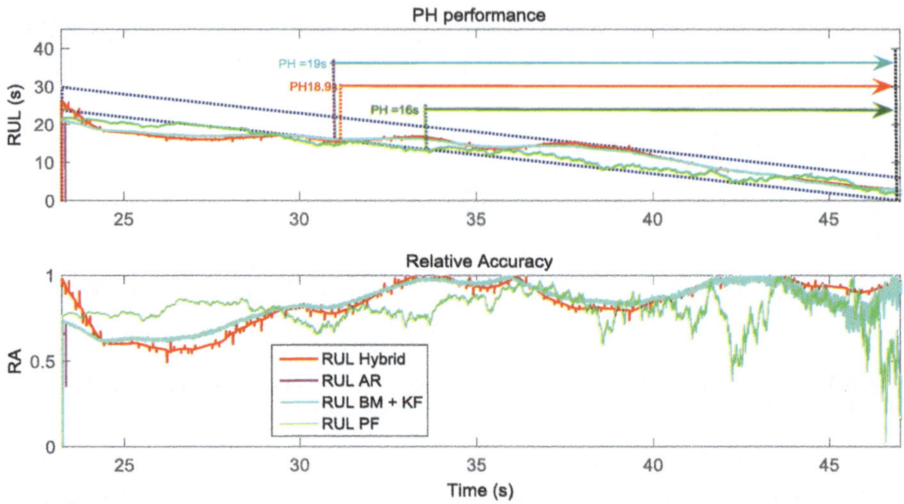

Fig. 8.5 Practical illustration of the use of PH and RA metrics for the performance comparison of the RUL estimation methods

Table 8.2 Performance comparison of the considered RUL estimation methods by universal metrics

	PH	RAmin	RAmax	RAmean
RUL Hybrid	18.9 s	0.5	1	0.8509
RUL AR	19 s	0.6207	1	0.8504
RUL BM + KF	19 s	0.2702	1	0.8487
RUL PF	16 s	0	1	0.7155

In addition, these two metrics can be used to compare the performances of different methods in a given context. A practical example is given in Fig. 8.5, where the PH and RA are calculated for four prognostic methods applied to the RUL prediction on the mechanical transmission system presented in [14]. In this paper, the analytical redundancy method is used for HIs generation and four trend modeling methods are applied for RUL estimation: an auto-regressive (AR) model, whose parameters are estimated using the least square methods, an updated Wiener process, whose drift parameter is updated using a Kalman filter, a first-order differential model whose parameter is updated using a particle filter (PF), and a deterministic model based on the calculation of the Euclidean distance [14]. The performance results of the considered RUL estimation methods are given in Table 8.2, which shows that the AR model and the Wiener model have the largest PH, thus giving the user more time to react, whereas the Wiener model is less stable since it presents a greater variability in its RA.

It would be also better to evaluate the RUL prediction accuracy using—accuracy and cumulative relative accuracy (CRA) proposed in [100, 101].

8.5 Discussion and Future Challenges

8.5.1 Discussion

After analyzing the methods described above, one can see that discrete Markov models are the most complex to implement because they require expert knowledge and rich databases on the previous operation of the system and its failures; the uncertainty brought by expert knowledge is often taken into account using fuzzy logic. The memoryless assumption, which is the main property of discrete Markov processes, and that they also share with continuous processes, is a major limitation in the use of these processes for the estimation of the RUL. To overcome it, the hidden semi-Markov models (HSMMs), that do not follow the unrealistic Markov chain assumption, to provide more powerful modeling and analysis capabilities for real fault prognosis problems.

Continuous Markov processes, especially the Wiener and Gamma processes, are widely used in the literature as they are easy to implement and are well adapted to modeling the progressive dynamics of degradation phenomena. The updating of the parameters by increasingly powerful techniques such as maximum likelihood, the Kalman filter, and the particle filter makes it possible to adapt the estimation to the possible changes in the rate of degradation and provide in part a solution to the limit related to the memoryless assumption. Research works have gone even further in modeling, drawing on the Cox model, by proposing a reliability function that takes into account the covariates representing changes in the operating modes of a system: the limit of this model is related to the fact that the evolution of the covariates must be known beforehand, which is difficult to obtain on systems such as energy and transport systems where covariates are environment variables that are not controlled.

The representation of degradation processes by adaptive differential models makes it possible to take into account the physical knowledge available on these phenomena for the choice of the order of the models. Continuous parameter updating adapts to the change in degradation rate, and structured and unstructured uncertainties are taken into account by the generation of normal operating thresholds and total failure thresholds. The main limitation of this type of model is related to problems of amplification of the noises generated by the successive derivations of the outputs, as well as the lack of physical knowledge about the degradation processes, which generally leads to an arbitrary choice of the order of the model. Geometric models are efficient and accurate, but require complex classification work to identify clusters, using the physical model for generating the useful databases for learning.

The choice of the HI modeling approach depends on the context of use, the complexity of the system. and the information available on its previous operating modes, especially for the definition of the structure of the model as well as for the identification of its parameters. Table 8.3 summarizes a set of criteria, not

Table 8.3 Trend modeling approaches for RUL estimation

	Choice of model structure	Estimation of model parameters	Data in degraded operation	Expert knowledge	Physical knowledge	Accuracy of estimation	Adaptation to setpoints change
Discrete Markov processes							
HMM	Expert knowledge	By training	Necessary	Necessary	Not necessary	Depends on prior knowledge	No
HSMM	Expert knowledge	By training	Necessary	Necessary	Not necessary		No
Continuous Markov processes							
Wiener process	Predefined	Maximum likelihood Kalman filter particle filter	Required to estimate parameters	Not necessary	Not necessary	More accurate when the rate of degradation is not very variable	Not very suitable
Updated Wiener process	Predefined	Maximum likelihood Kalman filter particle filter	Required to estimate parameters	Not necessary	Not necessary	Better adapted to frequent changes in degradation rate	Yes by updating the settings
Gamma process	Predefined	Maximum likelihood Kalman filter particle filter	Required to estimate parameters	Not necessary	Not necessary	More precise on processes of monotonous degradation and not very noisy	Yes by updating the current state

Reliability functions							
Modified Cox models	Predefined	Maximum likelihood Kalman filter particle filter	Required to estimate parameters	Required to estimate parameters	Knowledge of covariates is necessary	More precise on monotonous degradation and not very noisy	Yes, is the most suitable model
Physical models							
Differential models	Chosen or deduced from physical knowledge	Least squares maximum likelihood Kalman filter particle filter	Not necessary if physical knowledge is available	Not necessary if physical knowledge is available	Necessary to choose the model order and parameters initialization	Accurate when physical knowledge is sufficient	Yes the model can consider the system setpoints
Geometric models							
Distance model	Depends on the size of the monitoring space	Not necessary	Not necessary	Not necessary	Not necessary	More accurate when classification is accurate	Yes By updating the speed and the direction of the degradation

exhaustive, that can be used as a basis for choosing the HI modeling methodology for the estimation of RUL.

8.5.2 Future Challenges

The general formulation of the Remaining Useful Life (RUL) of a system can be expressed in a general form, as a function of the time t, the current condition monitoring $CM(t)$ and the current health state $HS(t)$ (Eq. (8.54) below). However, in practice, the state of degradation is neither available nor measurable in the majority of cases, health indices must be deduced from the physical knowledge, expert knowledge, and available measurements [90]:

$$\mathrm{RUL} = g(t, HS(t), CM(t)) \tag{8.54}$$

- The use of the techniques initially developed for fault detection and isolation (FDI) to estimate the health state (HS) of the system is a good idea, but in the context of failure prognosis, the early detection becomes a major issue, as it is necessary to estimate the RUL well in advance to allow maintenance operators to plan their maintenance interventions. FDI techniques treat uncertainties in a probabilistic or deterministic manner to generate thresholds that provide a better compromise between false alarms and non-detections. In the context of failure prognosis, it is necessary to take into account also the Prognosis Horizon [100, 101]. In addition, the problems related to the occurrence of multiple faults, their interaction, their effect on the HS of the system remain.
- Condition monitoring (CM(t)), necessary for RUL estimation, is not always known especially in systems operating in a randomly variable environment, such as offshore wind turbines and transportation systems. The solution proposed in the literature consists in taking account of the known or controlled CMs by using, for example, the modified Cox model [89], and in compensating the lack of knowledge about the unknown CMs by an online update of the model parameters. This solution is effective in some application cases, but in the case where these CMs vary strongly and continuously, the RUL estimate will change considerably and continuously, which will prevent the use of the estimated RUL for planning the maintenance. The solution may be to associate the RUL estimate with risk analysis methods taking into account several operating and degradation scenarios [123, 125].

8.6 Conclusion

This review of horizontal approaches for RUL estimation has highlighted the diversity of methods proposed in the literature as well as their formal description and context of use. The analysis shows that the two major stages of the procedure can be synthesized independently, but that the context of use, the complexity of the systems as well as the history of available data and expertise are common elements that govern the relevance of the choice of the HI generation methods and the method of modeling its tendency for the estimation of the RUL. The diversity of methods for generating health indices can cover a wide range of applications. The PCA makes it possible to simultaneously reduce the size of the data and generate health indices from large databases with linear, bilinear, or nonlinear dependencies. When the instrumentation of the systems is not rich, the statistical, frequency, and time-frequency attributes can be extracted from the signals and, then, analyzed to make them HIs for the estimation of the RUL. When physical knowledge is relevant enough to take modeling assumptions, build and validate physical models, these latter are, then, associated with health index generation methods such as analytical redundancy, observers, and parameter estimation. In this paper, we have also presented the trend modeling methods that are the simplest to implement and that are adapted to the physical properties of degradation processes, such as the progressive aspect and the influence of the environment and operating modes of the systems. The choice of the model depends on the context of use, the physical knowledge available on degradation processes, the expert feedback, and available data. The RUL can be presented as a stochastic, probabilistic, or deterministic variable.

References

1. M. Abdel-Hameed, A gamma wear process. IEEE Trans. Reliab. **24**(2), 152–153 (1975)
2. K. Abid, M.S. Mouchaweh, L. Cornez, Fault prognostics for the predictive maintenance of wind turbines: state of the art, in *In Joint European Conference on Machine Learning and Knowledge Discovery in Databases*, pp. 113–125 (2018)
3. D. Adams, M. Nataraju, A nonlinear dynamical systems framework for structural diagnosis and prognosis. Int. J. Eng. Sci. **40**(17), 1919–1941 (2002)
4. J. Altmann, J. Mathew, Multiple band-pass autoregressive demodulation for rolling-element bearing fault diagnosis. Mech. Syst. Signal Process. **15**(5), 963–977 (2001)
5. D. An, N.H. Kim, J.H. Choi, Practical options for selecting data-driven or physics-based prognostics algorithms with reviews. Reliab. Eng. Syst. Saf. **133**, 223–236 (2015)
6. J. Bakker, J.V. Noortwijk, Inspection validation model for life-cycle analysis, in *Proceedings of the 2nd International Conference on Bridge Maintenance, Safety and Management (IABMAS)*, pp. 18–22 (2004)
7. P. Bangalore, L.B. Tjernberg, An artificial neural network approach for early fault detection of gearbox bearings. IEEE Trans. Smart Grid **6**(2), 980–987 (2015)
8. D. Banjevic, Remaining useful life in theory and practice. Metrika **69**, 337–349 (2009)
9. D. Banjevic, A.K.S. Jardine, Calculation of reliability function and remaining useful life for a Markov failure time process. IMA J. Manag. Math. **17**, 115–130 (2006)
10. P. Baraldi, G. Bonfanti, E. Zio, Differential evolution-based multi-objective optimization for the definition of a health indicator for fault diagnostics and prognostics. Mech. Syst. Signal Process. **102**, 382–400 (2018)
11. J. Barbot, M. Fliess, T. Floquet, An algebraic framework for the design of nonlinear observers with unknown inputs, in *46th IEEE Conference on Decision and Control*, pp. 384–389 (2007)

12. E. Bechhoefer, A. Bernhard, D. He, P. Banerjee, Use of hidden semi-Markov models in the prognostics of shaft failure, in *Annual forum proceedings – American Helicopter Society*, pp. 1330–1335 (2006)
13. B. Bellali, A. Hazzab, I.K. Bousserhane, D. Lefebvre, Parameter estimation for fault diagnosis in nonlinear systems by ANFIS. Procedia Eng. **29**, 2016–2021 (2012)
14. S. Benmoussa, M.A. Djeziri, Remaining useful life estimation without needing for prior knowledge of the degradation features. IEEE IET Sci. Meas. Technol. **11**(8), 1071–1078 (2017)
15. S. Benmoussa, B.O. Bouamama, R. Merzouki, Bond graph approach for plant fault detection and isolation: application to intelligent autonomous vehicle. IEEE Trans. Autom. Sci. Eng. **11**(2), 585–593 (2014)
16. S. Bhat, D. Bernstein, Geometric homogeneity with applications to finite-time stability. Math. Control Signals Syst. **17**, 101–127 (2000)
17. S. Bhat, D. Bernstein, Finite-time stability of continuous autonomous systems. SIAM J. Control. Optim. **38**(3), 751–766 (2005)
18. M. Blanke, M. Kinnaert, J. Lunze, M. Staroswiecki, *Diagnosis and Fault-Tolerant Control* (Springer, Berlin, 2006)
19. C. Byington, M. Roemer, T. Galie, Prognostic enhancements to diagnostic systems for improved condition-based maintenance, in *In Proceedings of IEEE Aerospace Conference* (2002)
20. F. Cadini, E. Zio, D. Avram, Model-based Monte Carlo state estimation for condition-based component replacement. Reliab. Eng. Syst. Saf. **94**(3), 752–758 (2009)
21. J. Chen, R. Patton, *Robust Model-Based Fault Diagnosis for Dynamic Systems* (Springer, Berlin, 1999)
22. A. Christer, W. Wang, J. Sharp, A state space condition monitoring model for furnace erosion prediction and replacement. Eur. J. Oper. Res. **101**(1), 1–14 (1997)
23. E. Cinlar, E. Osman, Z. Bazant, Stochastic process for extrapolating concrete creep. J. Eng. Mech. Div. **103**(6), 1069–1088 (1977)
24. J. Coble, Merging data sources to predict remaining useful life – an automated method to identify prognostic parameters. Ph.D. Diss., University of Tennessee (2010)
25. J. Coble, J. Hines, Identifying optimal prognostic parameters from data: a genetic algorithm approach, in *Annual Conference of the Prognostics and Health Management Society* (2009)
26. D. Cox, H. Miller, *The Theory of Stochastic Processes*, vol. 134 (CRC Press, Boca Raton, 1977)
27. J. Cruz-Victoria, R. Martinez-Guerra, J. Rincon-Pasaye, On nonlinear systems diagnosis using differential and algebraic methods. J. Frankl. Inst. **345**, 102–117 (2008)
28. S.A. Dahidi, F.D. Maio, P. Baraldi, E. Zio, Remaining useful life estimation in heterogeneous fleets working under variable operating conditions. Reliab. Eng. Syst. Saf. **3156**, 109–124 (2016)
29. M.J. Daigle, K. Goebel, A model-based prognostics approach applied to pneumatic valves. Int. J. Prognosis Health Manage. **2**, 84–99 (2011)
30. W. Danwei, Y. Ming, L. Chang, A. Shai, *Model-based Health Monitoring of Hybrid Systems* (Springer, Berlin, 2013)
31. O. Djedidi, M.A. Djeziri, N. MSirdi, Data-driven approach for feature drift detection in embedded electronic devices, in *IFAC Proceeding of the IFAC Symposium on Fault Detection, Supervision and Safety of Technical Processes (SAFEPROCESS)*, pp. 964–969 (2018)
32. O. Djedidi, M.A. Djezizi, S. Benmoussa, Failure prognosis of embedded systems based on temperature drift assessment, in *Proceedings of the International Conference on Integrated Modeling and Analysis in Applied Control and Automation*, pp. 11–16 (2019)
33. M.A. Djeziri, A. Aitouch, B.O. Bouamama, Sensor fault detection of energetic system using modified parity space approach, in *Proceeding of the IEEE Control and Decision Conference*, pp. 2578–2583 (2007)

34. M.A. Djeziri, S. Benmoussa, M. Ouladsine, B.O. Bouamama, Wavelet decomposition applied to fluid leak detection and isolation in presence of disturbances, in *IEEE Proceeding of the 18th Mediterranean Conference on Control and Automation (MED)*, pp. 104–109 (2012)
35. M. Djeziri, S. Benmoussa, L. Nguyen, N. MSirdi, Fault prognosis based on physical and stochastic models, in *Proceeding of the 2016 European Control Conference* (2016)
36. M.A. Djeziri, S. Benmoussa, R. Sanshez, Hybrid method for remaining useful life prediction in wind turbine systems. Renew. Energy (2017). https://doi.org/10.1016/j.renene.2017.05.020
37. M. Dong, D. He, A segmental hidden semi-Markov model (HSMM)-based diagnostics and prognostics framework and methodology. Mech. Syst. Signal Process. **21**, 2248–2266 (2007)
38. R. Engel, G. Kreisselmeier, A continuous-time observer which converges in finite time. IEEE Trans. Autom. Control **47**(7), 1202–1204 (2002)
39. S. Ferreiro, A. Arnaiz, B. Sierra, I. Irigoien, Application of Bayesian networks in prognostics for a new integrated vehicle health management concept. Expert Syst. Appl. **39**(7), 6402–6418 (2012)
40. M. Fliess, Some basic structural properties of generalized linear systems. Syst. Control Lett. **15**(5), 391–396 (1990)
41. M. Fliess, C. Join, H. Sira-Ramirez, Robust residual generation for linear fault diagnosis: an algebraic setting with examples. Int. J. Control. **77**, 1223–1242 (2004)
42. D. Frangopol, M. Kallen, J.V. Noortwijk, Probabilistic models for life cycle performance of deteriorating structures: review and future directions. Prog. Struct. Eng. Mater. **6**(4), 197–212 (2004)
43. L. Fridman, Y. Shtessel, C. Edwards, X.G. Yan, Higher order sliding-mode observer for state estimation and input reconstruction in nonlinear systems. Int. J. Robust Nonlinear Control **18**, 339–412 (2007)
44. D. Galar, U. Kumar, J. Lee, W. Zhao, Remaining useful life estimation using time trajectory tracking and support vector machines. J. Phys. Conf. Ser. **364**(1), 012063 (2012)
45. Z. Gao, C. Cecati, S.X. Ding, A survey of fault diagnosis and fault-tolerant techniques; part I: fault diagnosis with model-based and signal-based approaches. IEEE Trans. Ind. Electron. **62**(6), 3757–3767 (2015)
46. Z. Gao, C. Cecati, S.X. Ding, A survey of fault diagnosis and fault-tolerant techniques; part II: fault diagnosis with knowledge-based and hybrid/active approaches. IEEE Trans. Ind. Electron. **62**(6), 3768–3774 (2015)
47. M.C. Garcia, M.A. Sanz-Bobi, J.D. Pico, SIMAP: intelligent system for predictive maintenance: application to the health condition monitoring of a wind turbine gearbox. Comput. Ind. **57**(6), 552–568 (2006)
48. J. Gertler, Fault detection and isolation using parity relations. Control. Eng. Pract. **5**, 653–661 (1997)
49. J.C. Gomez-Mancilla, L. Palacios-Pineda, V. Nosov, Software package evaluation for Lyapunov exponent and others features of signals evaluating the condition monitoring performance on nonlinear dynamic system. J. Energy Power Eng. **9**(5), 443–551 (2015)
50. D. Gucik-Derigny, Contribution au pronostic des systemes à base de modles : theorie et application. Ph.D. Thesis, Universite Paul Cezanne Aix-Marseille (2011)
51. D. He, R. Li, J. Zhu, Plastic bearing fault diagnosis based on a two-step data mining approach. IEEE Trans. Ind. Electron **60**(8), 3429–3440 (2013)
52. W. He, Q. Miao, M. Azarian, M. Pecht, Health monitoring of cooling fan bearings based on wavelet filter. Mech. Syst. Signal Process **64–65**, 149–161 (2015)
53. Y. Hu, P. Baraldi, F.D. Maio, E. Zio, A particle filtering and kernel smoothing-based approach for new design component prognostics. Reliab. Eng. Syst. Saf. **134**, 19–31 (2015)
54. N. Huang, Z. Shen, S. Long, M. Wu, H. Shih, Q. Zheng, H. Liu, The empirical mode decomposition and the Hilbert spectrum for nonlinear and nonstationary time series analysis. Proc. R. Soc. London A Math. Phys. Eng. Sci. **454**(1971), 903–995 (1998)
55. R. Isermann, *Fault Diagnosis Systems* (Springer, Berlin, 2006)
56. M. James, Finite time observers and observability, in *Proceedings of IEEE Conference on Decision and Control* (1990)

57. A. Jardine, D. Lin, D. Banjevic, A review on machinery diagnostics and prognostics implementing condition-based maintenance. Mech. Syst. Signal Process. **20**(7), 1483–1510 (2006)
58. M. Jha, G. Dauphin-Tanguy, B. Ould-Bouamama, Particle filter based hybrid prognostics for health monitoring of uncertain systems in bond graph framework. Mech. Syst. Signal Process. **75**, 301–329 (2016)
59. M. Kan, A. Tan, J. Mathew, A review on prognostic techniques for non-stationary and non-linear rotating systems. Mech. Syst. Signal Process. **62**, 1–20 (2013)
60. D.C. Karnopp, D. Margolis, R. Rosenberg, *Systems Dynamics: A Unified Approach.* (John Wiley, Hoboken, 1990)
61. Y. Khellil, G. Graton, M.A. Djeziri, M. Ouladsine, Fault detection and isolation in marine diesel engines, a generic methodology, in *IFAC Proceeding of the 8th IFAC Symposium on Fault Detection, Supervision and Safety of Technical Processes (SAFEPROCESS)*, pp. 964–969 (2012)
62. J. Lawless, M. Crowder, Covariates and random effects in a gamma process model with application to degradation and failure. Lifetime Data Anal. **10**(3), 213–227 (2004)
63. B.P. Leao, T. Yoneyama, G. Rocha, K.T. Fitzgibbon, Prognostics performance metrics and their relation to requirements, design, verification and cost-benefit, in *2008 International Conference on Prognostics and Health Management*, pp. 1–8 (2008)
64. M. Lebold, M. Thurston, Open standards for condition-based maintenance and prognostic systems, in *Maintenance and Reliability Conference*, pp. 6–9 (2001)
65. J. Lee, J. Ni, D. Djurdjanovic, H. Qiu, H. Liao, Intelligent prognostics tools and e-maintenance. Comput. Ind. **57**(6), 476–489 (2006)
66. J. Lee, F. Wu, W. Zhao, M. Ghaffari, L. Liao, D. Siegel, Prognostics and health management design for rotary machinery systems – reviews, methodology and applications. Mech. Syst. Signal Process. **42**(1), 314–334 (2014)
67. C. Li, H. Lee, Gear fatigue crack prognosis using embedded model, gear dynamic model and fracture mechanics. Mech. Syst. Signal Process. **19**(04), 836–846 (2005)
68. G. Li, S.J. Qin, Y.J.D. Zhou, Reconstruction based fault prognosis for continuous processes. Control. Eng. Pract. **18**(10), 1211–1219 (2010)
69. C. Lim, D. Mba, Switching Kalman filter for failure prognostic. Mech. Syst. Signal Process. **52**, 426–435 (2015)
70. J. Lin, M. Zuo, Gearbox fault diagnosis using adaptive wavelet filter. Mech. Syst. Signal Process. **17**(6), 1259–1269 (2003)
71. F. Liu, Synthèse d'observateurs à entrées inconnues pour les systèmes non linéaires. Ph.D. Thesis, Université de Basse-Normandie (2007)
72. J. Luo, M. Namburu, K. Pattipati, L. Qiao, M. Kawamoto, S. Chigusa, Model-based prognostic techniques, in *Proceedings of IEEE Systems Readiness Technology Conference* (2003), pp. 330–340
73. M. Luschen, Derivation and application of nonlinear analytical redundancy techniques with applications to robotics. Ph.D. Thesis, Houston, TX (2001)
74. F.D. Maio, J. Hu, P. Tse, M. Pecht, K. Tsui, E. Zio, Ensemble-approaches for clustering health status of oil sand pumps. Expert Syst. Appl. **39**(5), 4847–4859 (2012)
75. F.D. Maio, F. Antonello, E. Zio, Condition-based probabilistic safety assessment of a spontaneous steam generator tube rupture accident scenario. Nucl. Eng. Des. **326**, 41–54 (2018)
76. S. Mallat, A theory for multiresolution signal decomposition: the wavelet representation. IEEE Trans. Pattern Anal. Mach. Intell. **11**(7), 674–693 (1989)
77. C. Marcelo, J.P. Fossatti, J.I. Terra, Fault diagnosis of induction motors based on FFT, in *Fourier Transform-Signal Processing*. InTech (2012)
78. A. Mathur, K. Cavanaugh, K. Pattipati, P. Willett, T. Galie, Reasoning and modeling systems in diagnosis and prognosis, in *Aerospace/Defense Sensing, Simulation, and Controls, International Society for Optics and Photonics* (2001), pp. 194–203

79. T. Menard, E. Moulay, W. Perruquetti, A global high-gain finite-time observer. IEEE Trans. Autom. Control **55**(6), 1500–1506 (2010)
80. P. Menold, Finite time and asymptotic time state estimation for linear and nonlinear systems. Ph.D. Thesis, Institute for Systems and Automatic Control, University of Stuttgart, Allemagne (2004)
81. P. Menold, R. Findeisen, F. Allgower, Finite time convergent observers for nonlinear systems, in *Proceedings of the IEEE Conference on Decision and Control* (2003)
82. P. Menold, R. Findeisen, F. Allgöwer, Finite time convergent observers for linear time-varying systems, in *Proceedings of the Mediterranean Conference on Control and Automation* (2003)
83. Q. Miao, V. Makis, Condition monitoring and classification of rotating machinery using wavelets and hidden Markov models. Mech. Syst. Signal Process. **21**(2), 840–855 (2007)
84. M. Moisan, O. Bernard, Semi-global finite-time observers for nonlinear systems. Automatica **44**(12), 3152–3156 (2008)
85. E. Moulay, W. Perruquetti, Finite-time stability and stabilization of a class of continuous systems. J. Math. Anal. Appl. **323**(2), 1430–1443 (2003)
86. E. Moulay, W. Perruquetti, Finite-time stability of nonlinear systems, in *42nd IEEE International Conference on Decision and Control* (2003), pp. 3641–3646
87. A. Muller, M. Suhner, B. Iung, Formalisation of a new prognosis model for supporting proactive maintenance implementation on industrial system. Reliab. Eng. Syst. Saf. **93**(2), 234–253 (2008)
88. A. Nagy-Kiss, G. Schutz, J. Ragot, Parameter estimation for uncertain systems based on fault diagnosis using Takagi–Sugeno model. ISA Trans. **56**, 65–74 (2015)
89. L. Nguyen, M.A. Djeziri, B. Ananou, M. Ouladsine, J. Pinaton, Degradation modelling with operating mode changes, in *IEEE International Conference on Prognostics and Health Management*, Austin, TX (2015)
90. L. Nguyen, M.A. Djeziri, B. Ananou, M. Ouladsine, J. Pinaton, Health indicator extraction for fault prognosis in discrete manufacturing processes. IEEE Trans. Semicond. Manuf. **28**(3), 306–317 (2015)
91. J.V. Noortwijk, A survey of the application of gamma processes in maintenance. Reliab. Eng. Syst. Saf. **94**(1), 2–21 (2009)
92. A. Oppenheim, R. Schafer, J. Buck, *Discrete-Time Signal Processing*, vol. 2 (Prentice-Hall, Englewood Cliffs, 1989)
93. M. Orchard, B. Wu, G. Vachtsevanos, A particle filtering framework for failure prognosis, in *World Tribology Congress III*, pp. 883–884 (American Society of Mechanical Engineers, New York, 2005)
94. Y. Pan, J. Chen, L. Guo, Robust bearing performance degradation assessment method based on improved wavelet packet and support vector data description. Mech. Syst. Signal Process **23**(3), 669–681 (2009)
95. C. Park, W. Padgett, Accelerated degradation models for failure based on geometric Brownian motion and gamma processes. Lifetime Data Anal. **11**(4), 511–527 (2005)
96. M.A. Patil, P. Tagade, K.S. Hariharan, S.M. Kolake, T. Song, T. Yeo, S. Doo, A novel multistage support vector machine based approach for li ion battery remaining useful life estimation. Appl. Energy **159**(1), 285–297 (2015)
97. T. Raff, F. Lachner, F. Allgower, A finite time unknown input observer for linear systems, in *Mediterranean Conference on Control and Automation* (2006)
98. M. Roemer, G. Kacprzynski, M. Kawamoto, S. Chigusa, Advanced diagnostics and prognostics for gas turbine engine risk assessment, in *IEEE, Aerospace Conference Proceedings*, pp. 345–353 (2000)
99. S. Sankaraman, Significance, interpretation, and quantification of uncertainty in prognostics and remaining useful life prediction. Mech. Syst. Signal Process. **52**, 228–247 (2015)
100. A. Saxena, J. Celaya, E. Balaban, B. Saha, S. Saha, K. Goebel, Metrics for evaluating performance of prognostic techniques, in *International Conference on Prognostics and Health Management (PHM08)*, pp. 1–17 (2008)

101. A. Saxena, J. Celaya, B. Saha, S. Saha, K. Goebel, On applying the prognostic performance metrics, in *Annual Conference of the Prognostics and Health Management Society* (2009)
102. M. Schlechtingen, I.F. Santos, S. Achiche, Wind turbine condition monitoring based on SCADA data using normal behavior models. Part 1: system description. Appl. Soft Comput. **13**(1), 259–270 (2013)
103. X. Si, W. Wang, C. Hu, D. Zhou, Remaining useful life estimation: a review on the statistical data driven approaches. Eur. J. Oper. Res. **213**(1), 1–14 (2011)
104. R. Singleton, E. Strangas, S. Aviyente, Extended Kalman filtering for remaining useful life estimation of bearings. IEEE Trans. Ind. Electron. **62**(3), 1781–1790 (2015)
105. K.L. Son, M. Fouladirad, A. Barros, E. Levrat, B. Iung, Remaining useful life estimation based on stochastic deterioration models: a comparative study. Reliab. Eng. Syst. Saf. **112**, 165–175 (2013)
106. J. Sun, H. Zuo, W. Wang, M. Pecht, Prognostics uncertainty reduction by fusing on-line monitoring data based on a state-space-based degradation model. Mech. Syst. Signal Process. **42**(2), 396–407 (2014)
107. J. Sun, H. Zuo, W. Wang, M. Pecht, Prognostics uncertainty reduction by fusing on-line monitoring data based on a state-space-based degradation model. Mech. Syst. Signal Process. **45**(2), 396–407 (2015)
108. K. Tidriri, N. Chatti, S. Verron, T. Tiplica, Bridging data-driven and model-based approaches for process fault diagnosis and health monitoring: a review of researches and future challenges. Annu. Rev. Control. **42**, 63–81 (2016)
109. D. Tobon-Mejia, K. Medjaher, N. Zerhouni, The ISO 13381-1 standard's failure prognostics process through an example, in *IEEE Prognostics and Health Management Conference (PHM'10)* (2010), pp. 1–12
110. S. Tseng, C. Peng, Stochastic diffusion modeling of degradation data. J. Data Sci. **5**(3), 315–333 (2007)
111. S. Tseng, J. Tang, I. Ku, Determination of burn-in parameters and residual life for highly reliable products. Nav. Res. Logist. **50**, 1–14 (2003)
112. D. Van, C. Berenguer, Condition based maintenance model for a production deteriorating system, in *Conference on Control and Fault-Tolerant Systems* (2010)
113. X. Wang, Wiener processes with random effects for degradation data. J. Multivar. Anal. **101**(2), 340–351 (2010)
114. P. Wang, G. Vachtsevanos, Fault prognostics using dynamic wavelet neural networks. Technical Report. AAAI Technical Report (1999)
115. A.P. Wang, H. Wang, Fault diagnosis for nonlinear systems via neural networks and parameter estimation, in *2005 International Conference on Control and Automation*, vol. 1 (2005), pp. 559–563
116. W. Wang, M. Carr, W. Xu, K. Kobbacy, A model for residual life prediction based on Brownian motion with an adaptive drift. Microelectron. Reliab. **51**(2), 285–293 (2011)
117. W. Wu, J. Hu, J. Zhang, Prognostics of machine health condition using an improved ARIMA-based prediction method, in *2nd IEEE Conference on Industrial Electronics and Applications* (2007), pp. 1062–1067
118. J. Yan, M. Koc, J. Lee, A prognostic algorithm for machine performance assessment and its application. Prod. Plan. Control **15**(8), 796–801 (2004)
119. D. Yu, J. Cheng, Y. Yang, Application of EMD method and Hilbert spectrum to the fault diagnosis of roller bearings. Mech. Syst. Signal Process. **19**(2), 259–270 (2005)
120. Y. Zhang, R.H.H. Xiong, M.G. Pecht, Lithium-ion battery remaining useful life prediction with box–cox transformation and Monte Carlo simulation. IEEE Trans. Ind. Electron. **66**(2), 1585–1597 (2018)
121. Y. Zhang, R.H.H. Xiong, M.G. Pecht, Long short-term memory recurrent neural network for remaining useful life prediction of lithium-ion batteries. IEEE Trans. Veh. Technol. **67**(7), 5695–5705 (2018)
122. E. Zio, Prognostics and health management of industrial equipment, in *Diagnostics and Prognostics of Engineering Systems: Methods and Techniques* (IGI Global, Pennsylvania, 2012), pp. 333–356. https://doi.org/10.4018/978-1-4666-2095-7

123. E. Zio, Challenges in the vulnerability and risk analysis of critical infrastructures. Reliab. Eng. Syst. Saf. **152**, 137–150 (2016)
124. E. Zio, F.D. Maio, M. Stasi, A data-driven approach for predicting failure scenarios in nuclear systems. Ann. Nucl. Energy **37**, 482–491 (2010)
125. Z. Zng, E. Zio, A classification-based framework for trustworthiness assessment of quantitative risk analysis. Saf. Sci. **99**, 1215–226 (2017)

Chapter 9
How Machine Learning Can Support Cyberattack Detection in Smart Grids

Bruno Bogaz Zarpelão, Sylvio Barbon Jr., Dilara Acarali, and Muttukrishnan Rajarajan

9.1 Introduction

The world's demand for electricity has been steadily growing due to several aspects of modern life, causing a push in industrial production and giving rise to new electricity-dependent technologies. At the same time, society has refused the idea of increasing the use of fossil fuels as power sources, given that they are responsible for several environmental problems we have faced. To cope with these challenges, power grids have been reshaped to become more resilient, reliable, and efficient. Renewable and alternative power sources have been increasingly adopted to reduce greenhouse gas emissions. The new power grids emerging from this modernisation process are named smart grids [1–3].

To reach their goals, smart grids rely on advanced control and communication technologies. Although these technologies have been used to make power grids more reliable, they are also responsible for introducing new vulnerabilities. Smart grids are complex and large-scale systems, composed of multiple domains involving customers, utilities, operators, and service providers. Attackers can target any part of these systems, from smart meters at customer premises to core devices at transmission networks or power plants. As smart grids are highly interconnected, an attack on a particular point, which at first sight does not seem to be significant, can escalate quickly to a massive disruption of the whole system [3–5].

B. B. Zarpelão (✉) · S. Barbon Jr.
Computer Science Department, State University of Londrina, Londrina, Paraná, Brazil
e-mail: brunozarpelao@uel.br; barbon@uel.br

D. Acarali · M. Rajarajan
School of Mathematics, Computer Science and Engineering, City, University of London, London, UK
e-mail: dilara.acarali@city.ac.uk; r.muttukrishnan@city.ac.uk

M. Sayed-Mouchaweh (ed.), *Artificial Intelligence Techniques for a Scalable Energy Transition*, https://doi.org/10.1007/978-3-030-42726-9_9

Cybersecurity measures must be set over the entire smart grid to ensure its reliability. Among all the available security solutions, attack detection systems are particularly important. As smart grids are complex and large, it is impossible to make sure that there are no security breaches in any part of the system. Researchers discover new vulnerabilities on a daily basis even in long-used devices, and well-known vulnerabilities may remain unpatched due to the lack of sufficient resources to cover such a huge attack surface. Therefore, attack detection systems are necessary to monitor the whole system continuously and alert the administrators when needed.

Attackers have evolved along with the defence tools. They are usually able to bypass or evade existing attack detection systems, and capable of developing smarter attacks that can adapt to new security measures. Machine learning is a promising solution for creating attack detection systems capable of dealing with these advanced adversaries in smart grids, having been successfully applied to detect attacks in other domains.

In this chapter, we present a survey about the application of machine learning for attack detection in smart grids. Our goal is to enable a better understanding of the attack types that affect smart grids, the aspects that drive detection systems development (detection methods, data collection, and system distribution), and how machine learning algorithms are employed in this context. Finally, we discuss open issues related to the current usage of machine learning-based detection in smart grids and point out some paths to address them.

The rest of the chapter is organised as follows. Section 9.2 presents an overview of smart grids to build a foundation for the rest of the study. Section 9.3 discusses the types of attacks that affect smart grids, while Sect. 9.4 shows the main aspects of detection systems. Section 9.5 details the foundations of the machine learning algorithms used for attack detection in smart grids. Section 9.6 presents the surveyed solutions, and Sect. 9.7 discusses their open issues and possible improvements. Finally, Sect. 9.8 presents the concluding remarks.

9.2 Smart Grids Overview

Smart grids are the convergence of power grids and Information and Communication Technology (ICT). They have been developed as a response to the growing demand for electrical power and the rise of renewable energy sources. In this context, ICT tools are used to improve the management and control of the whole cycle of power generation, transmission, and distribution, making sure that multiple power sources are explored and faults and outages are significantly reduced even with the system under constant pressure [1–3].

Power grid operation is divided into generation, transmission, and distribution. Energy is generated in power plants of different kinds (e.g. nuclear, thermal, wind, hydroelectric, or solar) and transmitted over long distances through high-voltage transmission lines to electrical substations. From electrical substations, energy is distributed to end customers, according to their demand. As these systems spread over wide geographical areas, they are structured in a hierarchical fashion. A control centre monitors the power grid activity to ensure that multiple parameters like voltage, frequency, and current are within the expected range. Situational awareness is a key term to define the central control mission. Additionally, the power grid has some protection mechanisms, like breakers and relays, that take action when a fault occurs to keep the system up and avoid significant damage. Protection mechanisms can operate automatically or under the central control command. Summing up, power grids are huge and complex systems that operate under strict requirements and are monitored continuously to prevent outages that might have serious consequences [6–9].

In smart grids, ICT is used to enable two-way communication between the control centre and different parts of the power grid. Data about the power grid state are collected in real-time from all over the system, providing controllers with updated information that can be used to respond to unexpected behaviour, make demand predictions, and coordinate multiple power sources, among other tasks related to management and control. Most of these needs always existed in power grids. However, the reality has changed in recent years, making more sophisticated ICT solutions necessary to cope with rising challenges. For instance, renewable energy sources like wind power or solar power may generate energy intermittently, as less wind, cloudy weather or some other natural and unavoidable condition may affect their generation potential. In this sense, ICT solutions can help to forecast these occurrences and coordinate the use of these sources accordingly [6–9].

According to a conceptual model proposed by NIST (National Institute of Standards and Technology) [1], smart grids are organised into seven domains: customer, markets, service provider, operations, generation, transmission, and distribution. The customer domain encompasses electricity end-users. Smart grids include some differentials, such as dynamic pricing and generation of electricity by end-users, which add more complexity to the customer's role. For this reason, customers need a two-way communication interface with the grid, named ESI (Energy Services Interface). This interface sets the boundary between the customer and the utility and is usually deployed at the meter or local management system. Customers can have smart devices, which interact with the smart grid to provide details of their consumption and other energy parameters, while receiving commands from service providers that deliver management services.

The market domain consists of the operators responsible for commercialising grid assets, from bulk electricity suppliers to retailers that supply electricity to end-users. Organisations that deliver services such as billing, account management, and maintenance and installation to customers and utilities make up the service provider domain. The operations domain encompasses those who are responsible for ensuring that the smart grid's operations run smoothly. Their activities include grid monitoring and control, fault management, grid estimates calculation, analytics, planning, and maintenance. All of these tasks are performed from a control centre, which hosts some management systems, such as the EMS (Energy Management

System), dedicated to generation and transmission processes, and the DMS (Distribution Management System), responsible for distribution processes.

Electricity generation is the key process of the generation domain. Several energy sources such as nuclear fission, flowing water, wind, and solar radiation can be used to create electricity. The generation domain has a plant control system, which is used to monitor and control the power generation. It must report performance measures continuously, so the operators can predict possible issues and mitigate their effects. The transmission domain encompasses all the actors and functions needed to transmit the electrical power produced in the generation domain to the distribution domain. The transmission domain has the essential responsibility of balancing electricity generation and load. Any disturbance in this delicate balance can affect the grid frequency, leading to power outages or other kinds of damages to the system. The distribution domain is responsible for interconnecting the transmission domain and the customer domain. It also informs the operations domain about the power flow situation.

All of these domains have to interact and cooperate to reach their goals, and several technologies are available to support this need. Smart meters are deployed at the customers' side to measure their energy consumption and gather other management information in real-time (typically every 30 min) to report to other domains. These meters are part of a communication infrastructure referred to as AMI (Advanced Metering Infrastructure). In addition to smart meters, an AMI includes data concentrators for aggregating data collected from smart meters, and head-end systems, which are responsible for connecting smart meters and data concentrators to management information systems. Together, smart meters and AMIs behave as typical IoT (Internet of Things) systems, adding the many particularities of this paradigm [10].

SCADA (Supervisory Control and Data Acquisition) systems are also used to support data exchange among these domains. These systems are made up of three main components, RTUs (Remote Terminal Unit), MTUs (Master Terminal Unit), and HMI (Human Machine Interface). RTUs are deployed close to or at devices that are remotely controlled. MTUs (Master Terminal Units) are responsible for sending requests periodically to RTUs, asking for data about the monitored device, in a process referred to as polling. The polling frequency can range from multiple requests per second to one request every few minutes, depending on the importance of the monitored device. MTUs can also send commands to RTUs asking them to act over the controlled system. Human operators interact with these components through HMIs. In smart grids, RTUs can be deployed in the generation, transmission, and distribution domains. The control centre's management systems provide human operators with HMIs to monitor and control these RTUs.

Another solution to collect measurements from the transmission domain is the PMU (Phasor Measurement Unit). In a smart grid, PMUs are deployed at transmission substations to collect current and voltage phasor information. They operate at very high sampling rates, and are, therefore, able to collect many more measurements per second than a common RTU. All PMU measurements are timestamped, and GPS (Global Positioning System) devices are needed to synchronise measurements from PMUs at different locations. PDCs (Phasor Data Concentrators) aggregate data from PMUs, perform quality checks, and then forward these measurements to EMSs, where the collected data is analysed for state estimation, monitoring, control, and protection.

Although smart meters, RTUs, and PMUs have some particularities that make them unlike one another, all of these devices share at least a common characteristic: they generate continuous data streams. Therefore, management and control systems for smart grids, which consume these data, have to be designed to handle continuous streams. This means they have to be able to learn incrementally, to manage the constant inflow of huge amounts of data, and to cope with real-time changes in the statistical distributions underlying the collected data.

Communication networks underpin all of these domains. As smart grid networks have to connect a great number of endpoints over wide geographical areas, they are organised hierarchically. At the customer end, there are HANs (Home Area Network), which connect smart devices within the customer's premises to the smart grid structure, enabling the energy usage management at the customer level. HANs are connected to the distribution system's networks, referred to as FAN (Field Area Network). These networks connect components such as RTUs in the distribution domain and smart meters to control centres. Networks in the distribution domain are also named NAN (Neighbour Area Network). Finally, WANs (Wide Area Network) connect distant sites, making up a backbone for the integration of the networks that compose smart grids. They are responsible for connecting the transmission and generation domains to the control centre and for transmitting information like PMU measurements and RTU readings. Also, they establish a communication path between FANs and control centres, which are usually separated by long distances.

Figure 9.1 presents an overview of the seven domains along with devices and communication networks that compose a smart grid.

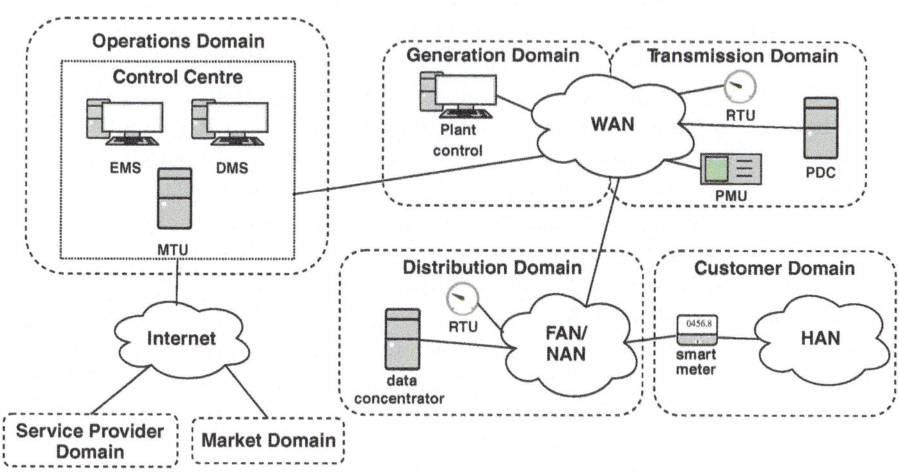

Fig. 9.1 Overview of main smart grid components and their relationships

9.3 Smart Grid Attacks

Smart grids are complex systems consisting of various specialised components working collaboratively to exchange sensitive data, process inputs, and make decisions, all in real-time. This combined complexity and sensitivity produces many vulnerabilities that can be exploited by malicious individuals. Furthermore, the accurate and sustained functionality of power infrastructures are non-negotiable; power is in constant demand. These issues are further exasperated by the vulnerabilities of wireless network technologies, and the presence of many potential access points (i.e. smart meters) [8]. All of this makes the smart grid a highly attractive target for those looking to cause large-scale disruption. A successful attack on the grid hinders everything in the affected region, as experienced in Ivano-Frankivsk in Ukraine in 2016, where thousands of people were left without electricity [11]. Despite the level of disruption caused in that incident, attacks on smart grids theoretically and feasibly have much a larger damage potential.

This section provides a taxonomy of smart grid attacks, along with detailed explanations of each category. In keeping with the basic principles of cybersecurity, the CIA triad is used to divide attacks into three main categories based on what they threaten: confidentiality, integrity, and availability. Each one is then further divided to distinguish between attack aims and methods. It should be noted that some categories inevitably have overlaps as attacks often interleave in complex campaigns. The outline given here aims to highlight individual malicious actions taken against the smart grid. In comparison to the taxonomy presented in [8], we consider "data attacks" and "device attacks" to fall under the integrity category, as the aim of both is to compromise the integrity of the grid network. Meanwhile, privacy attacks are directly analogous to attacks on confidentiality, and network availability attacks are captured in the same way.

Another important consideration is that attacks on smart grids can be considered over two planes: the cyber and the physical [8, 12–14]. This is because the grid is a digitised system that regulates and manages a physical utility. Hence, Cyber-Physical Threats (CPTs) are defined as attacks where a malicious action in the cyber plane has repercussions in the physical (and vice versa) [12, 13]. Examples of this include acts of remote sabotage (like the Stuxnet incident [13]), manipulation of the grid topology, and damage to hardware [13]. In [12], this idea is combined with big data concepts to categorise attacks by (a) data on the cyber plane, (b) data on the physical plane, and (c) metadata combining the cyber and physical planes. While Wu et al. [13] focus on manufacturing systems and Wang et al. [12] only consider false data injections (discussed in detailed in Sect. 9.3.2), the principles of CPTs can be applied across the attack spectrum. Therefore, the cross-plane nature of smart grid threats should be noted for the rest of the attacks discussed in this section.

Figure 9.2 presents the categories and attacks that are discussed in the rest of this section.

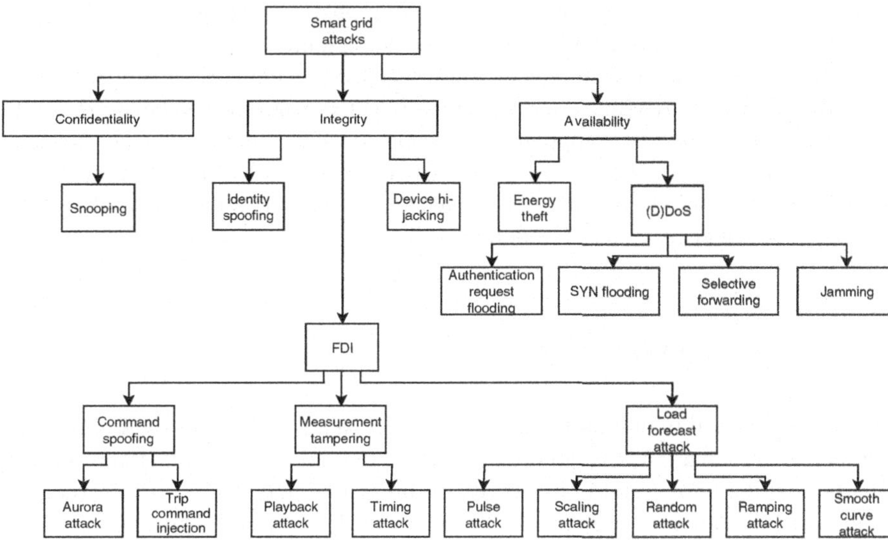

Fig. 9.2 Overview of the discussed attacks

9.3.1 Attacks on Confidentiality

Confidentiality is the quality of maintaining the privacy of data. By their nature, smart grids collect vast quantities of data that must be transmitted and processed in a secure manner. Recent regulations such as GDPR (General Data Protection Regulation) enshrine the privacy rights of users in law. Furthermore, grid devices generate very rich data, including user profiles, energy measurements, service specifications, telemetry details, and hardware information. Hence, the threat surface against confidentiality covers the whole of the smart grid infrastructure.

An example of a privacy-targeting attack is snooping. This is where malicious individuals aim to gain access to or visibility of data belonging to others. In the smart grid context, the communications between appliances, smart meters, and controllers are vulnerable to this. The power usage profiles of appliances are captured in readings and measurements made by smart meters; collectively these readings form a usage profile for the customers themselves [8]. An adversary may wish to capture this information to infer the activities and behaviours of users [8], which they may then use to plan intrusions or physical attacks on households/premises that appear to be unoccupied [8]. Similarly, such user profiling may form the basis of energy theft attacks (discussed in Sect. 9.3.3) to determine which accounts to steal electricity from with the least risk of detection [15].

Meanwhile, similar methods may also be used to infer the current topology of the smart grid. In their study of false data injections, Huang et al. [16] found that intelligence regarding the grid's structure could be mined from measurement data. To achieve this, the adversary requires some understanding of the grid's stochastic behaviour [4], and a length of time over which to observe readings—a single set of measurements taken at one point in time is not sufficiently revealing [16]. However, using linear Independent Component Analysis (ICA) techniques, Huang et al. [16] were able to demonstrate that measurements collected over time can be used to build a model of the grid. Their approach was based on the principle that the physical topology and the load change independently. In other words, the variation in load (which changes more frequently) can be analysed given the relative stability of the topology (which changes less frequently) [4].

9.3.2 *Attacks on Integrity*

Integrity is the quality of maintaining the intended states of systems (and/or the data within them) so that those systems can continue to serve their intended purposes. In smart grids, the preservation of integrity ensures the timely and accurate exchange of the data signals used to make decisions about delivery and distribution. It also ensures that all grid components are truthful about their identities. This is crucial to the correctness of measurements, given that meters and sensors are numerous and distributed widely over geographic regions.

One way to damage integrity is to spoof the identities of grid components. This is where someone other than the legitimate device fraudulently claims to be that component, thus allowing an adversary to interact with the system under false pretences. For example, smart meters may be spoofed to send fake data to controllers [12, 15]. Similarly, spoofed devices can send incorrect timestamps to PMUs, disrupting grid synchronisation [17]. Another method is device hi-jacking. This is slightly different to spoofing because while the compromised device can be wielded by a potential attacker, its identity is still intact. The primary version of this attack is the recruitment of grid devices into a botnet. A bot binary is injected into devices via a virus or a worm [18]. This binary then automatically executes and connects to a remote command and control (C&C) network from which it receives attack instructions. Adversaries may also harvest data from devices via the same C&C network. Botnets, which provide foundations for other types of attack, are a known threat to WSNs (Wireless Sensor Network) and IoT networks [11]. A prominent example is the Mirai botnet, which hijacked IoT routers and cameras and was used to launch massive-scale DDoS attacks in multiple countries [11].

The biggest threat to smart grid integrity comes from False Data Injection (FDI) attacks. As the name suggests, this involves the introduction of maliciously crafted data into sensitive communication streams, with the aim of manipulating system outputs. Hence, FDI attacks are mainly targeted at data-reliant management processes [2, 4, 19, 20]. They may be considered analogous to man-in-the-middle attacks. Some possible scenarios explored in literature include attacks on state estimation systems [19], the EMS [4], AMIs [21], SCADA systems [22], local systems with clustered measurement hierarchies [23], and in wind farms [2]. Generally, attackers engaging in FDI will compromise a subset of grid components but will not have visibility of the whole grid given its complexity and size [4]. However, due to the hierarchical structure, an FDI at one point in the network is still capable to causing widespread repercussions. Additionally, Anwar et al. [21] found that the impact of an FDI changes based on the characteristics of the targeted nodes, while Chen et al. [4] suggested that sophisticated script-based FDIs can learn the best injection approach through trial and error. This shows that even with incomplete information, this type of attack has great potential for damage and disruption.

FDI attacks are typically modelled using the formulation $z = Hx + a + n$ [2, 22, 24], where z is a set of measurements, x is the state vector (or bus voltage phase [24]), n is the measurement error or environmental noise [2, 24], and H is a Jacobian matrix of measurements that describe the current grid topology [2, 22]. Together these values determine the state of the grid. Then the attack vector a (describing the fake data and variables targeted) is added [2, 22, 24]. Attacks can be classified as normal or stealthy. For the latter, it is assumed that adversaries have some visibility of H, which allows them to set up their attacks intelligently so as to avoid threshold-based detection (i.e. residual test) mechanisms [2, 16, 19, 20, 22, 25]. Additionally, they may aim to manipulate the state variables corresponding to the targeted measurement variables to avoid noticeable anomalies [21]. An alternative approach is given by Chen et al. [4], who modelled FDIs as partially-observed Markov decision processes (POMDP), where the focus is on attackers (who have limited target visibility) aiming to learn the optimal setup [4].

The rest of the integrity-based attacks discussed in this section is specific sub-categories of FDI as identified in the smart grid literature.

Command spoofing is the intersection of identity spoofing and FDI attacks; fake data—styled as commands—is injected into the network, claiming to come from legitimate sources. Aurora is a type of command spoofing attack that targets the circuit breakers used to determine grid topology and the generators that they serve [26]. Specifically, fake control signals are sent to the breakers, instructing them constantly open and close at a high speed [5, 17]. Eventually, this causes the associated generators to desynchronise from the rest of the grid [17]. If a critical level is reached, this attack can cause physical damage to the generators [17], knocking them offline. Depending on the degree of physical damage and the positions of the affected breakers and generators, Aurora attacks can result in a significant drop in a smart grid's functional capacity and efficiency.

Another example of command spoofing is trip command injection attacks. These target protection relays (devices designed to respond to faults in power transmission lines) with fake relay trip commands, causing circuit breakers at the ends of transmission lines to open [5]. When this happens, additional strain is placed upon secondary transmission lines, as the system tries to meet demand [5]. Given the hierarchical nature of smart grid infrastructure, this can then result in cascading failures [5] and large-scale power outages. An alternative involves the disabling of relays so that faults do not trigger trip commands at all [5]. Meanwhile, fault replay attacks combine fake trip commands with fake transmission line faults [5]. To achieve this, measurements are altered to look like real-life faults either via some hijacked devices or through data injections [5]. These false readings cause controllers to make incorrect management and distribution decisions.

Attacks on integrity may be directed at specific devices in the grid. An example is against PMUs, as explained by Wang et al. [12]. These devices are used to measure and synchronise phasor values collected from distributed sensors and meters; these measurements are then used to perform state estimations. FDIs can be applied directly to PMU data to manipulate the resulting state estimations [12]. Examples of how this may be achieved include playback attacks (where captured data is played in reverse order, giving incorrect readings) [12] and time attacks (where captured data is sampled at varying rates, distorting the true readings) [12].

Attacks may also target specific functionalities. For example, load forecast attacks hinder the grid's ability to determine where to distribute power and the correct load [27]. This is achieved using data injections designed specifically to distort these forecasts, applied continuously for the duration of the attack [27]. In implementation, there are several variations defined by Cui et al. [27]. Pulse attacks change the forecast values at regular intervals to be higher or lower than the true reading [27]. Scaling attacks tamper with values by multiplying them by a scaler [27]. Random attacks insert randomly-generated positive values [27]. Ramping attacks use a ramping function to either increase values over time ("up-ramping") or increase and decrease values repeatedly ("up and down-ramping") [27]. Finally, smooth curve attacks change forecasts' start and end points [27]. Given that each of these approaches causes a different impact on controller behaviours, adversaries are able to fine-tune attacks to suit their specific goals.

9.3.3 Attacks on Availability

Availability is the quality of maintaining the accessibility and functionality of a system to a satisfactory degree at all times. Power is a basic utility, and so power grids are fundamental parts of urban and rural infrastructure. Furthermore, smart grids require efficient feeds of real-time data and a high level of responsiveness from all components (e.g. controllers, synchronisers, smart meters, and sensors) for accurate decision making. In other words, components must have high availability for the grid's internal functionality.

The primary attack against availability is Denial-of-Service (DoS). This is where an attacker generates lots of traffic to overwhelm the capacity of target devices, causing them to crash and hence, rendering the services they provide unavailable. When this flood of traffic comes from multiple distributed sources, it is known as a distributed DoS (DDoS). Smart grids are highly susceptible to such attacks because they (a) house a large consumer device population [18], (b) consist of many low-power devices, and (c) have a hierarchical infrastructure [11]. This indicates a large potential attack surface of low-capacity devices, and many centralised control points to target. Devices may be compromised physically, have their identities spoofed, or engage in DDoS as part of a botnet [18]. As DoS attacks inject lots of malicious data into the network, they may also be considered a form of FDI [28]. Typically, attacks take place between sensors and smart meters [18], or between smart meters and system controllers [3].

For example, sensors may be manipulated to send streams of malicious authentication requests to meters [18]. In the process of trying to verify request details, the capacity of the meter is exhausted, and so the authentication service is knocked offline [18]. A similar attack between meters and controllers would disrupt the collection of measurement data, causing controllers to make the wrong decisions about power management and distribution [3]. Smart grids are also vulnerable to typical application layer DDoS attacks like SYN flooding [29]. This is where 3-way TCP handshakes are intentionally left half-open (because the client never responds the server's SYN/ACK message), consuming server resources and causing their backlog queues to fill up so that new, legitimate requests are automatically dropped [29]. Choi et al. [29] demonstrated such an attack on the multicast-based communications of smart grid IEDs (Intelligent Electronic Device). Other documented examples of DDoS in the smart grid are those originating from buffer overflow attacks (where program code is tampered with) [29] and selective forwarding (where packets relating to a particular service are consistently dropped) [28].

Smart grids are based on wireless sensor technology, which uses broadcasting on open channels to enable the easy exchange of data between geographically distributed devices [30, 31]. This makes grids vulnerable to a special type of DDoS attack known as jamming, where attackers add random noise signals to wireless channels to corrupt the traffic exchanged between grid components [3, 31]. As with standard DDoS, this attack can disrupt traffic between appliances and meters or between meters and controllers [31]. In both cases, the accurate gathering of measurement data is denied, and where this causes controllers to make incorrect calculations about load, large-scale outages and mismanagement may result [31]. Additionally, jamming attacks may be easier to perform than conventional DDoS because they do not require a base of compromised devices to launch them [31].

To perform the attack, a jammer device or program selects a channel and then injects it with random noise [3, 30]. This is similar to the injection of fake requests into the network in DDoS. Generally, there are four jammer types identified in WSN literature, sorted into two categories. In the first category are "oblivious" jammers, i.e. those which operate only based on current channel availability [30]. These are static jammers (which always use the same channel) and random jammers (which switch channels randomly over time) [30, 31]. The second category consists of "intelligent" jammers, i.e. those that use historical data to make complex decisions [30]. These are myopic jammers (which learn users' channel usage patterns) and Multi-armed Bandits (MABs) (which use machine learning to predict user behaviours) [30]. In some cases, myopic and MAB jammers may be considered as one [31]. As suggested in [30], jamming attacks may be kept hidden by avoiding the use of licensed channels.

The attacks discussed so far cause disruptions in power distribution services by affecting particular functionalities provided or required by the grid. However, the availability of power can also be attacked directly through resource exhaustion attacks and energy theft. In resource exhaustion, adversaries demand large quantities of electricity by sending many requests in quick succession [28]. This maximises the amount of power drawn from the grid and can eventually lead to the depletion of the available energy [28]. Such an attack can feasibly be launched at the appliance level by malicious "consumers" using energy-inefficient or high-consumption devices [28]. Meanwhile, energy theft involves the consumption of power without providing proper compensation for this service [15]. There are three types of energy theft: those launched by malicious consumers, those launched by industry insiders, and those conducted by organised criminals [15]. Malicious consumers may tamper with their appliances and meters to avoid making payments as due [15]. Industry insiders (i.e. utility company employees) may manipulate internal records and readings [15], either for their own benefit or as part of a larger campaign. Lastly, organised crime syndicates may use both of the previous methods to syphon energy from paying customers to sell illegally [15] or for further criminal activity.

Resource exhaustion and energy theft tend to occur alongside attacks on integrity (like FDI, tampering, and spoofing) and attacks on confidentiality (like user profiling). For example, the energy theft process requires a disruption of the communications to and from smart meters. This prevents the grid from learning consumers' energy usage levels. Then smart meters can be spoofed, and fake readings be sent to controllers [15]. To prevent tracing, existing audit logs and records may also be deleted from the meters [15]. Meanwhile, criminals targeting other consumers can use profiling techniques to infer their usage patterns from sensor data, allowing them to plan out their attacks.

9.4 Attack Detection

The area of attack detection has been driven by some core concepts that must be considered when this kind of solution is developed or deployed. The first one is the detection method, which is essential because it defines which situations the detection system sees as an attack. Alongside this, the types of data collected and the system distribution are also pillars of attack detection. They can impact various functions, including the system's processing performance and the attacks that can be detected. Figure 9.3 outlines these concepts and the related techniques, which are discussed in the context of smart grids in this section.

9.4.1 Detection Methods

Smart grid defence incorporates classical intrusion detection methods, broadly categorised as signature-based and anomaly-based. In the former, systems use templates derived from historic attack instances to recognise new instances of the

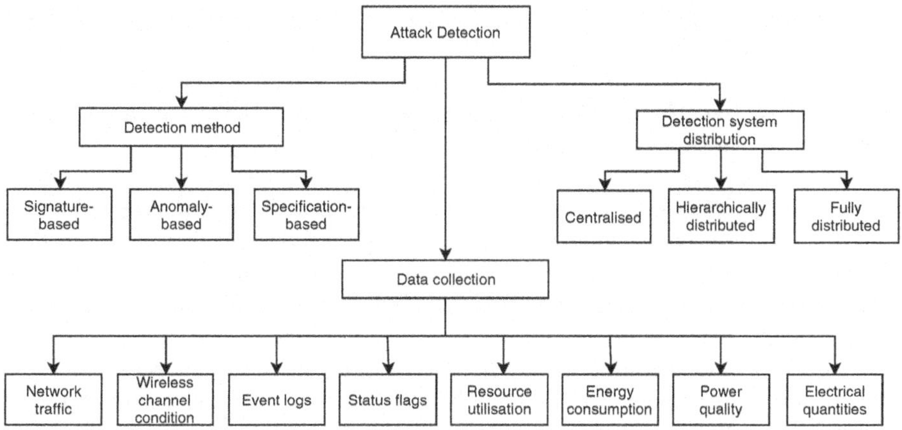

Fig. 9.3 Overview of attack detection core concepts and related techniques for smart grids

same or similar attacks. Typically, each attack on a system or network consists of a particular sequence of actions. These activities, observed in succession, can, therefore, be used to identify malicious activity. Signatures may be stored in a database and updated as new attacks are discovered. Popular attacks with generally well-understood methodologies (such as DDoS and FDI) may have strong patterns that make it easy and appropriate to model them [28]. Furthermore, one-to-one matching makes signatures very effective at detecting particular attacks. However, previously-unseen attacks will be missed, causing false negatives.

In contrast, anomaly-based detection observes the network or system for activity that deviates from a pre-defined norm. These norms may be derived from empirical baselines or heuristics. Given their malicious nature, many attacks fall outside the standard behavioural profile for systems or services, which results in unusual fluctuations in activity. Unlike the use of signatures, anomaly analysis is not limited by knowledge of historic incidents and is, therefore, capable of identifying day-zero attacks. Smart grids provide a wealth of data that can be used to generate complex and detailed activity profiles. However, not every anomaly corresponds to an attack attempt, and so a large number of false positives may be generated. This is especially true for smart grid sensors which are influenced by environmental factors. For example, high temperature readings may represent a sabotage attempt or may simply be caused by hot weather [32].

Novelty detection methods may be particularly useful for anomaly-based systems, once they can make the detection system reliable against previously-unseen attacks. The aim of these methods is to identify events that differ from the previously available data. New attack patterns, which were not present in the model induction, are placed out of the cluster of known patterns, and categorised as a novelty [33]. Novelty detection with multiple known classes is also widely applied in data stream classification, meaning that new classes may appear and previously known classes may change [34]. The most traditional methods are based on clustering algorithms, such as k-means [35].

A more recent addition is specification-based detection. This is similar to anomaly-based approaches, but the baseline profile used is created to reflect the expected behaviours of a particular application or protocol [11, 26]. Each is given its own set of expected behaviours and the detection system flags up instances where related activity deviates from this set. This overcomes the limitation of signature-based detection because unknown attacks may be discovered. In addition to this, the granular definition of expected behaviour can improve upon the false positive rates of anomaly-based detection. This approach applies well to metering systems in smart grids where state monitoring is an integral process and there are known thresholds for safe operation [5, 11, 26]. Conversely, the need to generate multiple specifications may make specification-based detection unscalable in larger grid networks [5]. An alternative approach is to shift focus from expected behaviours to the characteristics of a medium to determine the best attack vector [36]. This can be difficult to achieve in complex cases but enables pre-emptive actions.

9.4.2 Data Collection

Attack detection is essentially a data-driven process. No matter which detection method is followed (anomaly, signature or specification-based), the attack detection system always gathers data from the protected system, analyses it, and determines whether there is an ongoing attack. Smart grids offer several data sources that attack detection systems can use.

Multiple features can be extracted from network traffic data [14, 29]. The contents of protocol headers and payloads, rate of packets of a particular type, number of malformed packets, time of packet round trips, average packet size, and average volume of bytes per second are all examples of information that can reveal some change as a consequence of an attack. Network packets can be gathered at different points in the smart grid's networks, but it is important to consider that this choice will define which types of attacks can be detected. For instance, if the attack detection system analyses the packets carrying measurements from smart meters to the control centre, it will be certainly able to detect attacks involving the smart meters or the AMI, like a DDoS. However, if only packet statistics are checked, but not the payload content, an FDI will be hardly detected. As many smart grid sites will be located in remote locations, wireless technologies are good candidates to connect these sites to the rest of the network. With this in view, measuring wireless channels' conditions may be useful to detect some attacks like jamming. A possible way of measuring a channel condition is to transmit control packets in selected channels and wait for ACK packets to analyse the channel performance [30]. Signal Strength Intensity (SSI) of smart meters and data concentrators can also be checked to detect jamming attacks [28]. If the sensed SSI is much higher than an expected value, it can indicate a rogue or compromised device trying to jam legitimate transmissions.

Data about the status and events related to different devices in the smart grids can also be a good source of attack indicators. A power grid has several specialised devices (e.g. breakers) deployed across the system to control its operation and make sure that it is running correctly. Status flags collected periodically from these devices can reveal unexpected behaviours linked to command spoofing attacks, for example [26]. Several logs can also be processed to identify events that can help to uncover an attack episode [5]. Relay logs provide information about events involving breakers, while the control panel log can show whether there was scheduled maintenance for a particular grid component. Logs from Snort, a signature-based intrusion detection system for TCP/IP networks, can also be used to detect the presence of packets with some particular characteristics in the system. For example, it is possible to create a rule in Snort to trigger an alarm every time a packet carrying a command to change a breaker status is detected. Lastly, utilisation of hardware resources such as memory and CPU can also be monitored at the protected devices, as DoS/DDoS attacks may cause sudden changes in these measurements [29].

The data sources presented so far are particularly related to the operation of smart grids' infrastructure, such as breakers, relays, and networking devices. They encompass network packets, status flags, event logs, and resource utilisation data related to the daily routine of these devices. Smart grids also have another rich source of insights for attack detection: domain-specific data. Data about energy consumption and electrical quantities are already collected in real-time at multiple points of the grid for management and control purposes, and attacks (such as FDI and energy theft) can cause subtle but detectable changes in their behaviour.

FDI attacks typically affect the system state estimation. Therefore, measurements such as active and reactive power, current flow, voltage magnitude, and phase angles that feed the state estimation are used to detect these attacks [2–5, 12, 16, 19, 20, 22–26]. SCADA systems are usually employed to collect these measurements [4, 19, 22]. RTUs at different points in the power grid transmit these measurements every 2–5 s [19] to targets like the control centre. PMUs are also applied to collect this kind of data and send it to the control centre [5, 12, 23, 26]. They are much more precise than SCADA systems, reaching a sampling rate of 2880 samples per second [12]. However, this precision comes with a high computational cost, which poses a great challenge to management and control activities, including attack detection [5].

Alongside PMUs and SCADA systems, AMIs are also dedicated to collecting domain-specific data from smart grids for management and control. The main elements of AMIs are the smart meters, which are deployed at the customer end to send data related to energy consumption, power quality, and pricing to the utility provider [11]. Among them, energy consumption data has been used to detect energy theft [15], DoS [11], and FDI attacks [21]. The latter case is of particular interest here because it shows that AMI data improves state estimation, which usually takes energy consumption forecasts as input, instead of real consumption data. However, if the grid has an AMI, real-time energy consumption data is available and, consequently, consumption forecasts can be replaced by real data during state estimation. It is worth noting that due to the huge number of customers and frequent collections, control centres face significant difficulties in storing and processing smart meter data [15].

There are yet other data sources that, despite being used infrequently, can also be helpful in attack detection. Some attacks can be directed to load forecasting data, which is important in enabling operators to foresee the system's conditions and get ready for upcoming events. Hence, data that is typically used by feed load forecasting processes, such as historical load data, weather data (e.g. temperature and humidity), and time data (e.g. time of the day and day of the week), becomes useful for attack detection [27].

9.4.3 Detection System Distribution

Attack detection is a process composed of data collection, system profiling, detection, and response. As smart grids are huge and hierarchically-arranged systems, data collection is usually distributed. In the case of detection based on network traffic data, packets must be sniffed at multiple points of the several networks that compose a smart grid communication infrastructure. Unlike traditional enterprise networks, where there is often a single point of connection to an external network (e.g. the Internet) which is monitored for attack detection, smart grid networks present a wider attack surface with several points to monitor. Likewise, when system logs are used as a data source, there are different critical systems to be monitored and, hence, multiple data collection points. Approaches based on SCADA systems or PMUs are naturally distributed in terms of data collection, as there are always several RTUs and PMUs deployed over the power grid to perform their primary function: controlling and monitoring the power grid.

The goal of system profiling changes according to the detection method. For anomaly and specification-based approaches, system profiling is responsible for building the notion of which activities are normal, while for signature-based ones it specifies what defines an abnormal activity. Then, during the detection task, data collected in real-time is analysed according to the knowledge built during the system profiling, and the response takes place when an attack is detected. The response can vary, from alerts that are presented in a management console to an action to mitigate the attack, such as blocking the attacker's access to the protected asset. For simplicity's sake, system profiling, detection, and response can be summarised with a single term: decision making.

In smart grids, decision making in attack detection can be centralised, partially (hierarchically) distributed, or fully distributed. For the rest of this chapter, when used to classify a detection approach architecture, the terms centralised, hierarchically distributed and fully distributed will refer to how decision making is performed.

In centralised architectures, all of the data collected is transmitted to the control centre, where decision making is performed. Attack detection approaches based on data used for state estimation and event logs are usually centralised because this data is already transmitted to the central control as part of other control and management tasks [3, 12, 16, 19, 22, 26]. In these scenarios, data processing may rely on tools such as Hadoop, which runs in distributed computing environments [12]. Nevertheless, decision making is still centralised, meaning that there are no multiple instances concurrently determining whether an attack is occurring. Indeed, the main challenge for centralised architectures is to cope with the huge volume of data to be stored and processed, as smart grids have multiple data collection points operating at high sampling frequencies.

Hierarchically distributed architectures are directly linked to the hierarchical topology of smart grids. As mentioned in Sect. 9.2, smart grid networks are organised in three levels: HAN, FAN (or NAN), and WAN. Hierarchically distributed architectures seek to spread the decision-making process across network levels and, at the same time, keep some degree of central coordination. In other words, they transfer part of the burden of storing and analysing huge amounts of data from a central point to multiple points, while ensuring that a central element supervises attack detection. Hierarchically distributed solutions can rely on network traffic data [14], as well as on domain-specific data [11, 28]. Detection system agents are placed to monitor the communication traffic or the data collected from smart meters in HANs, FANs, and the WAN. Attacks detected in a particular network level can be checked in the next level up. For example, an attack detected in a smart meter after analysing data collected in a HAN can trigger an alert, that is sent to the detection system of the FAN where this HAN is connected. Then, that FAN's detection system analyses these alerts before confirming them [28]. Likewise, when a detection agent cannot decide if an attack is occurring based on the data it has, the decision can be passed up to the next level [14].

Alerts are not the only information that detection agents in lower levels send to their counterparts in upper levels. In some cases, the lower level detection agent can forward high-level statistics (e.g. a measure of anomaly evidence [11]) to the upper level agent. It is important to note that devices in lower network levels can face difficulties in hosting computationally costly processes because they are usually resource-constrained. A smart meter, for instance, may not be able to host a detection agent that runs a machine learning classifier. On the other hand, lower level monitoring can offer more detailed data for attack detection, particularly in situations where devices at the lower level are the targets. Therefore, those responsible for designing hierarchically distributed architectures have to consider this trade-off between computational capacity and data granularity.

Fully distributed architectures are not frequently proposed because the absence of any central control is seen as incompatible with the level of reliability demanded from smart grids. However, a fully distributed solution may be applicable in some specific situations. For example, to defend jamming attacks, distributed agents can be responsible for sensing communication channels and pointing out which ones are free from jamming attacks and, consequently, more suitable for transmission [30, 31].

9.5 Machine Learning

Machine Learning (ML) is a field that emerged from artificial intelligence and involves the use of algorithms belonging to different categories, such as supervised, semi-supervised, and unsupervised learning. In this chapter, we use the term ML to refer to algorithms where computers learn how to process their inputs, without this being explicitly implemented. In other words, with ML, computers are able to perform a task by making use of inference or based on observed patterns, and not by relying only on instructions that specify clearly how it should be done. Due to the vast number of ML algorithms, we focus on the widely used and most accurate ones in the smart grid field.

An ML model is a mathematical model that receives the description of a given problem as an input and delivers a generated solution as an output. This model is constantly updated by a data-driven induction towards making reliable predictions or decisions. Most ML models are induced using supervised algorithms, which demand a dependent variable. In classification problems, the dependent variable, commonly referred to as a label, is linked to the problem class of a given sample. For instance, to induce a supervised classification model to be embedded into a smart grid attack detection system, various examples of the attacks to be detected are needed, alongside instances of non-malicious behaviour. These examples are presented to the algorithm along with their respective labels, which inform the example's class. Examples and labels are then used to build a model capable of classifying new instances.

In an attack detection scenario, the classification problem is usually modelled as a binary classification task, as it supports two opposite classes: normal or anomalous behaviour. However, in some cases, there are more than two classes to be predicted, defining a multi-class problem. Multi-class detection systems are generally focused on identifying specific cyberattacks (energy theft, jamming, DoS, and FDI), supporting efficient countermeasures to minimise their damage and to combat the attack source. The most widely used algorithms for cyberattack detection in smart grids, for both binary and multi-class problems, are Support Vector Machine (SVM) [37], Artificial Neural Networks (ANN) [38], k Nearest Neighbours (kNN) [39], Naive Bayes (NB) [40], and Random Forest (RF) [41].

SVM is an algorithm developed to find a hyperplane in high-dimensional space from training samples, while attempting to maximise the minimum distance between that hyperplane and any training sample according to its class. The model (hyperplane) obtained by SVM is used for predicting new unlabelled samples. The default hyperparameters of SVM are the regularisation parameter (C) and the kernel. Some kernels, such as radial basis function (RBF) and polynomial, require additional hyperparameters.

The usage of ANN has been boosted by the advent of deep learning approaches. Its inducing architecture is based on connected artificial neurons used to simulate the learning process of a biological brain. From the shallow (Perceptron and Multilayer Perceptron) to the deep learning structure (Long Short-term Memory and Convolutional Neural Networks), ANN has several architectures and hyperparameters to be defined.

Unlike SVM and ANN, kNN is a simple machine learning method, which predicts new samples based only on the distance between a given sample and the training pattern. Using the *k* hyperparameter as the number of relevant neighbours, kNN classifies a new sample based on its closest training examples in the feature space. Another simple ML algorithm is NB, grounded on the assumption of independence among features for modelling a classifier. Although the conditional independence premise is rarely true in most real-life applications, NB generates competitive models in practice. The reason for this is that an NB classifier will be successful as long as the actual and estimated distributions agree on the most probable class, regardless of feature independence.

RF is an algorithm based on classification trees. More precisely, it is an ensemble of decision trees created through bagging strategy, which combines multiple random predictors to generate its final result. RF presents some important advantages, such as the ranking of features and the reduced possibility of overfitting. Furthermore, as hyperparameters, it requires only the number of decision trees and the number of variables available for splitting at each tree node.

In addition to classification, another important application of supervised ML algorithms is grounded on regression problems. Instead of using categorical outputs (i.e. dependent variables), regression problems require the prediction of continuous values, e.g. power flow in smart grids. In this scenario, the attack detection relies on a threshold-based strategy and statistical control techniques, such as Cumulative Sum (CUSUM) [11, 16, 17]. CUSUM follows the premise that an attack modifies the typical behaviour of the evaluated stream. Detection based on CUSUM is usually performed by computing some stream signatures such as mean value, root mean square value, peak values, amplitude probability density function, rate of signal change variations, and zero crossings per unit time. When one or more signatures are changed, the cumulative sum is computed for detecting an increase in the mean value of a sequence of independent Gaussian random variables. If the CUSUM's value exceeds a threshold, an attack is characterised. The success of attack detection depends on proper hyperparameters setup, which are related to the tolerance interval, the probability of false alarms, and the detection delay from the observed stream.

In several cases, the dependent variable is not completely available due to the cost or complexity of its extraction. Attack detection in smart grids is an example, as it typically suffers from scarcity of labelled data. In smart grid scenarios, there are several security behaviours that should be simulated, studied, and stored to produce labelled data for supervised learning. Thus, when it is challenging to obtain labelled data, we can employ semi-supervised approaches [42].

Unsupervised clustering is a third category of machine learning approaches applied to smart grid attack detection. More precisely, it is tailored for scenarios with a total absence of labels [43]. The most used algorithm of this category is k-means, which, coupled with a heuristic algorithm (e.g. Particle Swarm Optimisation or PSO), is able to assume the likely number of clusters (*k* value) required to properly distinguish attacks from normal situations. The k-means algorithm follows the premise that all evaluated instances can be associated with one of the *k* clusters. As the instances are grouped according to their behaviour, attack and normal instances will be clustered into different partitions. More recently, Bayesian clustering has also been used to address smart grid problems. For instance, Dasgupta et al. [44] made use of techniques from elastic shape analysis along with a Bayesian approach to cluster and evaluate electricity consumption curves according to their shapes.

Supervised, semi-supervised, and unsupervised categories cover most of the algorithms applied to cyberattack detection on smart grids. However, the current call for real-time (online) detection and high predictive performance paves the way for more recent paradigms of machine learning. On the one hand, dealing with real-time detection, we have the Hoeffding Tree (HT), an incremental decision tree for high-speed data streams classification [45]. On the other hand, focused on leveraging high predictive performance, we have Deep Learning (DL) methods [46].

HT is a supervised ML algorithm designed to induce models online in an incremental way (i.e. instance-by-instance) based on anytime learning as required for data stream processing. Therefore, the usage of HT for attack detection consists of the induction and classification of data flows from smart grids without *apriori* knowledge, i.e. there is no offline phase to train a model. Unlike offline learning, which assumes that all training data needed to create a model is already available, online learning assumes that new data can arrive at any time, which can make a model outdated [47]. Like in traditional ML processing, data stream mining can be performed by supervised and unsupervised algorithms. Considering high-speed stream scenarios, which may be the case for smart grids, unsupervised approaches have been reported as more feasible. In [48], an online unsupervised clustering algorithm was used for load profiling. The proposed approach takes advantage of the stream structure of the data, keeping the identified profiles updated in accordance with newly collected data. It is worth mentioning that the kernel of the proposed solution is based on the k-means algorithm.

In recent years, DL methods have drawn academic and industrial attention. These methods are grounded on discovering the intricate structure of inputs to learn representations of data with various levels of abstraction. Among all DL methods, some deep variants of multi-layer perceptron (MLP) were used to detect FDI. In [19], deep MLP was applied to identify attacks on smart grids using active power-flows, active power-injections, reactive power, and voltage measurements as features to induce the DL model.

All the algorithms, methods, and categories mentioned so far in this section are applied as unmixed or hybrid approaches when addressing attack detection in smart grids. Most of the works surveyed are designed as a pipeline composed of steps such as pre-processing, feature selection, and ML predictive algorithms. A hierarchical overview of ML algorithms and their combination is presented in Fig. 9.4.

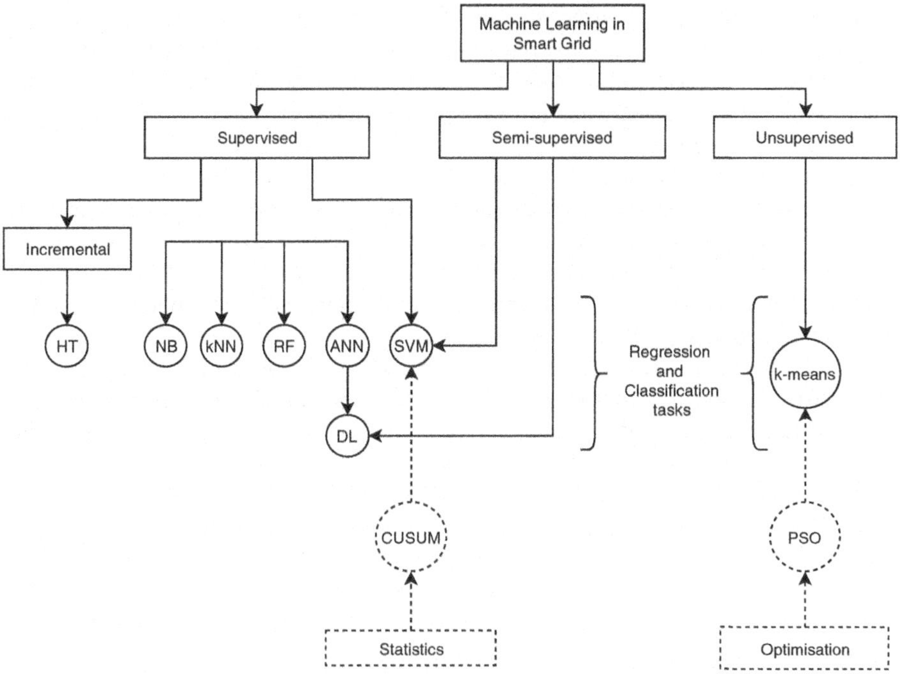

Fig. 9.4 Overview of machine learning algorithms (Hoeffding Tree, k Nearest Neighbours, Random Forest, Artificial Neural Networks, Deep Learning and Support Vector Machine) and their correlations with Statistics (Cumulative Sum) and Optimisation (Particle Swarm Optimisation) techniques

9.6 Existing Solutions

Solutions proposed in the literature for cyberattack detection in smart grids are diverse, which reveals the many decisions a researcher has to make when developing a new approach. Which attacks will be addressed, which data will be collected, how to distribute the system, and how to combine machine learning techniques are among the questions that must be answered. This section presents a literature review of proposals to tackle cyberattacks in smart grids using machine learning, providing a discussion of how the different authors addressed these issues.

FDI is the most addressed threat in works on attack detection in smart grids. Its high potential of disrupting smart grid operations is probably the leading cause for this concern. To identify both standard and stealthy FDIs, Huang et al. [16] contributed a centralised anomaly detection scheme applied to state estimation data. They define Gaussian-based vectors for observation and for an unknown data injection (commencing at a random time) with the aim of identifying the change in the observation vector's distribution from the idle state to the attack state. Based on the time of detection, the average run length (ARL) is calculated. Then, the detection time and a threshold h are used in multi-thread processing (with a linear solver and Rao tests) to solve the problem recursively. The alarm is raised when a cumulative score reaches h. Based on tests on a 4-bus environment, the authors reported that the best detection is achieved at higher ARLs, and that the value of h influences the timeliness of detections.

Qiu et al. [49] addressed FDI attacks as part of their investigation into the application of cognitive radio networks for smart grids. They proposed the centralised use of Independent Component Analysis (ICA) to overcome FDI attacks, characterising them as instances of high interference. A data transmission matrix Z is defined, which contains a matrix X of source signals originating from smart meters. The aim is to fill out X with signal estimations. To do so, the attacker's signals must first be filtered out. This is achieved using the statistical properties of signals, with Principal Component Analysis (PCA) used to deal with differing power levels in the interference. Through simulations, the authors were able to demonstrate that ICA can effectively separate different signals.

Esmalifalak et al. [20] also used PCA to tackle FDI attacks but in a different way. The authors proposed two methods to detect stealth FDI attacks. The data used to detect the attacks consisted of measurements for state estimation, which is collected from multiple points in the power grid. These data can present some redundancies, and the number of dimensions in the detection problem is linked to the power system size. For instance, a 118 bus system used in this work for tests generated 304 dimensions. In the two methods, to avoid the curse of dimensionality, the authors employed PCA. Their underlying assumption is that normal data is generated according to physical laws, while tampered data is not, so these data should be separated in the projected space. The first detection method was based on a statistical anomaly detection technique, which made use of a threshold learnt from historical data. It is centralised and relies on data sent to the control centre. The second method was built upon distributed SVM, which, unlike the first method, requires labelled data from both classes (normal and attack) for training. To test their system, they used an IEEE 118-bus test system.

Ozay et al. made extensive use of machine learning classifiers to detect FDIs in two different works [23, 24]. In the first one [24], they proposed a centralised signature-based method for state estimation data. Supervised learning is used to classify samples as either "secure" or "attacked". Three machine learning algorithms were used: kNN, SVM, and sparse logic regression (SPR). kNN sorts feature vectors into neighbourhoods based on Euclidean distance. SVM identifies hyperplanes for the binary splitting of samples. SPR uses Alternating Direction Method of Multipliers (ADMM) with distributions for labels-to-samples matching. Testing on IEEE 9-, 30-, 57-, and 118-bus systems revealed that both kNN and SVM are negatively impacted by data sparsity, unlike SPR [24]. The authors suggested that kNN is suited for smaller systems, and SVM and sparse logistic regression for larger ones.

In [23], Ozai et al. presented a thorough study exploring multiple families of machine learning algorithms, including supervised, semi-supervised, and online machine learning. In their evaluations, the authors employed feature-level fusion and ensemble methods. IEEE 9-, 57-, and 118-bus test systems were used again in the experiments. The authors pointed out that semi-supervised algorithms were more robust against sparse data than supervised ones. Also, feature-level fusion and ensemble methods were shown to be robust against changes in system size. Lastly, the performance of online classifiers was comparable to batch ones.

Yan et al. [25] also presented a comparative study exploring multiple machine learning classifiers. In their work, three supervised algorithms, namely SVM, kNN, and eNN, were used to detect FDI attacks on measurements for state estimation. The authors considered balanced and imbalanced cases, and analysed the impact of the magnitude of false data in the detection performance. Tests were based on the IEEE 30-bus test system, with the SVM classifier obtaining the best overall results.

In [12], the authors described the additional challenge of detecting FDIs in the vast amounts of data collected in smart grids. Based on experiments using a 6-bus power network in a wide area measurement system environment, these authors proposed a Margin Setting Algorithm (MSA). The proposed algorithm was compared to the SVM and ANN algorithms in a binary classification scenario for detecting playback and time attacks. Results demonstrated that the MSA achieved minimal errors and better accuracy than traditional machine learning algorithms with conventional hyperparameters.

Unlike the works presented so far, Hink et al. [26] approached FDI detection with three different classification schemes. They aimed to study the performance of multiple machine learning algorithms in distinguishing power system disturbances as malicious or natural. In the first classification scheme, each type of event was modelled as a class, meaning that it was a multi-class problem with 37 distinct classes. The second scheme took into account three classes: malicious event, non-malicious event, and no event; the latter corresponds to data related to normal operations. The last classification scheme had two classes: attack or normal. Seven machine learning algorithms were tested, namely OneR, NNge, RF, NB, SVM, JRipper, and AdaBoost. Although the results varied significantly for different classification schemes, the authors pointed out the combination of Adaboost, JRipper, and the 3-class model as the best solution among the studied ones.

Neural networks were also used as a feasible solution for FDI detection. Hamedani et al. [2] made use of Reservoir Computing (RC), an energy-efficient computing paradigm grounded on neural networks. The proposal was implemented by combining DFN (Delayed Feedback Network) and MLP to support spatio-temporal pattern recognition. Considering wind turbines as the major source of electrical power generation, collected measurements (i.e. temporal data) were encoded as feature vectors to be the input of the binary classification task, which distinguished instances between normal or under attack. Simulation results showed DFN + MLP could detect attacks under different conditions, such as different magnitudes and number of compromised meters, overcoming the performance of single MLP and SVM algorithms.

Another solution based on neural networks for detecting FDI was proposed by He et al. [22]. In this work, CDBN (Conditional Deep Belief Network) was explored to extract high-dimensional temporal features for recognising the differences between the patterns in data compromised by FDI attacks and in normal data. The system architecture, composed of five hidden layers, obtained the best results when comparing three different numbers of layers. In comparison with SVM (Gaussian kernel) and ANN (MLP with a single hidden layer), CDBN achieved superior accuracy, followed by MLP and SVM. The authors claimed that they performed online attack detection, but it is important to mention that the training and updating were performed offline.

Following a different path, the proposal by Adhikari et al. [5] is based on pure online modelling. The authors proposed the use of the Hoeffding Tree coupled with a mechanism to handle concept drift when classifying binary and multi-class power system events and cyberattacks. A total of 45 classes of cyber-power issues were addressed using a combination of attributes from power and network transactions (such as voltage, current, and frequency), and logs from Snort. The authors put effort into tuning all the algorithms and deployed real-time analysis with a high level of accuracy. The main advantages of the proposed method are related to consuming less memory than traditional batch processing, as well as providing real-time analysis to classify a broad number of power system contingencies and cyberattacks. However, HT is a supervised machine learning algorithm, which depends on a labelling step. This becomes a pitfall for real-time applications.

Semi-supervised learning methods are an attractive alternative to ease the need for the labelling step. In [50], cyber-physical attacks on power systems were addressed using Reinforcement Learning (RL), more precisely a Q-learning semi-supervised algorithm. A contingency analysis system was proposed to handle sequential attacks in power transmission grids, such as blackout damage and hidden line failures. Based on simulated study cases with IEEE 5-bus, RTS-79, and 300-bus, it was possible to discover a more vulnerable target sequence in sequential attacks. Furthermore, when varying the blackout size and topology of attacks, the proposed solution was capable of reducing the number of successful attacks by excluding failed attack sequences.

Chen et al. [4] improved on the usage of the Q-learning algorithm in their proposal to enable the online learning of non-malicious and attack behaviour. Focused on detecting FDI attacks that affect the normal operation of a power system regulated by automatic voltage controls (AVCs), the authors proposed to model the attacks as a POMDP. An FDI mitigation method was developed, consisting of offline and online modules capable of detecting multiple attacks. The experiments performed on an IEEE 118-bus system assessed the scalability of the proposed solution and its ability to provide insights about attacks and their impact in the whole power system. The main contribution was the study of the RL usage, providing theoretical assumptions about scalability and feasibility of FDI detection, as well as further results from a mitigation system. However, the paper lacks real-life cases (very sparse attacks) and considers a naive virus spreading strategy.

Deep learning was the method used by Ashrafuzzaman et al. [19] to deal with FDI attacks. The experiment was carried out using a simulated IEEE 14-bus system. Four different architectures of deep learning models based on MLP were compared to Gradient Boosting Machines (GBM), Generalised Linear Models (GLM), and Random Forest (RF). The authors explored 122 measurement features to find the 20 most important with RF importance. As an outcome, deep MLP structures obtained the best accuracy results. Also, the use of a smaller set of selected features resulted in training time speed-up. Lastly, the deep learning training cost was mentioned as a challenge that needs to be handled for speeding up the process. To obtain more confidence, the authors planned to use real-life datasets as future work.

Instead of detecting FDI attacks, Anwar et al. [21] just clustered AMI nodes according to their vulnerability to such attacks. Their idea is that some nodes, due to their inter-dependency to other ones, can cause more damage to the entire system when attacked. Therefore, these nodes should be identified to be better protected. To cluster the AMI nodes, the authors applied the k-means algorithm combined with CF-PSO over the nodes' voltage stability index. Three clusters were defined: one for the least vulnerable nodes, other for the nodes with moderate vulnerability, and the last one for the most vulnerable. Experiments were performed in a 33-bus and a 69-bus test systems.

Alongside FDI, DoS attacks are among the top concerns regarding smart grids cybersecurity. Fadlullah et al. [18] proposed a centralised Bayesian approach for early DoS detection. The DoS attack is modelled as an attacker with access to one or more smart meters (via a worm), which they use to generate many fake authentication requests to saturate the network and strain target devices. The system uses Gaussian process regression to create an attack forecast based on the current state of communications. A composite covariance function is used to analyse trends, and samples are taken to create a set of real observations. The method was tested in a simulated BAN (Building Area Network), with 50% of smart meters vulnerable to worm infection. The authors found the forecasting system to be effective with both long and short training times, noting that the BAN gateway can be impacted at different times depending on attack particularities.

Comparative studies on machine learning algorithms were also carried out for DoS detection. To achieve this, Choi et al. [29] simulated a SYN flood attack and a buffer overflow attack on the bay and the station levels of the grid. PCs were implemented to emulate IEDs, with the GOOSE protocol's publisher-to-subscriber multicast feature used to spread attack commands via a router. The data generated was then collected, and a set of traffic-based metric attributes extracted (consisting of both normal and attack state information). The authors used Weka's machine learning library to process the data using various algorithms including Bayes classifiers, neural networks, SVM, lazy classifiers, Voting Feature Intervals (VFI), rule-based classifiers, RF, and decisions trees. They reported that for both attack types, the use of key attributes improved detection ratings, and that decision trees produced the best results overall.

Yilmaz and Uludag [11] explored the online classification paradigm to develop the MIAMI-DIL (Minimally Invasive Attack Mitigation via Detection Isolation and Localization) approach. It focuses on detecting DoS attacks against nodes on the distribution and customer domains such as data concentrators, smart meters, and smart appliances. Their approach is based on an online and non-parametric detector named ODIT, which combines features from GEM (Geometric Entropy Minimization) and CUSUM. ODIT is applied at different levels of the network, so an anomaly evidence score is computed for smart appliances (HAN level), smart meters, and data concentrators (FAN level). Anomaly evidence scores from different levels are gathered to decide whether any node in the system is under attack. If so, a mitigation procedure is carried out, which isolates the node involved in the detected attack.

As smart grid networks rely on wireless networks due to their capacity of covering long distances and reaching remote spots, jamming attacks are also a significant threat. Su et al. [31] and Niu et al. [30] proposed a distributed jamming-avoidance strategy where the efficient use of channels for secondary users (SU) is defined as a POMDP. Each SU uses the MAB algorithm to generate a set of possible strategies (i.e. channels it can sense), weighted by availability. It then selects a random strategy to try, calculating the distribution for the channel set. Estimated and actual success rates are then compared to update weightings. Simulations revealed that the more sophisticated the jammer, the more difficult the problem. However, the authors reported that over time, SUs could achieve a highly unified view of channel availability and were less likely to be affected by jammed channels. It is important to note that these works do not propose a jamming attack detection scheme, but a solution to avoid channels under this kind of attack.

As load forecasting is helpful to improve the smart grid operation and planning, attacks against this activity can lead operators to make wrong decisions. Cui et al. [27] employed some classical machine learning algorithms in a three-stage anomaly detection approach to address this issue. In the first stage, the data is reconstructed to deliver a suitable forecast based on feature selection. Afterwards, the attack template is detected via k-means clustering. Finally, in the third step, the identification of the occurrence of a cyberattack is performed using Naive Bayes algorithm and dynamic programming. Five different attack templates were studied: pulse attack, scaling attack, ramping attack, random attack, and smooth curve attack, with the latter ones being the more difficult to detect. The authors discussed the importance of feature selection in enhancing the accuracy of attack prediction. They also discussed the impact of adversaries in the anomaly detection model and detection performance, highlighting this topic as an important challenge for future works in cybersecurity.

Some works proposed schemes to address multiple types of attacks. Kurt et al. [3] explored RL and POMDP to detect FDI, DoS, and jamming attacks. They also claimed that the proposed solution would allow new unknown attack types to be detected. The authors implemented a framework to track slight deviations of measurements from normal system operation. To evaluate the results, the proposal was compared to a Euclidean detector and a Cosine-similarity detector. The best results were achieved by the RL proposed detector, followed by the Euclidean detector and the Cosine-similarity detector. The proposed solution achieved satisfactory results but, throughout the experiments, the authors had to handle several hyperparameters to tune the algorithms appropriately. This might present a challenge to this method's wide-scale adoption. Furthermore, when discussing the results, some concerns were raised about the memory cost, and the possibility of improving performance with a DL algorithm was suggested.

Sedjelmaci and Senouci [28] also targeted multiple types of attack. They proposed a hierarchically distributed system to detect FDI, DoS, and energy theft by analysing data collected from smart meters. The system is composed of three agents, with one for each hierarchy level. The LLIDS (Low Level IDS) is deployed at smart meters, the MLIDS (Medium Level IDS) is embedded in data concentrators, and the control centre hosts the HLIDS (High-Level IDS). Firstly, a rule-based system analyses collected data. This system has a specific threshold for each kind of attack, FDI, DoS, or energy theft. When the rule-based system detects a threshold violation, it passes the analysis on to the IDS agent of the next upper level. Then, an SVM classifier is used to confirm whether the anomaly is an attack.

Works such as [32] and [14] do not specify the types of attacks they are aiming at. They build their approaches to detect anomalies or unusual behaviours that can signal an attack, but do not focus on any specific threat. Kher et al. [32] developed a hierarchical sensor model for anomaly detection using sensor data, covering both the lower node and the upper cluster head levels. The proposed protocol initially has all nodes in sleep mode (for synchronisation). Clusters are formed through the exchange of “Hello” messages (at the lower level), a cluster head is selected, and multiple cluster heads establish linear links with each other (at the upper level). Data from each cluster is integrated before being sent up the chain. Data received at towers is integrated again before being sent to the base station. This integrated data can then be analysed for anomalies. The authors used Weka-implemented supervised learning for this purpose, and reported that the decision tree classifier (J48) achieved the best performance compared to other algorithms like ZeroR, decision table, RF, and ADTree [32].

Zhang et al. [14] proposed a distributed IDS system that uses intelligent analysis modules (AMs) sitting across HAN, NAN, and WAN layers. AMs at each level work with other modules to form a self-contained IDS on that grid layer. At the NAN and WAN levels, the lower level IDS is used together with the local IDS, such that the overall system is formed hierarchically. For difficult decisions, data may be sent to higher layers for further analysis. Either unsupervised SVM (with a Gaussian radial basis function) or Artificial Immune System (AIS) algorithms (with a focus on clustering) are used for detection. Clonal selection algorithms CLONALG and AIRS2Parallel were tested, and the authors found that SVM had better overall performance, especially for remote-to-user (R2L) and user-to-root (U2R) attacks. They suggested that the detection accuracy of the AIS algorithms could be improved with a larger sample of attack data.

Table 9.1 presents a summary of all the reviewed works and their main characteristics in chronological order. In some cases, the reviewed work does not define the data source used or how the solution would be distributed. In these cases, the table presents “–” for the undefined feature.

Table 9.1 Surveyed solutions and their main features

Reference	Year	Data source	Distribution	Attack types	ML algorithms
[18]	2011	–	–	DoS	Other[a]
[49]	2011	–	–	FDI	Other[a]
[14]	2011	Network traffic	Hierarchically distributed	Anomalies	SVM
[32]	2012	–	–	Anomalies	RF
[29]	2012	Network traffic, CPU, and memory utilisation	–	DoS	SVM, ANN
[31]	2012	Wireless channel condition	Fully distributed	Jamming	Other[a]
[24]	2012	Measurements for state estimation	–	FDI	kNN, SVM
[16]	2013	Measurements for state estimation	Centralised	FDI	CUSUM
[20]	2014	Measurements for state estimation	Centralised and distributed	FDI	SVM
[26]	2014	PMU measurements and devices status flags	Centralised	FDI	RF, SVM
[30]	2015	Wireless channel condition	Fully distributed	Jamming	Other[a]
[21]	2015	Smart meter data (energy consumption)	–	FDI	PSO
[23]	2015	PMU measurements	–	FDI	ANN, SVM, kNN
[25]	2016	Measurements for state estimation	–	FDI	SVM, kNN
[11]	2016	Smart meter data (energy consumption)	Hierarchically distributed	DoS	CUSUM
[28]	2016	Smart meter data, amount of requested energy by customer, signal strength intensity	Hierarchically distributed	FDI, DoS, energy theft	SVM
[12]	2017	PMU measurements	Centralised	FDI	SVM, ANN
[5]	2017	PMU measurements, relay logs, central control logs, snort logs	Centralised	FDI	HT
[50]	2017	–	–	Sequential topology attack	RL
[22]	2017	Measurements for state estimation (SCADA)	Centralised	FDI	SVM, ANN
[19]	2018	Measurements for state estimation (SCADA)	Centralised	FDI	RF
[4]	2018	Measurements for state estimation (SCADA)	–	FDI	Other[a]
[3]	2018	Measurements for state estimation	Centralised	FDI, DoS, jamming	RL
[2]	2018	Measurements for state estimation	–	FDI	SVM, ANN
[27]	2019	Load data, weather info, temporal info	–	Load forecast attack	k-Means

[a]Refers to pure statistical approaches (e.g. maximum likelihood estimation) or probabilistic approaches (e.g. Markov models), outside of machine learning boundaries

9.7 Open Issues

Considerable progress has been made in smart grids scenarios when applying machine learning for cyberattack detection. However, several key issues need to be handled to allow the development of feasible solutions capable of achieving suitable performance in real-life scenarios.

The main challenges are related to the real-time nature of the problem, the need for labels in supervised learning, demand for more comprehensive and human-friendly models, and solution scalability. Additionally, some inherent challenges of machine learning (such as hyperparameter tuning and the capacity of algorithms for dealing with imbalanced data) open new paths for further research in applied smart grid security.

Real-time classification algorithms are often demanded in the literature. This type of algorithm could be implemented by an offline induction and online classification, as in most of the current proposals. However, these solutions require some additional effort to leverage reliable models since they become obsolete when the smart grid behaviour changes, culminating in the concept drift problem. Also, the cost of feature extraction needs to be suitable to support a real-time classification. The algorithms that meet the real-time classification requirements are grounded on stream mining. Stream mining algorithms are able to induce a model online, which eliminates the offline step and keeps the model updated. Some algorithms such as Very Fast Decision Tree [45] and Strict Very Fast Decision Tree [47] are important examples of stream algorithms.

Even though online classification algorithms pave the way for more useful solutions, their requirement for labelled instances is a hindrance. In other words, it is impossible to label each instance on the smart grid data flow. Thus, it is necessary to rely on semi-supervised approaches or unsupervised algorithms. The DenStream algorithm [51] is an unsupervised algorithm for stream clustering. Based on three types of clusters, DenStream can point out the core, potential, and outlier behaviours, giving insights into the smart grid's behaviour.

Changes are expected in smart grid behaviour during an attack, and consequently, the recognition of these pattern deviations allows a machine learning model to detect the attack. For this reason, the predictive performance of detection systems has been the main focus of current systems, avoiding false positives and improving the computational complexity of the designed solutions. However, some concerns on how the attacks happen, the importance of features used to describe the event, and a user-friendly model for supporting attack comprehension are also relevant demands from industry. In addition to being highly accurate, a machine learning model needs to produce meaningful results and help operators to make better decisions through the usage of more descriptive modelling.

Meaningful results from descriptive models support suitable incident comprehension and mitigation. Thus, choosing an algorithm that matches certain model legibility criteria is necessary. However, the amount of data collected from a smart grid environment poses an additional constraint: scalability. A highly accurate algorithm that is able to output a user-friendly model considering the current smart grid scenario also needs to be scalable to handle huge amounts of data. Scalability is related to the parallelism inherent in an algorithm and can be measured according to its speed-up on a particular architecture [52]. Most of the current works are limited to experimental scenarios with controlled, finite, and synthetic data sets, which do not offer a close-to-reality challenge in terms of data volume.

Another problem related to synthetically-produced data is that they usually result in a balanced dataset, which promotes an unrealistically smooth model induction. Attack detection problems are usually imbalanced, since deviations caused by attacks are much less frequent than expected behaviour episodes. More precisely, the attack-related samples provided by a smart grid to induce a machine learning model are often much fewer than the non-malicious samples, making up an imbalanced dataset. Highly imbalanced problems generally present high non-uniform error, which compromises the overall performance when errors occur in the minority classes. There are several approaches to work around this issue. The most commonly used are based on undersampling the majority class or oversampling the minority one. Considering the possibility of losing important samples with the former approach, oversampling strategies, such as the Synthetic Minority Over-sampling Technique (SMOTE) [53], can be used to balance the original dataset and provide a reliable scenario for the machine learning algorithms.

However, if the synthetic data design is driven by simple constraints and deterministic behaviour, the performance achieved during the experiments can be biased by patterns that are easier to learn than they would be in real scenarios. Therefore, when applied to real-life scenarios, the solutions can demand a more complex pipeline or unfeasible modifications, which prevents their adoption. The same reasoning applies to the adversarial model design. In some works, researchers assume simple or very specific attack models, which can hinder the effective application of the proposed solutions in production environments.

The open issues mentioned so far are related to the application of machine learning to the smart grid domain. Nevertheless, the machine learning area has its own challenges, which are intrinsic to its algorithms and must be addressed regardless of the application domain. Hyperparameters tuning [54], temporal vulnerabilities [55], stream classification trade-offs [56], and more recent topics like adversarial machine learning attacks [57] are examples of these issues and pave the way for future studies.

Lastly, for some authors like [13, 17, 58], the use of multiple sources of data leads to improvements in detection performance. For example, the amalgamation of features extracted from computer network traffic and smart grid measurements can form a more robust feature vector, which covers a wider range of attacks. Working on the several possible combinations of smart grid data sources to assess how they enhance the range of detected attacks is another possible subject for future work.

9.8 Conclusion

As smart grids are critical infrastructures, cyberattacks against these systems have a high potential for causing large-scale disruption to electricity supplies. To assist in the fight against this threat, we have provided a study of how machine learning algorithms can be applied to detect attacks on smart grids. We outlined the possible attacks types, as well as the concepts that underpin the detection of such attacks. Then, we presented the machine learning algorithms that have been employed in proposed detection schemes. Following this discussion, a list of existing attack detection approaches based on machine learning was given, detailing how each one addressed the characteristics of this problem.

Some open issues were identified in the reviewed approaches, such as algorithms depending on a labelling process, approaches not prepared to deal with imbalanced datasets and real-time aspects of smart grids, algorithms producing poor descriptive models, experiments relying on poorly designed synthetic data, and testing with a limited range of attack behaviours. Among the recommendations for future work are suggestions to use stream mining algorithms and oversampling techniques, multiple data sources, and to invest more effort into producing more realistic data sets.

References

1. C. Greer, D.A. Wollman, D.E. Prochaska, P.A. Boynton, J.A. Mazer, C.T. Nguyen, G.J. FitzPatrick, T.L. Nelson, G.H. Koepke, A.R. Hefner Jr. et al., NIST framework and roadmap for smart grid interoperability standards, release 3.0. Technical Report (2014)
2. K. Hamedani, L. Liu, R. Atat, J. Wu, Y. Yi, Reservoir computing meets smart grids: attack detection using delayed feedback networks. IEEE Trans. Ind. Inf. **14**(2), 734–743 (2017)
3. M.N. Kurt, O. Ogundijo, C. Li, X. Wang, Online cyber-attack detection in smart grid: a reinforcement learning approach. IEEE Trans. Smart Grid **10**(5), 5174–5185 (2019)
4. Y. Chen, S. Huang, F. Liu, Z. Wang, X. Sun, Evaluation of reinforcement learning-based false data injection attack to automatic voltage control. IEEE Trans. Smart Grid **10**(2), 2158–2169 (2018)
5. U. Adhikari, T.H. Morris, S. Pan, Applying hoeffding adaptive trees for real-time cyber-power event and intrusion classification. IEEE Trans. Smart Grid **9**(5), 4049–4060 (2017)
6. W. Wang, Y. Xu, M. Khanna, A survey on the communication architectures in smart grid. Comput. Netw. **55**(15), 3604 – 3629 (2011) [Online]. Available: http://www.sciencedirect.com/science/article/pii/S138912861100260X
7. H. He, J. Yan, Cyber-physical attacks and defences in the smart grid: a survey. IET Cyber-Phys. Syst. Theory Appl. **1**(1), 13–27 (2016)
8. X. Li, X. Liang, R. Lu, X. Shen, X. Lin, H. Zhu, Securing smart grid: cyber attacks, countermeasures, and challenges. IEEE Commun. Mag. **50**(8), 38–45 (2012)
9. G. Loukas, *Cyber-Physical Attacks: A Growing Invisible Threat* (Butterworth-Heinemann, Oxford, 2015)
10. F. Samie, L. Bauer, J. Henkel, *Edge Computing for Smart Grid: An Overview on Architectures and Solutions* (Springer, Cham, 2019), pp. 21–42 [Online]. Available: https://doi.org/10.1007/978-3-030-03640-9_2

11. Y. Yilmaz, S. Uludag, Mitigating IoT-based cyberattacks on the smart grid, in *2017 16th IEEE International Conference on Machine Learning and Applications (ICMLA)* (IEEE, Piscataway, 2017), pp. 517–522
12. Y. Wang, M.M. Amin, J. Fu, H.B. Moussa, A novel data analytical approach for false data injection cyber-physical attack mitigation in smart grids. IEEE Access **5**, 26022–26033 (2017)
13. M. Wu, Z. Song, Y.B. Moon, Detecting cyber-physical attacks in cyber manufacturing systems with machine learning methods. J. Intell. Manuf. **30**(3), 1111–1123 (2019)
14. Y. Zhang, L. Wang, W. Sun, R.C. Green II, M. Alam, Distributed intrusion detection system in a multi-layer network architecture of smart grids. IEEE Trans. Smart Grid **2**(4), 796–808 (2011)
15. R. Jiang, R. Lu, Y. Wang, J. Luo, C. Shen, X.S. Shen, Energy-theft detection issues for advanced metering infrastructure in smart grid. Tsinghua Sci. Technol. **19**(2), 105–120 (2014)
16. Y. Huang, M. Esmalifalak, H. Nguyen, R. Zheng, Z. Han, H. Li, L. Song, Bad data injection in smart grid: attack and defense mechanisms. IEEE Commun. Mag. **51**(1), 27–33 (2013)
17. H. He, J. Yan, Cyber-physical attacks and defences in the smart grid: a survey. IET Cyber-Phys. Syst. Theory Appl. **1**(1), 13–27 (2016)
18. Z.M. Fadlullah, M.M. Fouda, N. Kato, X. Shen, Y. Nozaki, An early warning system against malicious activities for smart grid communications. IEEE Netw **25**(5), 50–55 (2011)
19. M. Ashrafuzzaman, Y. Chakhchoukh, A.A. Jillepalli, P.T. Tosic, D.C. de Leon, F.T. Sheldon, B.K. Johnson, Detecting stealthy false data injection attacks in power grids using deep learning, in *2018 14th International Wireless Communications & Mobile Computing Conference (IWCMC)* (IEEE, Piscataway, 2018), pp. 219–225
20. M. Esmalifalak, L. Liu, N. Nguyen, R. Zheng, Z. Han, Detecting stealthy false data injection using machine learning in smart grid. IEEE Syst. J. **11**(3), 1644–1652 (2014)
21. A. Anwar, A.N. Mahmood, Z. Tari, Identification of vulnerable node clusters against false data injection attack in an AMI based smart grid. Inf. Syst. **53**, 201–212 (2015)
22. Y. He, G.J. Mendis, J. Wei, Real-time detection of false data injection attacks in smart grid: a deep learning-based intelligent mechanism. IEEE Trans. Smart Grid **8**(5), 2505–2516 (2017)
23. M. Ozay, I. Esnaola, F.T.Y. Vural, S.R. Kulkarni, H.V. Poor, Machine learning methods for attack detection in the smart grid. IEEE Trans. Neural Netw. Learn. Syst. **27**(8), 1773–1786 (2015)
24. M. Ozay, I. Esnaola, F.T.Y. Vural, S.R. Kulkarni, P.H. Vincent, Smarter security in the smart grid, in *2012 IEEE Third International Conference on Smart Grid Communications (SmartGridComm)* (IEEE, Piscataway, 2012), pp. 312–317
25. J. Yan, B. Tang, H. He, Detection of false data attacks in smart grid with supervised learning, in *2016 International Joint Conference on Neural Networks (IJCNN)* (IEEE, Piscataway, 2016), pp. 1395–1402
26. R.C.B. Hink, J.M. Beaver, M.A. Buckner, T. Morris, U. Adhikari, S. Pan, Machine learning for power system disturbance and cyber-attack discrimination, in *2014 7th International Symposium on Resilient Control Systems (ISRCS)* (IEEE, Piscataway, 2014), pp. 1–8
27. M. Cui, J. Wang, M. Yue, Machine learning based anomaly detection for load forecasting under cyberattacks. IEEE Trans. Smart Grid **10**(5), 5724–5734 (2019)
28. H. Sedjelmaci, S.M. Senouci, Smart grid security: a new approach to detect intruders in a smart grid neighborhood area network, in *2016 International Conference on Wireless Networks and Mobile Communications (WINCOM)* (2016), pp. 6–11
29. K. Choi, X. Chen, S. Li, M. Kim, K. Chae, J. Na, Intrusion detection of NSM based DoS attacks using data mining in smart grid. Energies **5**(10), 4091–4109 (2012)
30. J. Niu, Z. Ming, M. Qiu, H. Su, Z. Gu, X. Qin, Defending jamming attack in wide-area monitoring system for smart grid. Telecommun. Syst. **60**(1), 159–167 (2015)
31. H. Su, M. Qiu, H. Wang, Secure wireless communication system for smart grid with rechargeable electric vehicles. IEEE Commun. Mag. **50**(8), 62–68 (2012)
32. S. Kher, V. Nutt, D. Dasgupta, H. Ali, P. Mixon, A detection model for anomalies in smart grid with sensor network, in *2012 Future of Instrumentation International Workshop (FIIW) Proceedings* (IEEE, Piscataway, 2012), pp. 1–4

33. J.L. Viegas, S.M. Vieira, Clustering-based novelty detection to uncover electricity theft, in *2017 IEEE International Conference on Fuzzy Systems (FUZZ-IEEE)* (IEEE, Piscataway, 2017), pp. 1–6
34. A.E. Lazzaretti, D.M.J. Tax, H.V. Neto, V.H. Ferreira, Novelty detection and multi-class classification in power distribution voltage waveforms. Expert Syst. Appl. **45** 322–330 (2016)
35. E.R. Faria, J. Gama, A.C. Carvalho, Novelty detection algorithm for data streams multi-class problems, in *Proceedings of the 28th Annual ACM Symposium on Applied Computing* (ACM, New York, 2013), pp. 795–800
36. R. Kaviani, K.W. Hedman, A detection mechanism against load-redistribution attacks in smart grids (2019). Preprint. arXiv:1907.13294
37. C. Cortes, V. Vapnik, Support-vector networks. Mach. Learn. **20**(3), 273–297 (1995)
38. S. Haykin, *Neural Networks: A Comprehensive Foundation* (Prentice Hall PTR, Englewood Cliffs, 1994)
39. T.M. Cover, P. Hart et al., Nearest neighbor pattern classification. IEEE Trans. Inf. Theory **13**(1), 21–27 (1967)
40. I. Rish et al., An empirical study of the Naive Bayes classifier, in *IJCAI 2001 Workshop on Empirical Methods in Artificial Intelligence*, vol. 3, no. 22 (2001), pp. 41–46
41. L. Breiman, Random forests. Mach. Learn. **45**(1), 5–32 (2001)
42. O. Chapelle, B. Scholkopf, A. Zien, Semi-supervised learning (Chapelle, O. et al., eds.; 2006) [Book reviews]. IEEE Trans. Neural Netw. **20**(3), 542–542 (2009)
43. S. Zanero, S.M. Savaresi, Unsupervised learning techniques for an intrusion detection system, in *Proceedings of the 2004 ACM Symposium on Applied Computing* (ACM, New York, 2004), pp. 412–419
44. S. Dasgupta, A. Srivastava, J. Cordova, R. Arghandeh, Clustering household electrical load profiles using elastic shape analysis, in *2019 IEEE Milan PowerTech* (IEEE, Piscataway, 2019), pp. 1–6
45. J. Gama, R. Fernandes, R. Rocha, Decision trees for mining data streams. Intell. Data Anal. **10**(1), 23–45 (2006)
46. Y. LeCun, Y. Bengio, G. Hinton, Deep learning. Nature **521**(7553), 436 (2015)
47. V.G.T. da Costa, A.C.P. de Leon Ferreira, S. Barbon Jr., et al., Strict very fast decision tree: a memory conservative algorithm for data stream mining. Pattern Recogn. Lett. **116**, 22–28 (2018)
48. G. Le Ray, P. Pinson, Online adaptive clustering algorithm for load profiling. Sustain. Energy Grids Netw. **17**, 100181 (2019)
49. R.C. Qiu, Z. Hu, Z. Chen, N. Guo, R. Ranganathan, S. Hou, G. Zheng, Cognitive Radio Network for the smart grid: experimental system architecture, control algorithms, security, and microgrid testbed. IEEE Trans. Smart Grid **2**(4), 724–740 (2011)
50. J. Yan, H. He, X. Zhong, Y. Tang, Q-Learning-based vulnerability analysis of smart grid against sequential topology attacks. IEEE Trans. Inf. Forensics Secur. **12**(1), 200–210 (2016)
51. F. Cao, M. Estert, W. Qian, A. Zhou, Density-based clustering over an evolving data stream with noise, in *Proceedings of the 2006 SIAM International Conference on Data Mining* (SIAM, Philadelphia, 2006), pp. 328–339
52. D. Nussbaum, A. Agarwal, Scalability of parallel machines. Commun. ACM **34**(3), 57–61 (1991)
53. N.V. Chawla, K.W. Bowyer, L.O. Hall, W.P. Kegelmeyer, SMOTE: synthetic minority over-sampling technique. J. Artif. Intell. Res. **16**, 321–357 (2002)
54. R.G. Mantovani, A.L. Rossi, J. Vanschoren, B. Bischl, A.C. Carvalho, To tune or not to tune: recommending when to adjust SVM hyper-parameters via meta-learning, in *2015 International Joint Conference on Neural Networks (IJCNN)* (IEEE, Piscataway, 2015), pp. 1–8
55. J. Gama, I. Žliobaitė, A. Bifet, M. Pechenizkiy, A. Bouchachia, A survey on concept drift adaptation. ACM Comput. Surv. **46**(4), 44 (2014)
56. V.G.T. da Costa, E.J. Santana, J.F. Lopes, S. Barbon, Evaluating the four-way performance trade-off for stream classification, in *International Conference on Green, Pervasive, and Cloud Computing* (Springer, Berlin, 2019), pp. 3–17

57. A. Kurakin, I. Goodfellow, S. Bengio, Adversarial machine learning at scale (2016). Preprint. arXiv:1611.01236
58. R.S. de Carvalho, S. Mohagheghi, Analyzing impact of communication network topologies on reconfiguration of networked microgrids, impact of communication system on smart grid reliability, security and operation, in *2016 North American Power Symposium (NAPS)* (IEEE, Piscataway, 2016), pp. 1–6

Part III
Artificial Intelligence for Control of Smart Appliances and Power Systems

Chapter 10
Neurofuzzy Approach for Control of Smart Appliances for Implementing Demand Response in Price Directed Electricity Utilization

Miltiadis Alamaniotis and Iosif Papadakis Ktistakis

10.1 Introduction

Advancements in information technologies have led the efforts in the so-called fourth revolution of technology. Society, cities, cars, and homes—to name a few—are aspects of the modern world that have been significantly benefited and transformed by the use of the new wave of information technologies. The new and widely used terms of "smart grid," "smart city," and "smart homes" are coined to those advancements, and more particularly to the coupling of current infrastructure with information networks [1].

Of profound significance toward moving to a smart world is the use of artificial intelligence (AI). AI offers the necessary tools to automate the processes involved in processing data, extracting information, and making decisions [2]. Furthermore, AI compensates for physical constraints imposed in decision-making: for instance, humans cannot monitor incoming data and make decision 24/7—this is physically impossible. Driven by the above motives, AI is expected to infiltrate in various aspects of our human activities and improve the quality of our daily lives [3].

One of the domains that is expected to significantly benefit from the use of AI technologies is the energy domain. In particular, the use of AI has already significantly contributed to the development of smart grid technologies and will

M. Alamaniotis (✉)
Department of Electrical and Computer Engineering, University of Texas at San Antonio, San Antonio, TX, USA
e-mail: miltos.alamaniotis@utsa.edu

I. P. Ktistakis
ASML Inc., Wilton, CT, USA
e-mail: iosif.papadakis@wright.edu

M. Sayed-Mouchaweh (ed.), *Artificial Intelligence Techniques for a Scalable Energy Transition*, https://doi.org/10.1007/978-3-030-42726-9_10

play a significant role in the implementation and deployment of a truly smart grid in the near future [4]. One of the areas that will be greatly influenced by the use of artificial intelligence is the demand response programs. Smart grid technologies will enable the use of price signals for shaping the electricity demand in a way that satisfies the physical constraints of the delivery systems [5].

In the traditional delivery grid, demand was satisfied by controlling the generation, i.e. usually by operating as many generation units as needed to satisfy the demand. In demand response program the level of control has been shifted from generation units to consumers: demand response aspires in shaping the demand in such a way to meet the generation capacity [6, 7].

One of the ways to conduct demand response is by utilization of price signals. Market operator utilizes the available generation capacity in order to determine prices that will drive the consumers to reduce their demand at levels that meets the generated amount. These programs target the individual consumer's demand but in practice their aggregation is the one that is mainly shaped by the demand response program [8]. To make it clearer, the broadcasted prices are sent to all market consumers, which subsequently decide the amount of consumption that will be reduced in order for each consumer to stay within the limits of its desired budget [9]. In this example, though the individual consumers make decisions over their consumption, their aggregated behavior is the one observed by the system operator.

In this chapter, we present a new approach for demand response at the appliance level. We assume that the demand response of each consumer is determined by the operation of the electric appliances [10]. In particular, we assume that the consumer consumption is the sum of the individual appliance consumption. Therefore, controlling the consumption of each appliance impacts the overall consumption of the individual consumer and subsequently the power flow at the grid level. The proposed approach implements a neurofuzzy approach to control the amount of time that the appliance shall operate in the next duty cycle [11]. The neuro part is comprised of a neural network and more particularly by an extreme learning machine (ELM) [12], while the fuzzy part is comprised of a set of fuzzy rules [11]. The ELM is used for making predictions of the future electricity prices and then using this prediction together with the current conditions to make a decision over the operational time [13] in the next duty cycle of the appliance.

The structure of the paper is as follows. In the next section, the ELM is introduced and a brief introduction to fuzzy sets and inference is provided. Section 10.3 presents the neurofuzzy method, while Sect. 10.4 describes its application for demand response in smart appliances. The last section, i.e., Sect. 10.5, concludes the paper and highlights its main points.

10.2 Background

10.2.1 Extreme Learning Machines

Artificial neural network (ANN) is the widest used artificial intelligence tool [2]. ANN has been used in classification and regression problems, with numerous applications. The key "ingredient" in ANN is the utilization of a set of parameters known as weights. The ability of weights to be evaluated based on a set of training data (i.e., known outputs for known inputs) is the main strength of the ANN [11], given that they can be used to approach the underlying function that associates the inputs to the outputs. There are several types of neural networks, spanning a wide range of architectures and forms. One type of ANN, which attracts attention recently, is the extreme learning machine [14].

ELM is a special case of ANN comprised of two layers (it should be noted that the input layer is not considered as a computing layer). In particular, ELM is a feedforward neural network whose architecture contains a single hidden layer, an input and an output layer [14]. Based on that architecture, there are two sets of parameters in the ELM architecture: the set of weights between the input and the hidden layer, and the set of weights between the hidden layer and the output layer. For visualization purposes, a general architecture of ELM is provided in Fig. 10.1.

The essential feature of ELM that distinguishes it from the rest types of the neural networks is the way the weights are evaluated [15]. As it was mentioned above there are two types of weights in the ELM architecture, with each of them being evaluated based on a different procedure. Regarding the weights of the first layer (i.e., hidden layer in Fig. 10.1), a set of values is randomly assigned to the weights. With regard

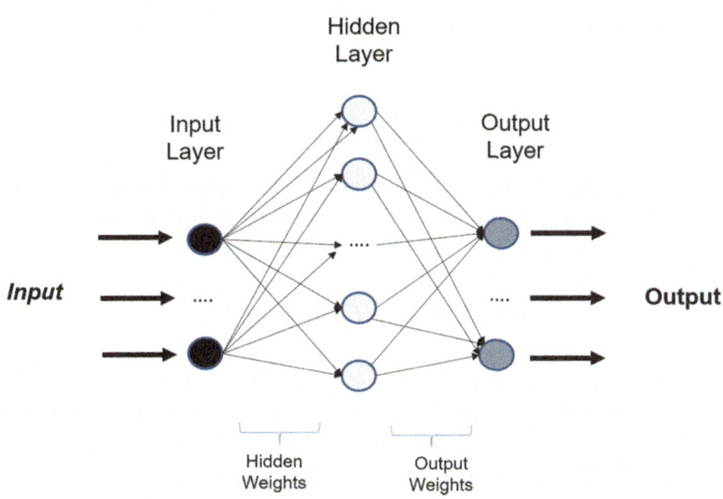

Fig. 10.1 General architecture of ELM

to the output weights (see Fig. 10.1), weight values are taken as the solution to a simple linear optimization problem. The objective function is evaluated based on the difference between the available training data and the output of the hidden layer. A detailed description of the ELM structure and training process may be found in [15] and (Tang et al. 2006).

Analytically, an ELM can be expressed as:

$$f(x) = \sum_{i=1}^{N} w_i h_i(x), \tag{10.1}$$

where w are the output weights, N is the number of hidden nodes, and $h_i()$ expresses the output of the hidden node i. Going further, the hidden layer output mapping gets the following vector form:

$$\mathbf{h}(\mathbf{x}) = [G(h_1(\mathbf{x}), \ldots, h_L(\mathbf{x}))], \tag{10.2}$$

where $\mathbf{x}$ is the input vector. Moreover, by assuming that there is an availability of L datapoints, then the hidden layer matrix becomes

$$\mathbf{H} = \begin{bmatrix} \mathbf{h}(\mathbf{x}_1) \\ \ldots \\ \mathbf{h}(\mathbf{x}_N) \end{bmatrix} = \begin{bmatrix} G(a_1, b_1, \mathbf{x}_1) & \ldots & G(a_N, b_N, \mathbf{x}_1) \\ \ldots & & \ldots \\ G(a_1, b_1, \mathbf{x}_L) & \ldots & G(a_N, b_N, \mathbf{x}_L) \end{bmatrix} \tag{10.3}$$

with α_i, b_i being the parameters of the hidden node i. By consolidating the training data targets in a matrix T, then it is obtained:

$$T = [\mathbf{t}_1 \ \ldots \ \mathbf{t}_L]^T \tag{10.4}$$

that can be used to set the training problem of the output layer. In particular, the evaluation of the output weights is the solution to the following problem:

$$\min : \|w\|_p^{\sigma_1} + C\|\mathbf{H}\beta - \mathrm{T}\|_q^{\sigma_2}, \tag{10.5}$$

where $\sigma_1 > 0$, $\sigma_2 > 0$, p, $q = 0$, 1/2, 1, …,+∞. The values of σ_1, σ_2, p, q define the type of problems—various combinations determined different problem—like regression, classification, clustering. By assuming a single hidden layer with sigmoid functions then the training process is reduced to:

$$\hat{Y} = \mathbf{W}_2\sigma\,(\mathbf{W}_1 x) \tag{10.6}$$

with $\mathbf{W}_1$ being the matrix of the hidden layer weights, σ the activation function, and $\mathbf{W}_2$ the output weights. The solution to (10.6) is taken by a least square fit—for known target in Y—with the solution taking the form:

$$\mathbf{W}_2 = \sigma(\mathbf{W}_1\mathbf{X})^{+}\mathbf{Y}, \tag{10.7}$$

where the symbol + denoted the pseudoinverse matrix of the quantity $\mathbf{W}_1\mathbf{X}$. In the case of ill-posed pseudoinverse matrix, resulting by noisy data, then a regularization techniques are used to secure that the pseudomatrix is not ill-posed [11].

At this point, it should be noted that the main strength of the ELM is its very short training time. This mainly results because the weights of the first layer are randomly evaluated and are not further changed. Hence, this very short training time is a characteristic that promotes ELM in real time application scenarios, especially at cases where adaptiveness—via training—is needed. Regarding the current work, it should be emphasized that the demand response at the level of appliance should be fast, and thus, making the use of ELM suitable for this type of applications.

10.2.2 Elements of Fuzzy Inference

In the classical set theory, an object in the world of the problem either belongs to a set or not. In other words, objects are imposed to a binary logic that can be analytically expressed by

$$f_A(x) = \begin{cases} 1, \ if \ \mathrm{x} \in A \\ 0, \ if \ \mathrm{x} \notin A \end{cases}, \tag{10.8}$$

where A denotes the set at hand, and x denotes the object. To make it clearer, classical set theory excludes the concurrent participation of an object in multiple sets. In practice, there are several complex problems that demand the participation of an object in more than one sets: hence, an object does not strictly belong to one set but to multiple ones at the same time [11].

Fuzzy logic comes to extend the classical set theory by allowing the partial participation of an object to more than one set. The essential idea is that every object belongs to a set with some degree [11]. The degree value is determined by the membership function as given below:

$$\mu_A(x) = \left\{ d, \ if \ \mathrm{x} \in A, \right. \tag{10.9}$$

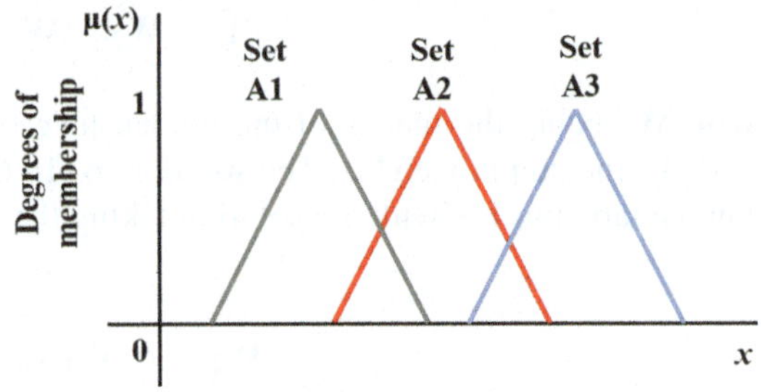

Fig. 10.2 Example of membership function for fuzzy sets A1, A2, and A3 of the variable *X*

where the degrees of membership take a value in the range:

$$d \in [0\ 1] \tag{10.10}$$

with 0 denoting the object does not belong to set *A*, and 1 that it fully belongs to the set *A*. The above schema allows an object to be a member with different degrees of membership to more than one set. A set that is expressed via a membership function (i.e., Eq. (10.9)) is called fuzzy set.

For visualization purposes, an example of fuzzy sets is given in Fig. 10.2, where it is clear that the membership functions of fuzzy sets are allowed to overlap. The overlapped parts assign the associated objects to more than one set with degrees of membership determined by the respective membership function. It should be noted that in the example of Fig. 10.2 the membership functions have a triangle shape, with the peak being equal to 1.

The fuzzy set representation allows the development of inference mechanism using empirical IF/THEN rules. In particular, spanning the input space with a group of fuzzy sets, and the output space with another group of fuzzy sets allows the correlation among the input–output sets using rules. For instance, if the input variable is spanned by three sets *I1*, *I2*, *I3* and the output with *O1*, *O2*, *O3*, respectively, then a possible example of fuzzy rules is the following:

- IF Input is *I1*, THEN Output is *O1*
- IF Input is *I2*, THEN Output is *O2*
- IF Input is *I3*, THEN Output is *O3*,

where the input is the condition and the output is the consequence. By using linguistics terms for inputs and output, then the fuzzy rules closely mimic the human way of thinking—i.e., association of linguistic terms via empirical rules—as given in the example here: IF Temperature is High, THEN Season is Summer. Furthermore, fuzzy logic allows the IF/THEN rules to have several conditions and several consequence parts. Multiple conditions and consequences are connected via the logical operators AND and OR depending on the semantics of the problem at hand [16]. It should be noted that in a fuzzy inference system there may be more than one rule that may be fired because of the overlap among the fuzzy sets. In that case, the output of the system is also a fuzzy set that goes under defuzzification. Defuzzification transforms a fuzzy set into a single value; the most popular method of defuzzifying a fuzzy set is the mean of area method [11], whose analytic formula is given by

$$y^{\mathrm{defuz}} = \frac{\sum_{n=1}^{N} y_n \mu_{\mathrm{out}}(y_n)}{\sum_{n=1}^{N} \mu_{\mathrm{out}}(y_n)} \tag{10.11}$$

with y_n being the elements of the output fuzzy set and N stands for the population of set elements.

Therefore, the association of two or more variables via fuzzy rules allows the development of fuzzy inference systems that are able to make decisions based on complex scenarios [17]. The use of fuzzy rules allows those decisions to be computational inexpensive and fast [18].

10.3 Methodology

Demand response programs target to driving consumers to morph their demand in such a way that is secure and safe for the power grid infrastructure [19]. Given that electricity demand is heavily associated with the number of electric appliances that operate at every time instance, then it is rational to approach the demand response as an appliance control problem [20]. Controlling the operation of each appliance in a given operational cycle, implicitly implements a demand response program [21]. The demand response program in a smart home can be seen as the aggregated operation of the set of smart appliances [22, 23].

In this work, we assume that each smart appliance can be autonomous and make optimal decisions over its operation [24]. Thus, each appliance utilizes the information from the smart grid and promotes decisions without any human intervention [25]. The proposed methodology is based on two basic principles:

- Demand response is connected to the electricity prices [26],
- Forecasting of prices is a mechanism that allows decision-making [27].

The proposed methodology implements a two-stage data processing process. In the first stage the ELM is utilized to make predictions over the variables of interest to the operation of the appliance, while the fuzzy inference utilizes the forecasting values as well as the current ones to make a decision over the operational time of the appliance for the next cycle [10]. The individual steps that comprise the proposed demand response methodology of an electric appliance are depicted in Fig. 10.3, where we can observe the individual steps taken to determine the operational cycle of the appliance.

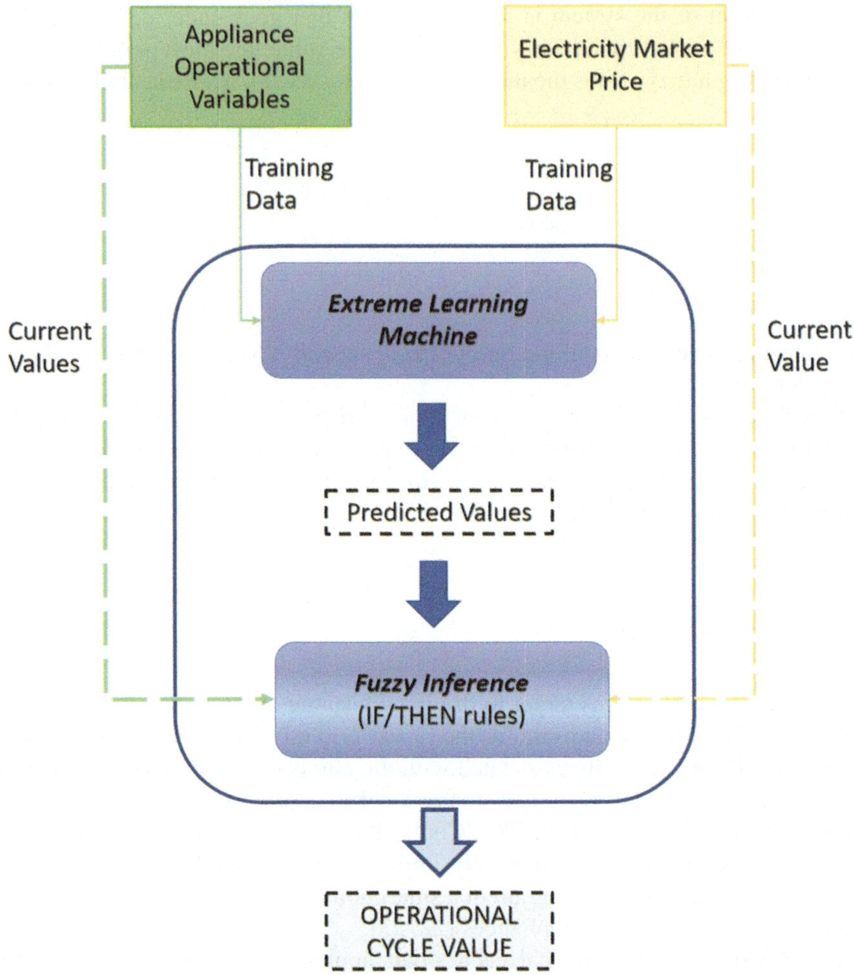

Fig. 10.3 Block sketch of the proposed neurofuzzy methodology for deciding the operational cycle of the appliance

The first step contains the determination of the operation variables of the electric appliance. This step is crucial in order to make the decision. For instance, the operational variable of an HVAC system refers to the determination of the variable temperature that expresses the temperature of the environment that needs to be controlled [28]. Another example may include the coffee maker and the operational variable is the temperature of the coffee that needs to be retained for an interval of time. Therefore, every appliance will determine its own set of operational variables.

Essential in our methodology is the values of the electricity market price. To that end, the ELM component provides an anticipation of the near future prices. The goal is to utilize the current price together with the anticipated ones to facilitate the

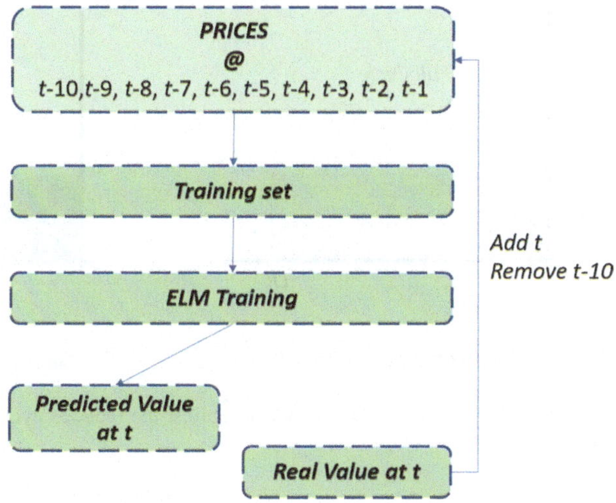

Fig. 10.4 ELM training process for price forecasting based on a rolling window of ten values

decision-making. It is a common sense that predicting whether the prices will be lower or higher in the near future is an essential factor in scheduling our electricity consumption. As seen in Fig. 10.3, ELM is utilized for price forecasting by using the most recent ten electricity prices as training set. Notably, the training dataset is based on a rolling window of length ten and therefore it is different at each time. The training process for ELM with regard to prices is depicted in Fig. 10.4.

The predicted values by ELM are forwarded to a fuzzy inference system together with the current operational values and the current price value. The goal of the fuzzy inference system is to utilize the predicted and the current values and provide a final decision over the operational time of the appliance for the next cycle. Therefore, the fuzzy inference system is comprised of two types of rules:

1. IF Variable(t) is A1, THEN *Operation Time* is B2
2. IF Variable(t) AND Variable($t + 1$) is A2, THEN *Operation Time* is B3

where we observe that the rule of type (1) utilizes only the current information, while rules of type (2) utilize the current as well as the predicted values. The number of rules depends on the electric appliance at hand and the demand response strategy that the user wants to follow.

The presented neurofuzzy system provides the operational time of the appliance for the next cycle. By assuming that the full length of an operational cycle of an appliance is denoted as OC^f then the neurofuzzy system aims at determining a value in the range $[0\ OC^f]$ as is shown in Fig. 10.5. The value defined by the neurofuzzy system is named *reduced operational cycle* and is denoted as OC^r.

Overall, the neurofuzzy system outputs the value OC^r that defines the length of the operational time of the appliance in the next cycle. The proposed methodology is mainly price driven given that the market electricity prices are taken into consideration to make a decision. However, the proposed methodology also considers the

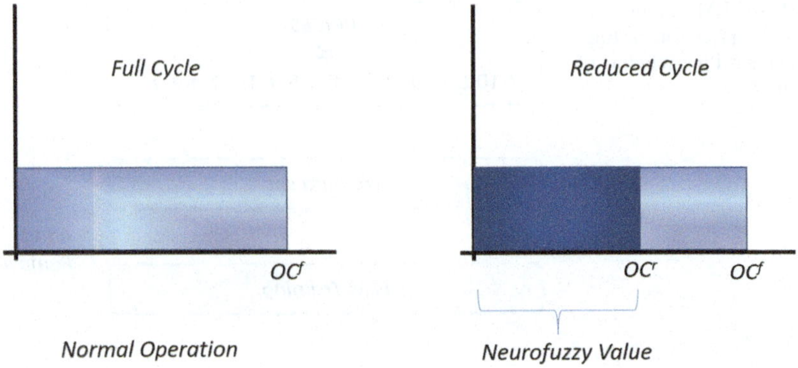

Fig. 10.5 Full operational cycle notion against the reduced cycle determined by the neurofuzzy method

future value of operational variables as part of its decision-making process. It should be noted that the ELMs for predicting the operational variables are already trained in the available data and they do not use rolling window as is the case for prices.

The final decision is made by a fuzzy inference mechanism, whose input contains the current and the predicted values by the ELM. Lastly, it should be noted that both ELM and fuzzy inference have very short execution times, thus making the presented methodology suitable for cases that decision can be made in very short-term time intervals.

10.4 Application to HVAC System

In this section, the presented methodology is applied for controlling an HVAC system. The HVAC system is deployed in a smart house and though the house digital infrastructure is connected to the information of the grid. Therefore, the HVAC system is able to receive the electricity price values at any time.

The HVAC data are simulated using the GridLAB-d software and the interface that has been developed by Nasiakou et al. [29]. The operational variables used for the HVAC are the following:

- Actual temperature
- Minimum desired temperature
- Maximum desired temperature
- Time for reaching the minimum desired temperature from the current temperature

where the last variable is a time driven variable and could not be simulated via GridLAB-D software. Furthermore, the "time for reaching the min desired value" depends on several randomly varying factors that cannot be accurately predicted

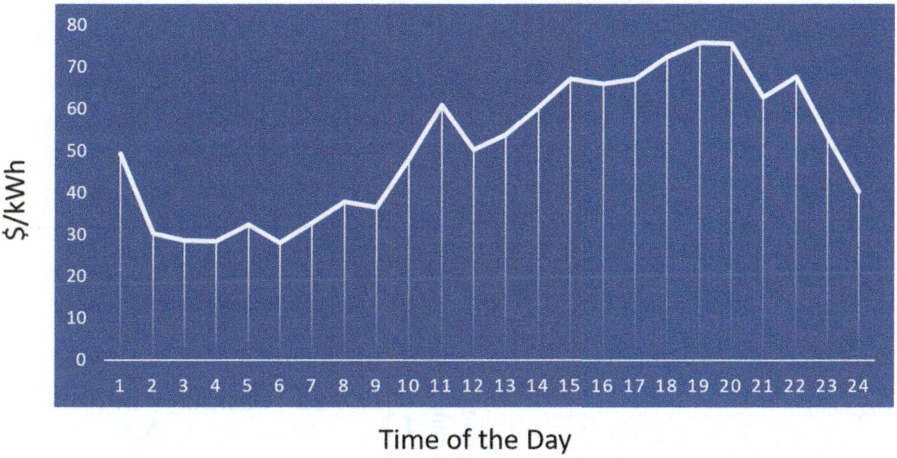

Fig. 10.6 Price signal utilized for testing the proposed demand response methodology for the HVAC system

or known a priori (e.g., leaving the room door open for long time). Therefore, to simulate data for this variable we created a randomizer by using Matlab. To that end, the training dataset was created based on the randomizer and contained a set of 300 values.

The price signal used in this manuscript for assessing the demand response cost benefit was randomly selected from the available data summer values from the New England ISO [30]. The resolution of the signal was taken in hourly intervals for a whole day (price signal changed every hour) and its curve is depicted in Fig. 10.6.

With regard to the neurofuzzy methodology we adopted 5 ELMs: The first four were assigned to one of the four operational variables, while the fifth was used for price forecasting. Their architecture was as follows:

- Operational parameter ELM
  - 1 input neuron
  - 10 hidden neurons
  - 1 output neuron
- Price signal ELM
  - 1 input neuron
  - 2 hidden neurons
  - 1 output neuron.

The training data for the operational variables (the first 4 ELMs) were obtained from the GridLAB-D and includes simulated HVAC every 5 min for 3 days—for the three parameters, namely, actual temperature, minimum desired temperature, and maximum desired temperature,—while the fourth variable training dataset was populated by the developed randomizer as we mentioned earlier.

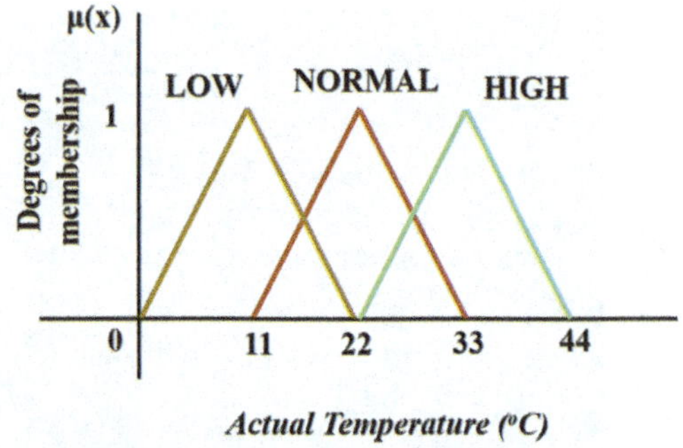

Fig. 10.7 Fuzzy sets for variable *actual temperature*

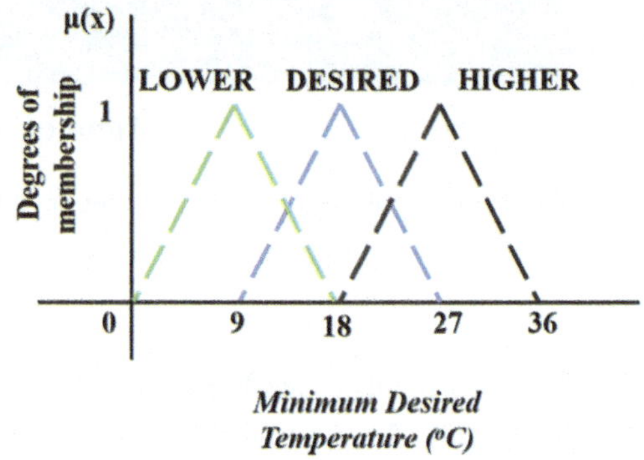

Fig. 10.8 Fuzzy sets for variable *minimum desired temperature*

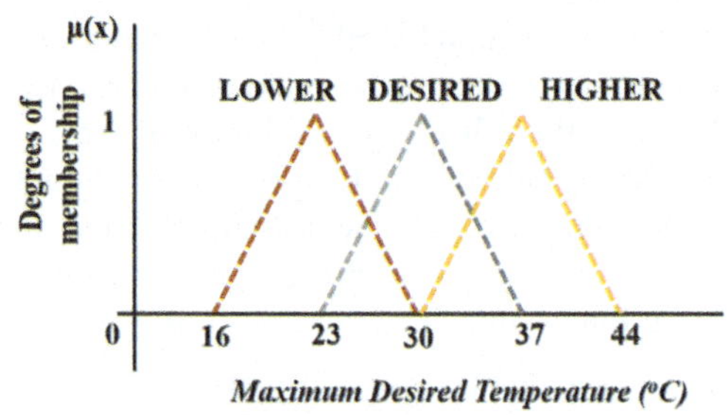

Fig. 10.9 Fuzzy sets for variable *maximum desired temperature*

The fuzzy inference mechanism gets ten inputs (the current values and the predicted values of the operational parameter and electricity price) and one output (i.e., the operational time of appliance). The fuzzy sets adopted for fuzzifying each of the operational parameters is given in Figs. 10.7, 10.8, 10.9, and 10.10, for price in Fig. 10.11 and for the output operational time in Fig. 10.12, respectively. It should be noted, that both the current and the predicted values use the same fuzzy sets.

In the current work, the HVAC cycle has been set to 8 min, and this is the reason that the operational time variable lies in the range of [0–8] min. However, this range can be modified based on the appliance and the current needs. The strength of fuzzy sets is that their range can be easily modified without affecting the computational cost of the system. Furthermore, it should be mentioned that the predicted values of the operational variables refer to the end of the cycle (every 8 min).

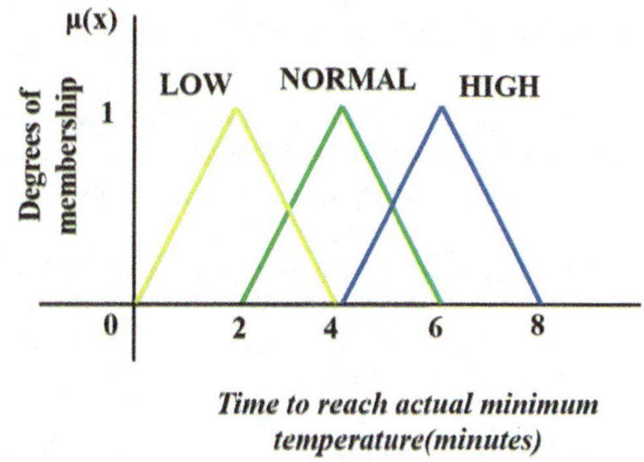

Fig. 10.10 Fuzzy sets for variable *time to reach minimum temperature*

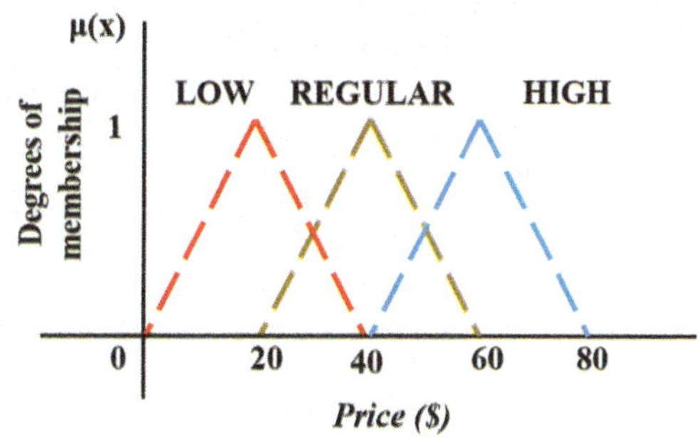

Fig. 10.11 Fuzzy sets for *electricity price*

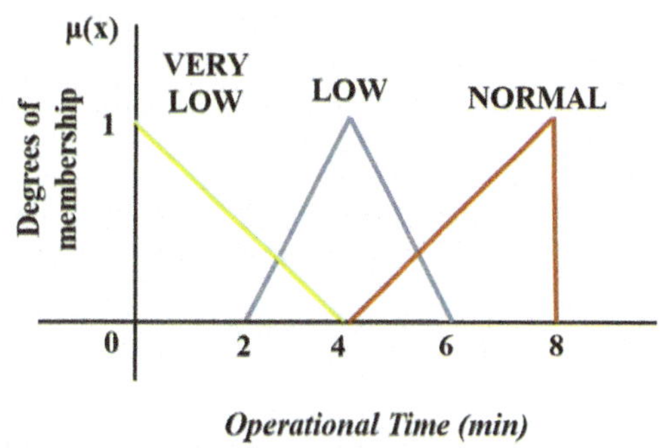

Fig. 10.12 Fuzzy sets for variable operational time (output value)

Regarding the fuzzy rules, the developed inference mechanism is comprised of 40 IF/THEN rules, a subset of which is provided in Table 10.1. The number of rules may vary depending on the appliance and the modeler's choices in demand respond strategy. The specific set of rules was developed based on our own experience and the response strategy we wanted our methodology to follow. Different modelers, which follow different response strategies, may derive different number and type of rules. For instance, the second rule is fired, when the both the current and predicted prices are between 0 and 20 dollars, and the min_temp is between 9 and 27 °C, providing as an output the fuzzy set VERY LOW. Then the defuzzification method will select a value from 0 to 4 min. In case more than one rule are fired then the output will be a fuzzy set that combines the respective output fuzzy sets using the

Table 10.1 Fuzzy inference rules

Rules
IF price is LOW AND predicted price is LOW, THEN operational time is NORMAL
IF price is LOW AND predicted price is LOW AND MIN_TEMP is DESIRED, THEN operational time is VERY LOW
IF price is LOW AND predicted price is HIGH AND MIN_TEMP is DESIRED, THEN operational time is LOW
IF price is HIGH AND predicted price is HIGH AND MIN_TEMP is LOWER AND ACTUAL_TEMP is HIGH, THEN operational time is LOW
IF price is HIGH AND predicted price is REGULAR AND MAX_TEMP is LOWER AND ACTUAL_TEMP is HIGH, THEN operational time is LOW
IF price is REGULAR AND predicted price is HIGH AND MIN_TEMP is NORMAL AND ACTUAL_TEMP is NORMAL, THEN operational time is VERY LOW
IF price is REGULAR AND predicted price is HIGH AND MIN_TEMP is NORMAL AND ACTUAL_TEMP is NORMAL, THEN operational time is VERY LOW
IF price is HIGH AND predicted price is HIGH AND predicted MIN_TEMP is NORMAL AND ACTUAL_TEMP is HIGH, THEN operational time is VERY LOW
IF price is REGULAR AND predicted price is LOW AND MIN_TEMP is NORMAL AND predicted MIN_TEMP is NORMAL, THEN operational time is VERY LOW
IF price is HIGH AND predicted price is HIGH AND MAX_TEMP is NORMAL AND TIME_to_MIN is LOW, THEN operational time is LOW
IF price is HIGH AND predicted price is REGULAR AND MIN_TEMP is LOWER AND Predicted_Max-Temp is NORMAL and ACTUAL_TEMP is HIGH and TIME_to_MIN is HIGH, THEN operational time is NORMAL

AND operator (as explained in previous section), and the defuzzification will select a value from the synthesized fuzzy set. If the synthesized set includes the VERY LOW and LOW sets, then the defuzzification will select a value between 0 and 6 min.

Next, the neurofuzzy methodology for the HVAC system is tested for 3 days. Based on the 8 min cycle that we have selected, for every day the appliance has to make 450 decisions. The values obtained for the 3 days are given in Fig. 10.13 in the form of curve that expresses OC^r against # of cycle.

The values of the operational cycles of the HVAC system presented in Fig. 10.13 provide economical benefit to the user. At this point, we need to mention that we compare the benefit attained by the neurofuzzy methods with two cases: (a) in the first case the HVAC operates fully (8 min) for all cycles (#450 cycles), and (b) in this case the HVAC operates fully in 50% of the cycles (#225 cycles). The results are

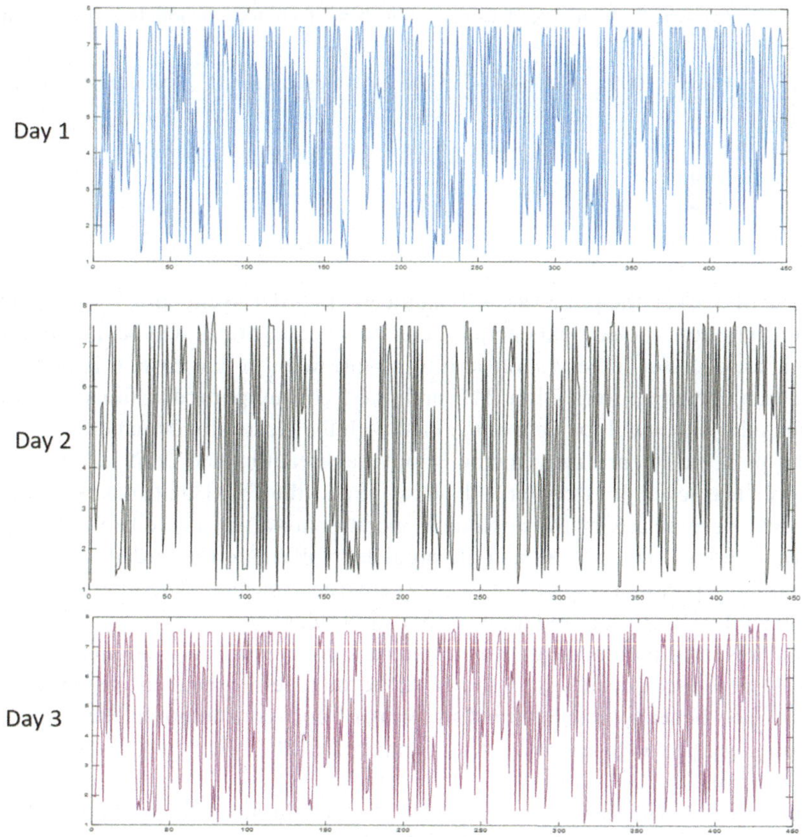

Fig. 10.13 Reduced operational cycle values for the three test days

provided in Table 10.2, where we observe an average of about \$6.5 reduction in cost of appliance operation by using the neurofuzzy methodology as compared to case 1 and an approximate reduction of \$2 compared to case 2. Furthermore, we observe that our methodology provided lower cost for all three tested cases. In addition, our methodology presented a more sophisticated operational way by adapting to current operation conditions; the latter was attained by considering several variables in determining the operational time as compared to the simple on/off for a whole cycle.

10.5 Conclusion

In this chapter, a new intelligent methodology for demand response at the electric appliance level was introduced. The methodology assumes that the electric appliance is part of a smart home that participates in a price directed electricity market. In

Table 10.2 Test results with regard to economic benefit attained (amounts expressed in US Dollars $)

# Day	Case 1 (100% cycles used)	Case 2 (50% cycles used)	Neurofuzzy method
Day 1	$ 19	$ 14	$ 13.5
Day 2	$ 19	$ 15.2	$ 12.1
Day 3	$ 19	$13.6	$ 12.9
Average	$ 19	$ 14.26	$ 12.8

this type of market, the price varies dynamically in order to encourage or discourage the consumption of electrical energy. In the ideal case, prices will be announced by the market operator in very short-term time intervals. In that case, it is impossible for human consumer to monitor 24/7 and make consumption decisions. Therefore, automated systems will be able to respond with the electricity demand.

The presented neurofuzzy method implements automated demand response at the appliance level. It is comprised of two parts: the ELM that makes prediction over the future values of the appliance operational values as well over the future electricity prices. The current and the predicted values are forwarded to the second methodology part that implements a fuzzy inference system. The inference is done with the aid of fuzzy sets and a set of IF/THEN rules.

The presented methodology was tested on simulated HVAC systems, comprised of 480 operational cycles, and shown to be more efficient than a simple on/off system. The benefit was quantified in terms of cost consumption: the neurofuzzy methodology provided lower cost as compared to the other two cases taken into consideration.

Future work will focus on extensive testing of the neurofuzzy methodology in other electrical appliances beyond the HVAC system. Furthermore, more advanced neural networks will be studied such as deep neural networks.

References

1. N. Bourbakis, L.H. Tsoukalas, M. Alamaniotis, R. Gao, K. Kerkman, Demos: a distributed model based on autonomous, intelligent agents with monitoring and anticipatory responses for energy management in smart cities. Int. J. Monit. Surveill. Technol. Res. **2**(4), 81–99 (2014)
2. S.J. Russell, P. Norvig, *Artificial Intelligence: A Modern Approach* (Pearson Education Limited, Malaysia, 2016)
3. T. Nam, T.A. Pardo, Conceptualizing smart city with dimensions of technology, people, and institutions, in *Proceedings of the 12th Annual International Digital Government Research Conference: Digital Government Innovation in Challenging Times* (ACM, New York, 2011), pp. 282–291
4. L.H. Tsoukalas, R. Gao, From smart grids to an energy internet: assumptions, architectures and requirements, in *2008 Third International Conference on Electric Utility Deregulation and Restructuring and Power Technologies* (IEEE, Piscataway, 2008), pp. 94–98

5. M. Alamaniotis, R. Gao, L.H. Tsoukalas, Towards an energy internet: a game-theoretic approach to price-directed energy utilization, in *International Conference on Energy-Efficient Computing and Networking* (Springer, Berlin, 2010, October), pp. 3–11
6. V. Chrysikou, M. Alamaniotis, L.H. Tsoukalas, A review of incentive based demand response methods in smart electricity grids. Int. J. Monit. Surveill. Technol. Res. **3**(4), 62–73 (2015)
7. P. Siano, Demand response and smart grids—A survey. Renew. Sust. Energ. Rev. **30**, 461–478 (2014)
8. M. Alamaniotis, Morphing to the mean approach of anticipated electricity demand in smart city partitions using citizen elasticities, in *2018 IEEE International Smart Cities Conference (ISC2)* (IEEE, Piscataway, 2018), pp. 1–7
9. M. Alamaniotis, N. Gatsis, L.H. Tsoukalas, Virtual budget: integration of electricity load and price anticipation for load morphing in price-directed energy utilization. Electr. Power Syst. Res. **158**, 284–296 (2018)
10. M. Alamaniotis, L.H. Tsoukalas, Multi-kernel anticipatory approach to intelligent control with application to load management of electrical appliances, in *2016 24th Mediterranean Conference on Control and Automation (MED)* (IEEE, Piscataway, 2016), pp. 1290–1295
11. L.H. Tsoukalas, R.E. Uhrig, *Fuzzy and Neural Approaches in Engineering* (Wiley, New York, 1996)
12. G.B. Huang, Q.Y. Zhu, C.K. Siew, Extreme learning machine: theory and applications. Neurocomputing **70**(1–3), 489–501 (2006)
13. M. Alamaniotis, A. Ikonomopoulos, A. Alamaniotis, D. Bargiotas, L.H. Tsoukalas, *Day-Ahead Electricity Price Forecasting Using Optimized Multiple-Regression of Relevance Vector Machines* (IET Conference Publications, Cagliari, 2012)
14. J. Tang, C. Deng, G.B. Huang, Extreme learning machine for multilayer perceptron. IEEE Trans. Neural Netw. Learn. Syst. **27**(4), 809–821 (2015)
15. G.B. Huang, Q.Y. Zhu, C.K. Siew, Extreme learning machine: a new learning scheme of feedforward neural networks. Neural Netw. **2**, 985–990 (2004)
16. T.J. Ross, *Fuzzy Logic with Engineering Applications* (Wiley, Chichester, 2005)
17. Z.S. Xu, Models for multiple attribute decision making with intuitionistic fuzzy information. Int. J. Uncertainty Fuzziness Knowledge-Based Syst. **15**(03), 285–297 (2007)
18. L.H. Tsoukalas, A. Ikonomopoulos, R.E. Uhrig, Neuro-fuzzy approaches to anticipatory control, in *Artificial Intelligence in Industrial Decision Making, Control and Automation* (Springer, Dordrecht, 1995), pp. 405–419
19. J.S. Vardakas, N. Zorba, C.V. Verikoukis, A survey on demand response programs in smart grids: Pricing methods and optimization algorithms. IEEE Commun. Surveys Tutorials **17**(1), 152–178 (2014)
20. M. Alamaniotis, I.P. Ktistakis, Fuzzy leaky bucket with application to coordinating smart appliances in smart homes, in *2018 IEEE 30th International Conference on Tools with Artificial Intelligence (ICTAI)* (IEEE, Piscataway, 2018), pp. 878–883
21. A.S. Meliopoulos, G. Cokkinides, R. Huang, E. Farantatos, S. Choi, Y. Lee, X. Yu, Smart grid technologies for autonomous operation and control. IEEE Trans. Smart Grid **2**(1), 1–10 (2011)
22. Z. Chen, L. Wu, Y. Fu, Real-time price-based demand response management for residential appliances via stochastic optimization and robust optimization. IEEE Trans. Smart Grid **3**(4), 1822–1831 (2012)
23. D. Setlhaolo, X. Xia, J. Zhang, Optimal scheduling of household appliances for demand response. Electr. Power Syst. Res. **116**, 24–28 (2014)
24. C.B. Kobus, E.A. Klaassen, R. Mugge, J.P. Schoormans, A real-life assessment on the effect of smart appliances for shifting households' electricity demand. Appl. Energy **147**, 335–343 (2015)
25. M. Alamaniotis, N. Gatsis, Evolutionary multi-objective cost and privacy driven load morphing in smart electricity grid partition. Energies **12**(13), 2470 (2019)
26. N.G. Paterakis, O. Erdinc, A.G. Bakirtzis, J.P. Catalão, Optimal household appliances scheduling under day-ahead pricing and load-shaping demand response strategies. IEEE Trans. Ind. Inform. **11**(6), 1509–1519 (2015)

27. M. Alamaniotis, L.H. Tsoukalas, N. Bourbakis, Virtual cost approach: electricity consumption scheduling for smart grids/cities in price-directed electricity markets, in *IISA 2014, the 5th International Conference on Information, Intelligence, Systems and Applications*, (IEEE, Piscataway, 2014), pp. 38–43
28. E. Tsoukalas, M. Vavalis, A. Nasiakou, R. Fainti, E. Houstis, G. Papavasilopoulos, et al., Towards next generation intelligent energy systems: design and simulations engines, in *IISA 2014, the 5th International Conference on Information, Intelligence, Systems and Applications* (IEEE, Piscataway, 2014), pp. 412–418
29. A. Nasiakou, M. Alamaniotis, L.H. Tsoukalas, MatGridGUI—A toolbox for GridLAB-D simulation platform, in *2016 7th International Conference on Information, Intelligence, Systems & Applications (IISA)* (IEEE, Piscataway, 2016), pp. 1–5
30. New England ISO (2019) [Online] Available: https://www.iso-ne.com/isoexpress/web/reports/pricing/-/tree/final-5min-lmp-by-node. Accessed Aug 2019

Chapter 11
Using Model-Based Reasoning for Self-Adaptive Control of Smart Battery Systems

Franz Wotawa

11.1 Introduction

Self-adaptive systems are characterized by their ability to change their behavior by themself during operation. Ordinary systems are designed to fit a certain purpose where requirements are used that describe how a system must behave considering boundaries between the system and its environment. All changes of the behavior of such systems are foreseen during system design and appropriately implemented. Hence, behavioral changes are usually deterministic relying on the system's state and its current input values obtained from the environment. In truly autonomous self-adaptive systems this needs not to be the case anymore. Such systems change their behavior in a way that might not be predicted during design in order to react on certain input stimuli. For such systems usually we are using a monitoring system for assuring that these systems' behavioral changes do not lead to an unwanted situation during operation.

In this chapter, we want to discuss the foundations behind self-adaptive and autonomous systems but this time taking an approach that limits potential behavioral changes. In particular, we focus on model-based reasoning systems that take a model of themself and utilize this model for behavioral changes. In this way, the system and its decision are well-informed and the final decision can be explained using the system model and the current set of inputs. Such systems can react on internal and external faults, i.e., faults related to internal components, the perceptions system, or the actuators, and faults related to environmental interactions not expected or not foreseen during system development. Such adaptive systems are becoming of increasing interest due to application areas like smart cities or autonomous driving.

F. Wotawa (✉)
CD Lab for Quality Assurance Methodologies for Autonomous Cyber-Physical Systems, Institute for Software Technology, Graz, Austria
e-mail: wotawa@ist.tugraz.at

M. Sayed-Mouchaweh (ed.), *Artificial Intelligence Techniques for a Scalable Energy Transition*, https://doi.org/10.1007/978-3-030-42726-9_11

In autonomous driving, for example, it is necessary not only to detect a failure during operation but also to react in a safe way. In ordinary cars with or without automated driving functions this means to give control back to the driver, which cannot be achieved when dealing with truly autonomous driving, where there is no driver behind a steering wheel anymore. In this case there must be an emergency procedure to be executed in order to find a safe state and to reach it. When considering hardware faults, such a procedure must be able to compensate a fault or ideally to repair it before continuing driving for reaching a place where the car might stop safely. Please note that there are many situations like driving through a tunnel where emergency braking and finally stopping the car can hardly be considered reaching a safe state.

When considering the domain of smart battery systems, we are interested in two possible adaptations required at runtime. The first is due to changes in the required electrical characteristics, i.e., the maximum required current or voltage such a system should deliver, or to change the configuration of the battery system in order to replace empty battery cells with charged ones. The second is due to faults that might happen during operation. The charge of the battery might drop and reach a not acceptable limit, or a battery that was consider recharged might not work because it reached its lifetime expectation. As a consequence we need to implement two different functionalities in a smart battery system, i.e., (1) the functionality of re-configuring itself, and (2) to diagnose itself. Using model-based reasoning we are able to provide both configuration and diagnosis functionality.

Although both functionalities can be implemented based on the same foundations, it is inevitable to use different models of the same system. For the smart battery system, we will introduce a model that allows configuring the battery based on electrical requirements and available battery cells. Furthermore, we will introduce another model that is capable of providing diagnosis capabilities and to identify the causes of an observed misbehavior. Hence, instead of requiring the implementation of two different functions, we only need to change the models and can re-use existing implementations of model-based reasoning systems.

The intention behind this chapter is to show how model-based reasoning can be applied to smart battery systems. We outline the foundations and also cite related literature that is important to implement model-based reasoning systems. However, the given explanations should be sufficient to introduce the basic ideas and necessary prerequisites.

This chapter is organized as follows: First, we are going to introduce a smart battery system where cells can be arranged in parallel or sequentially in order to deliver more current, more voltage, or both. In the description we focus on basic principles but ignore details like how to switch from one configuration to another, or details about electrical issues that have to be considered when coming up with a real implementation of a smart battery. Second, we introduce the foundations behind model-based reasoning, show how to use it for diagnosis, and adopt it for configuration. In addition, we show how to use model-based reasoning in the context of smart batteries, where we use the smart battery as case study for configuration and diagnosis. Afterwards, we discuss related research. Finally, we conclude this chapter.

11.2 Smart Battery Description

In this section, we outline the concept of a smart battery comprising an arbitrary size of re-chargeable batteries and other components for providing the required voltage and current. The objective behind the smart battery is to allow its reconfiguration to meet the specific needs during operation. For this purpose, we first have a look at the schematics of a base module (or cell) depicted in Fig. 11.1. A base module either has one battery or wire that can be switched for connecting two consecutive outputs from $x_1, \ldots, x_{k+1}$. The idea is that we are able to place a battery or wire in a stack at a position $1, \ldots, k$ for a fixed value of k allowing to obtain a higher output voltage. For example, if we have $k = 5$ and we place five batteries at each of the positions 1 to 5 connecting lines x_1, x_2 with the first battery up to x_5, x_6 with the 5th one, we obtain an output voltage between x_1 and x_6 of five times the voltage of each individual battery. If we want to obtain more power, we can set two or more batteries on each of the lines. In addition, the system enables to disconnect or re-charge batteries that are not used. Hence, such a system enables re-charging each battery separately.

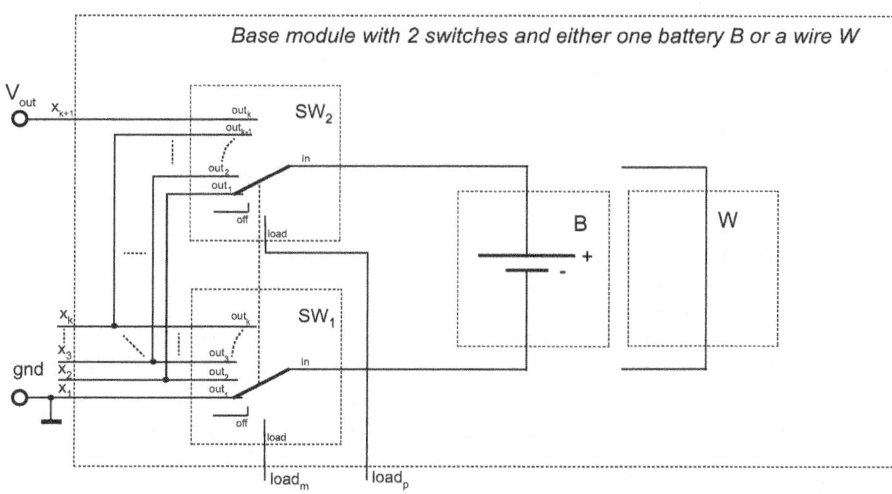

Fig. 11.1 The base module or cell for a smart battery stack of size k. Note that the base module can be either a battery cell or a wire cell. In the latter case we do not use the $load_m$ and $load_p$ ports, which are, therefore, not considered in any schematics representing a wire cell

Designing a smart battery, i.e., choosing the right value for the stack size k and the right number of batteries n, we have to consider the specification of a single battery used in a cell. The required specification includes at least the battery voltage v_B in Volt, its internal resistor value r_B in Ohm, and the available electrical power p_B in Ampere hours. The value of k can be determined by the maximum voltage v_{max} that should be delivered by the smart battery, i.e., $k = \frac{v_{max}}{v_B}$ rounded to the next larger integer value. The number of required batteries depends on the maximum current i_{max} to be requested and has to be computed using the internal resistor value and its corresponding voltage drop $i_{max} \cdot r_B$ that shall be allowed. If we allow a vd, e.g., 10%, voltage drop, the number of batteries to be put in parallel should be an integer value n_p larger than $\frac{(1-vd)\cdot v_B}{i_{max}\cdot r_B}$. The number of required batteries (and, therefore, cells) n has to be at least $n_p \cdot k$. It is worth noting that the number of batteries obtained should be increased also to allow re-charging batteries. Hence, the overall number of batteries n must be as large as required to deliver enough current and also to allow re-charging over time. In addition, we might also consider redundancies in order to cope with faults in a battery cell that should not be used anymore. Furthermore, we have also to add $k-1$ wire components (or wire cells) in order to allow the smart battery to deliver output voltage between v_B and $k \cdot v_B$.

In Fig. 11.2 we depict a smart battery comprising six battery cells, one wire cell, and having a stack size of 2, i.e., the maximum voltage is $2 \cdot v_B$. In the following, we discuss potential configurations of such a smart battery system. A configuration for a base module, i.e., a cell, corresponds to the value of the two switches, which are assumed to be synchronized. Hence, a cell can be in configuration (or state) off, $load$, or 1 to k. In case of configuration $1 \leq i \leq k$ the cell is connected with line x_i and x_{i+1}. The configuration of a smart battery system is a set of configuration for all the cells in the system. For example, for

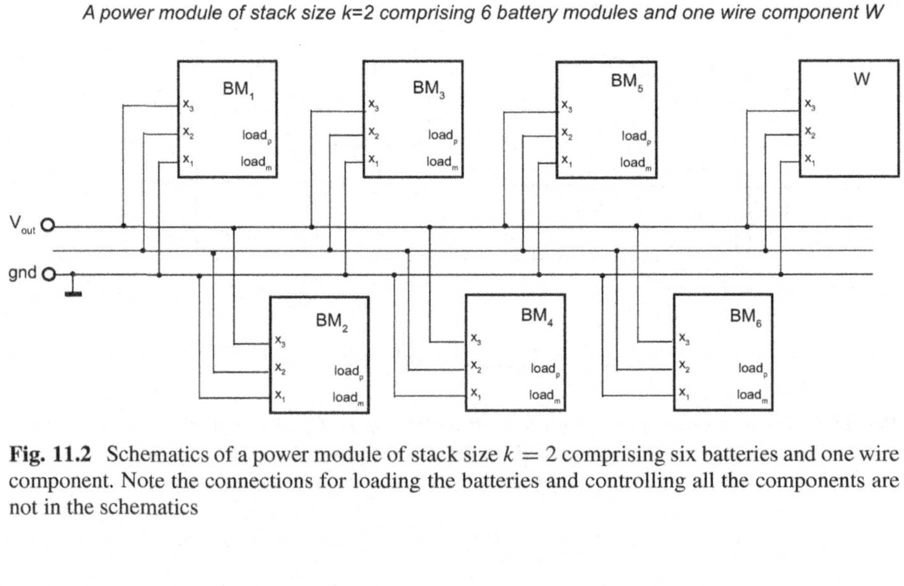

Fig. 11.2 Schematics of a power module of stack size $k = 2$ comprising six batteries and one wire component. Note the connections for loading the batteries and controlling all the components are not in the schematics

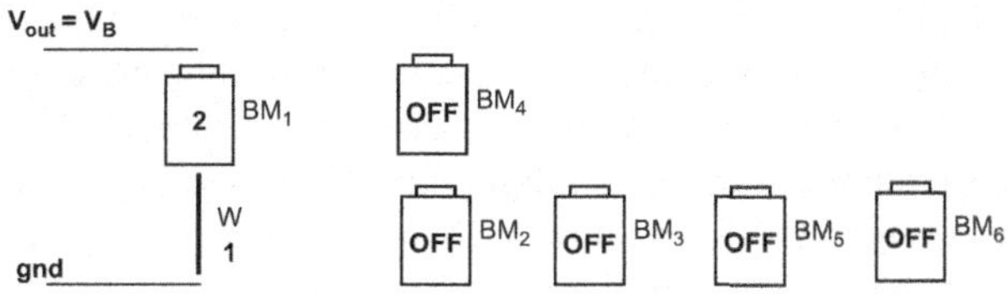

Fig. 11.3 Configuration 1: a valid configuration comprising battery cell BM_1 and the wire cell W with an output voltage of v_B

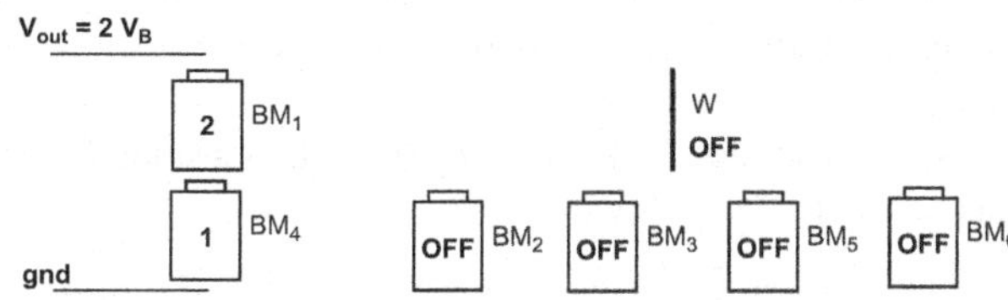

Fig. 11.4 Configuration 2: a valid configuration connecting battery cells BM_1 and BM_4 in series to deliver $2 \cdot v_B$ to the output

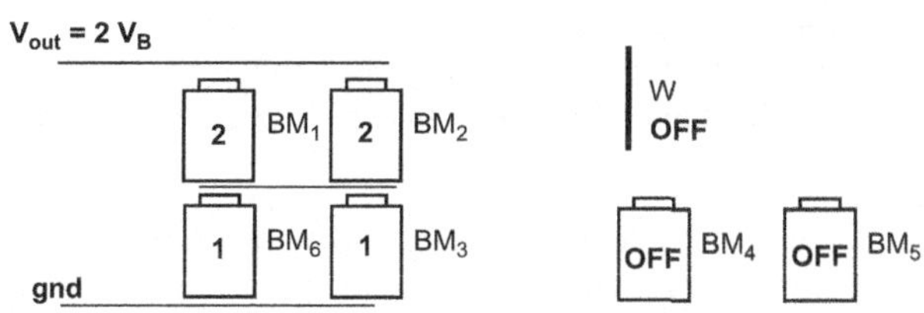

Fig. 11.5 Configuration 3: a valid configuration with two battery cells in parallel (BM_1, BM_2 and BM_6, BM_3) and both connected in series for delivering a higher current and an output voltage of $2 \cdot v_B$

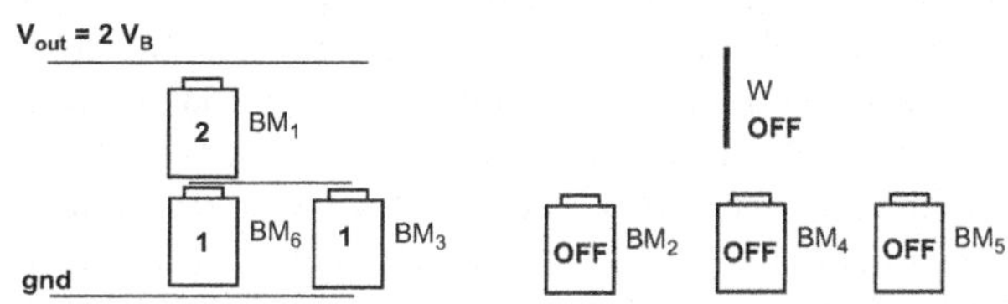

Fig. 11.6 Configuration 4: an invalid configuration that does not allow to deliver a higher current to the output

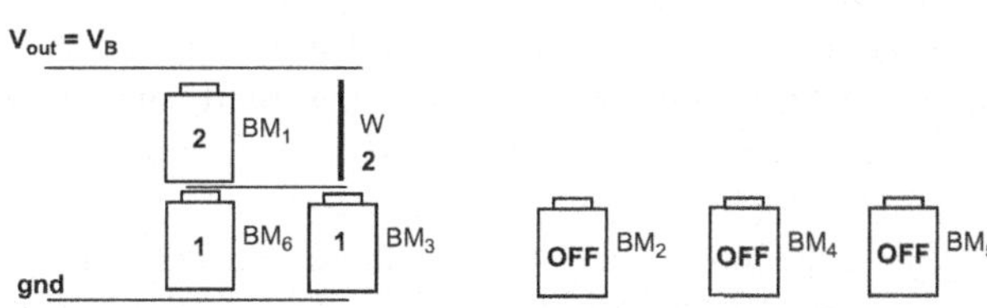

Fig. 11.7 Configuration 5: an invalid configuration causing battery cell BM_1 to be short circuited

the smart battery from Fig. 11.2 every configuration specifies a state for each of the 7 cells ($BM_1, \ldots, BM_6, W$). In Figs. 11.3, 11.4, 11.5, 11.6, and 11.7 we depict five different configuration. The first configuration shows a setting where we only have one battery cell attached delivering electrical power. The second configuration represents two batteries connected in series in order to increase the output voltage. The third configuration shows the same configuration but including batteries in parallel for increasing the current to be delivered to the output. The fourth and the fifth configuration represent invalid configurations that should not occur in practice. In Fig. 11.6 the configuration is missing one battery connected in parallel and cannot deliver more current as expected. The last configuration has a wire connected in parallel with a battery. This configuration would cause the battery to uncharge at the maximum current most likely leading to a severe damage of the battery cell, which cannot be used again. Hence, it is important to avoid configurations that are either not delivering specified properties or cause harm on side of batteries or the electronics.

As a consequence, the control module of a smart battery as outlined in this section has to assure the following:

1. Fulfill given electrical requirements like the maximum output current i_{out} and output voltage v_{out}.
2. Never connect wire cells in parallel with battery cells.
3. Always assure that the same number of battery cells are connected at each level of the stack.

Furthermore, we are interested in coming up with implementations of such a control module. In particular, we are going to discuss the use of model-based reasoning [8, 38] for obtaining valid configurations given a smart battery comprising a stack size of k, n battery cells, and $k - 1$ wire cells, and the given specification of the power to be delivered to the output. We make use of the smart battery system depicted in Fig. 11.2 to illustrate the approach.

In the following, we summarize the parameters, possible specifications, and system constraints given for configuring a smart battery system based on battery and wire cells. For the overall smart battery system we know the maximum stack size k, and the number of batteries n_B. Obviously, the number of battery cells n_B must be larger than or equal to k. For battery cells, we know their maximum supply voltage v_B, and current i_B. In order to use the full configuration space, i.e., all different output voltage levels, the number of wire cells must be equal to $k - 1$ allowing to have one and up to k batteries on the stack. Hence, a smart battery systems of stack size k has to have $b_1, \ldots, b_{n_B}$ battery cells, and $w_1, \ldots, w_{k-1}$ wire cells as base modules.

The behavior, i.e., the provided output voltage and current, depends on the batteries on the stack and the batteries in parallel, i.e., on the configuration of the system, and has to fulfill its requirements, i.e., the expected output voltage and current, which is limited to the following values: $v_{out} = m \cdot v_B$ for $m \in \{1, \ldots, k\}$ and $i_{out} = p \cdot i_B$ for $p \in \{1, \ldots, \lfloor \frac{n_B}{k} \rfloor\}$.[1] In order to formalize the behavior, we introduce a predicate *state* for cells indicating the state of the given cell. To distinguish battery cells from wire cells we use a function *bat* and *wire*,

[1]The function $\lfloor\rfloor$ rounds to the nearest smaller integer value.

respectively. Hence, $state(bat(b_1), 1)$ indicates that battery b_1 is in state 1, i.e., used in the first level of the battery stack. $state(wire(w_2), off)$ states that wire w_2 is not connected to any level of the stack, i.e., is in an off-state. A configuration now is an assignment of states to every battery and every wire where we have possible states from $D_B = \{1, \ldots, k, off, load\}$ for battery cells and $D_W = \{1, \ldots, k, off\}$ for wire cells. Formally, we write

$$\left\langle \begin{array}{c} state(bat(b_1, s_1)), \ldots, state(bat(b_{n_B}, s_{n_B})), \\ state(wire(w_1, t_1)), \ldots, wire(w_{k-1}, t_{k-1})) \end{array} \right\rangle$$

for a configuration where states $s_1, \ldots, s_{n_B}$ are from D_B and $t_1, \ldots, t_{k-1}$ from D_W.

In order to be valid a configuration has to fulfill constraints as already discussed. We now are able to state these constraints formally:

$$\begin{array}{c} v_{max} = m \cdot v_B \\ \leftrightarrow \\ \exists c_1, \ldots, c_m, c_{m+1}, \ldots, c_k : state(bat(c_1), l_1) \wedge \ldots \wedge state(bat(c_m), l_m) \wedge \\ state(wire(c_{m+1}), l_{m+1}) \wedge \ldots \wedge state(wire(c_k), l_k) \wedge \\ (\forall l_i, l_j \in \{1, \ldots, k\} : i \neq j \rightarrow l_i \neq l_j) \end{array} \tag{11.1}$$

$$\begin{array}{c} i_{max} = p \cdot i_B \\ \leftrightarrow \\ \forall l \in \{1, \ldots k\} : (state(bat(c_1^l), l) \wedge \ldots \wedge state(bat(c_p^l), l) \\ \vee state(wire(w_l), l)) \end{array} \tag{11.2}$$

$$\nexists l \in \{1, \ldots k\} : state((bat(c), l) \wedge state(wire(c'), l) \tag{11.3}$$

Equation (11.1) states that in a battery stack of size k we need m batteries and $k - m$ wires to assure that $v_{max} = m \cdot v_B$. Equation (11.2) is for stating that in each level of the stack where we have batteries, we need p of them to deliver the required current of $p \cdot i_B$. Equation (11.3) assures that in none of the levels of the stack we have wire and battery cells in parallel. Equations (11.1)–(11.3) are for assuring that the configured structure of the smart battery system provides the expected specification parameters. However, what is missing is assuring that the batteries themselves are able to provide their nominal voltage and current if working as expected, i.e., they are not empty. We formalize that a battery is empty using a predicate *empty* having the battery as parameter and being true if and only if the given battery cannot deliver electrical power for a predefined time.

$$\forall p \in \{1, \ldots k\} : \forall l \in \{1, \ldots k\} : state(bat(c_p^l), l) \rightarrow \neg empty(bat(c_p^l)) \tag{11.4}$$

In Eq. (11.4) we formalize this constraint stating that all batteries that are attached to the smart battery systems are not empty.

Computing a valid configuration for a given specification of the smart battery system is not trivial and has a high computational complexity. The whole search space is restricted by the number of cells and their particular domains of the configuration. Hence, the computational complexity of finding a valid configuration for a given specification is of order $O((k+2)^{n_B} + (k+1)^{k-1})$. For larger values of k the search for valid configuration becomes infeasible. However, we can simplify the search for a valid configuration a little bit. We only need to configure $m \times p$ batteries correctly. All $k - m$ levels on the stack where we do not have a battery can be filled with a wire. Hence, we are able to write a program selecting the batteries accordingly to the specification, and setting the wires afterwards. Such a program, however, is rather inflexible hardly able to handle cases where certain batteries have to be replaced or recharged during operation and are no longer available. We would need to capture all critical cases and have to verify that the battery system will always work as expected. The solution, we are going to discuss, avoids such troubles because it is delivering results only based on models and the available specification.

It is worth noting that in this section, we only have focussed on the configuration part of a smart battery system stating constraints for fulfilling certain (electrical) requirements. We have not discussed other issues like the concrete dynamic behavior, the necessity to provide reconfigurations over time in order to cope with batteries becoming empty, the re-charging of batteries so that we always have a certain number of spare batteries available, nor implementation details. The smart battery system as explained in this section, however, will be used as an example illustrating and explaining how AI methodology can be used for providing configurations of such a system.

In addition to the configuration challenge of dealing with finding the right setup of batteries for fulfilling given electrical requirements, we may face also a diagnosis problem during operation of such battery systems. For example, a battery run out of power when being used causing the given voltage to drop, or the required current cannot be delivered due to a faulty battery. Such a behavior can be identified using given sensor information, causing actions to bring the smart battery again into a good state. For this purpose, we have to identify first the cause of a detected misbehavior, i.e., the battery that is not working as expected anymore, and second, to re-configure the system using the health state of batteries as further information. This can be achieved by assigning faulty or empty batteries a certain fault state and constraining configuration search to only consider batteries that are not faulty.

Sensor in the context of diagnosis are used to obtain information, i.e., observations, about the current state of a system. Such sensor information for smart batteries may include measurements of voltage levels, currents, or even the temperature of a specific cell at particular points in time. From this information, we may obtain further estimates, e.g., the available power of a battery. For the smart battery case study, we make use of a simplified scenario comprising measuring the voltage drops between the different battery levels and the overall current delivered to the output at any particular time. In particular, we will use the scenario depicted in Fig. 11.8 where we consider two voltage sensors v_1 and v_2 and one current sensor i_{out}. Note that the name of the sensors correspond to the values they provide and we use these names when referring to their values.

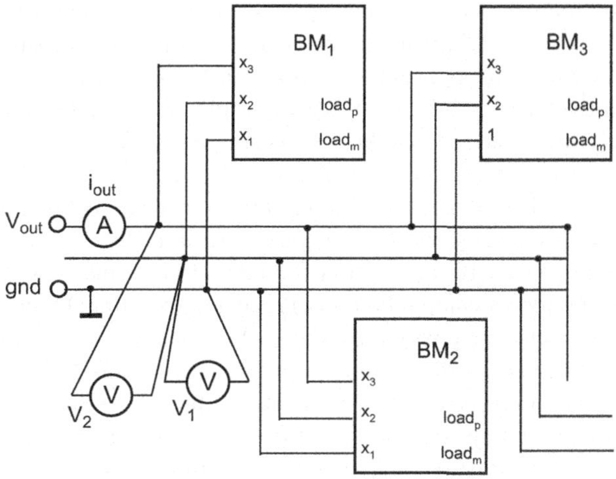

Fig. 11.8 The placement of the voltage and the current sensors (v_1, v_2, i_{out}) for our smart power module from Fig. 11.2

Obviously there must be some equations holding during operation in case the smart power module works as expected. For example, the measured current must be smaller or equal to the requested current, i.e., $i_{out} \leq i_{max}$. Because of using the same batteries we also assume that the measured voltages v_1 and v_2 are also always either 0 or equal to the voltage provided by a single battery v_B, i.e., $v_1 = 0 \vee v_1 = v_B$ and $v_2 = 0 \vee v_2 = v_B$. Finally, the sum of the two voltages measured must be equivalent to the voltage required, i.e., $v_1 + v_2 = v_{max}$. It is worth noting that in practice the equations would not hold because of differences in underlying component specification that has to fall within a known boundary. Hence, when implementing in practice equivalence usually is interpreted as being close within a predefined boundary. This can be achieved by mapping voltage values of 1.21 or 1.18 to the expected voltage level of 1.2 and use this value in an equation whenever needed. After introducing the basic foundations behind model-based reasoning in the next section, we will outline how we can use the behavior of the smart battery module components and the observations in order to compute diagnoses, i.e., the root causes of observed deviations between the expected behavior and the observed one.

11.3 Model-Based Reasoning

Before discussing the underlying principles behind model-based reasoning and its application to diagnosis and configuration, we discuss some necessary prerequisites. In particular, we introduce the basics behind first-order logic, we are going to use for modeling the systems and refer the interested reader to text books like [41] or [5] for more details. In first-order logic we have predicates p that basically represent either attributes, properties, or relationships between concepts. We might, for example, want to indicate that zero is an integer. This can be done using the predicate int having one parameter, i.e., $int(zero)$. Or we might want to say that Julia is taller than Bob using a predicate $taller$ taking two parameters, i.e., $taller(julia, bob)$. A predicate can be true or false in a particular world. For example, $taller(julia, bob)$ might be the case if there is a person Julia in this world who is taller than Bob.

In addition to predicates, we have logical operators like or ($\vee$), and ($\wedge$), not ($\neg$), implication ($\rightarrow$), or bi-implication (or logical equivalence $\leftrightarrow$). These basic operators have the same meaning than for classical propositional logic. Moreover, in first-order logic we also have variables like X. Note that we write variables starting with a capitalized letter to distinguish them from constants like $zero$, $julia$, or bob. Variables can—in principle—take arbitrary values. However, in most cases we restrict these values stating that they must be from a given set. Using variables in first-order logic we are able to come up with generalized rules. We may want to say that if a person A is taller than another person B who is also taller than a person C, then A must be taller than C. We can formalize this using quantifiers, i.e., either the for-all ($\forall$) or the exists ($\exists$) quantifier. The relationship between the size of persons can be formalized as follows:

$$\forall A : \forall B : \forall C : taller(A, B) \wedge taller(B, C) \rightarrow taller(A, C)$$

We might also write this in the form:

$$\forall A, B, C : taller(A, B) \wedge taller(B, C) \rightarrow taller(A, C)$$

Or alternatively, we might skip the $\forall$ part completely assuming all variables that are not having an exists quantification are all quantified. For example, let us formulate that a person that is taller than every other person is the tallest person:

$$\nexists B : taller(B, A) \rightarrow tallest(A)$$

In this formulae, we say that a person A is tallest if there exists no other person B that is taller.

We can use first-order logic for deriving new facts from known facts and rules. For example, we might now that $taller(julia, bob)$ and $taller(bob, jeff)$. Let us now consider the rule $taller(A, B) \wedge taller(B, C) \rightarrow taller(A, C)$. It would be great to use this rule and the known facts to derive that Julia should also be taller than Jeff. This can be done using the substitution of variables. If we “assign” $julia$ to A, bob to B, and $jeff$ to C, we come up with the rule $taller(julia, bob) \wedge taller(bob, jeff) \rightarrow taller(julia, jeff)$. From the semantics of the implication we know that if the left side is true, the right side must be true. We know that $taller(julia, bob) \wedge taller(bob, jeff)$ is true, from which we conclude that $taller(julia, jeff)$ must be true too. Hence, we are able to derive the fact $taller(julia, jeff)$ given the other facts, rules, and substitution.

Using first-order logic we are also able to derive contradictions, i.e., $\perp$. For example, we may measure the height of Julia and Jeff leading to the conclusion that Julia is not taller than Jeff anymore, i.e., $\neg taller(julia, jeff)$. When using the same facts as before and the new one, we obviously derive a contradiction because $\neg taller(julia, jeff) \wedge taller(julia, jeff)$ cannot be true at the same time. In this case, we say that the whole logical sentence

$$taller(julia, bob) \wedge taller(bob, jeff) \wedge \neg taller(julia, jeff) \wedge \\ taller(A, B) \wedge taller(B, C) \rightarrow taller(A, C)$$

is a contradiction or not satisfiable. In the following, we assume that there is a basic understanding of how to interpret first-order logic equations and that there exists theorem provers that allow to check for satisfiability or to derive new facts and knowledge. We are going to use first-order logic to write our models, i.e., representations of concrete systems, to be used for configuration and diagnosis.

11.3.1 Model-Based Diagnosis

The idea behind model-based reasoning arose in the 1980s of the last century. In AI expert systems based on logical rules were becoming more and more important and first issues emerged including reduced flexibility and maintainability of the used knowledge base. As an answer to these issues Davis and colleagues [9] introduced model-based diagnosis as a methodology for localizing root causes for a detected misbehavior based on the system structure and functionality. The methodology was further elaborated (see [8]) and formalized. Reiter [38], De Kleer and Williams [12], and later De Kleer and colleagues [11] provided a well founded theory behind model-based reasoning based on first-order logic. In Fig. 11.9 we depict the basic principles behind model-based reasoning (MBR). There we have the system on the left from which we obtain observations. On the right, we have a model of the system, which covers the structure and behavior of the system in a way such that we are able to deduce the expected behavior. When comparing the observations with the expected behavior, we may obtain deviations. These deviations together with the observations and the system model are used to compute diagnoses, i.e., components of the system that when faulty explain the deviations. We will see that this kind of reasoning can not only be applied to diagnosis but also to configuration.

Let us explain the basic idea and principle behind MBR using a simple example. In Fig. 11.10 we see a battery connected to two bulbs in parallel. From our experience, we know that both bulbs should be illuminated. Behind this prediction there are more or less hidden assumptions like that both batteries are working as expected and the battery itself delivers enough electricity. We often use such assumptions during the day. These assumptions can be classified as common sense knowledge that can be applied to a large extent when interacting with our environment. However, sometimes common sense cannot be applied anymore, e.g.,

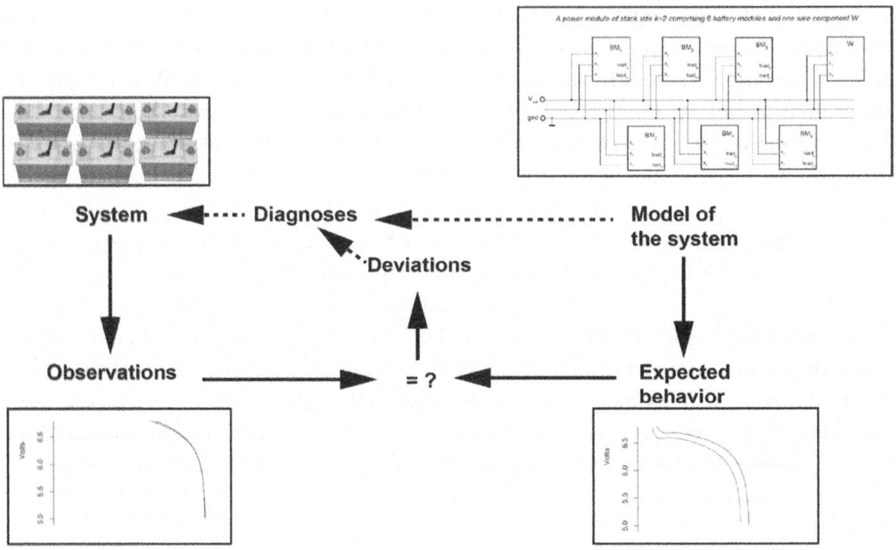

Fig. 11.9 The basic principle behind model-based reasoning

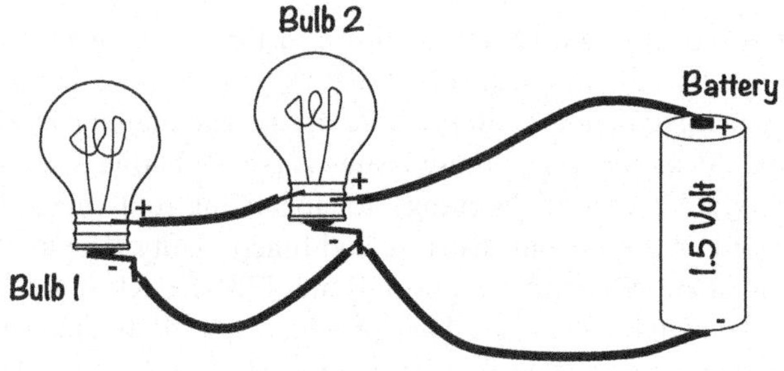

Fig. 11.10 A simple electrical circuit comprising two bulbs connected in parallel to a battery. There is a cable from the "+" pin of the battery to the "+" pin of the bulb and a cable connecting the "+" pins of the two bulbs. The other cables are connecting the "−" pins of the components in the same way

in case of faults. Let us consider that Bulb 1 in our example is glowing but Bulb 2 is not. In this case, obviously there is a contradiction with our common sense reasoning and we have to identify the reasons behind this unexpected behavior. How are we now able to find the cause behind this deviation from expectations? We are using the hidden assumptions. Assuming that Bulb 2 is not working perfectly explains why it is not glowing but Bulb 1 is. Assuming that the battery is empty, however, is not a valid solution, because in this case also Bulb 1 should not be illuminated. We know that a bulb can only glow in case of electricity applied.

So what can we take with us from the simple battery example? First, we are using basic physical principles, e.g., there must be electricity available to light a bulb. Second, we are relying on assumptions when providing predictions. These assumptions in case of diagnosis, describe the health state of components, e.g., a bulb is not broken. Third, we are more or less searching for a set of assumptions that may be retracted in order to explain a situation that contradicts our expectations. In the following, we discuss the basic definitions behind MBR, provide means for applying MBR for finding valid configurations, and show how MBR can be used in the context of our smart battery system.

We rely on Reiter's definitions of model-based diagnosis (MBD) [38] but modify some of them slightly. We start with defining a diagnosis problem, which comprises a model of the system, which is called a system description SD in the context of MBD, the components $COMP$ of a system, which can be considered working as expected or being faulty, and a set of observations OBS. In the system description, we provide a model of the system comprising the behavior of the components in case they are not faulty, and the structure of the system. The health state of a component is an assumption that is represented using a predicate. In the classical definitions of MBD, we use the predicate AB for stating a component to be abnormal or faulty. Hence, in SD we use the negation of AB, i.e., $\neg AB$ to specify a behavior of a component. For example, the behavior of the battery and the bulbs for our simple electrical circuit depicted in Fig. 11.10 can be stated using first-order logic as follows:

$$\neg AB(bat) \rightarrow electricity$$
$$\neg AB(bulb1) \wedge electricity \rightarrow glow(bulb1)$$
$$\neg AB(bulb2) \wedge electricity \rightarrow glow(bulb2)$$

In the first rule, we state that a battery bat is delivering $electricity$ in case of working appropriately. In the other two rules, we formalize that a bulb is glowing if it is working as expected and $electricity$ is available. These three rules are part of SD for the simple battery example. For this example, we have three components, i.e., $COMP = \{bat, bulb1, bulb2\}$. Observations OBS for this example might be $\{glow(bulb1), glow(bulb2)\}$ saying that both bulbs are lighting, or $\{glow(bulb1), \neg glow(bulb2)\}$ where the first bulb is lighting but the second is not. For the latter case we have a diagnosis problem, i.e., we want to identify the causes behind the observation that one bulb is working but the other is not. In MBD this means identifying a set of components that have to be assumed as not working correctly in order to explain observations. Formally, we are able to formulate the **diagnosis problem** as follows:

Definition 11.1 (Diagnosis Problem) A tuple $(SD, COMP, OBS)$ states a diagnosis problem where SD is the system description, $COMP$ is the set of system components, and OBS is a set of observations.

A solution for a diagnosis problem is a set of components explaining the given observations. In case of MBD, explaining observations is mapped to finding components that when assumed to behave incorrect remove all inconsistencies with the given observations. Components that are not in this set are assumed to work as expected. In the following definition, we state solutions for a diagnosis problem, i.e., the **diagnoses**, formally.

Definition 11.2 (Diagnosis) Given a diagnosis problem $(SD, COMP, OBS)$. A set $\Delta \subseteq COMP$ is a diagnosis (a solution) if and only if $SD \cup OBS \cup \{\neg AB(c) | c \in COMP \setminus \Delta\} \cup \{AB(c) | c \in \Delta\}$ is satisfiable (i.e., consistent).

In this definition $SD \cup OBS$ is a logical sentence comprising the model of the system, i.e., SD, and the observations OBS, $\{\neg AB(c) | c \in COMP \setminus \Delta\}$ is a logical sentence stating that all components that are not in the diagnosis have to work as expected, and $\{AB(c) | c \in \Delta\}$ is a logical sentence saying that all components in the diagnosis are behaving in an abnormal way. If this sentence is consistent, then Δ is a diagnosis, explaining which components to replace in order to retract all inconsistencies.

Let us consider our simple battery example again. Let SD be $\{\neg AB(bat) \rightarrow electricity, \neg AB(bulb1) \wedge electricity \rightarrow glow(bulb1), \neg AB(bulb2) \wedge electricity \rightarrow glow(bulb2)\}$, $COMP$ be $\{glow(bulb1), glow(bulb2)\}$, and OBS be $\{glow(bulb1), \neg glow(bulb2)\}$. Obviously assuming all components to be working as expected leads to an inconsistency. In this case $\Delta = \emptyset$ and we have $\{\neg AB(bat), \neg AB(bulb1), \neg AB(bulb2)\}$. From $\neg AB(bat)$ and $\neg AB(bat) \rightarrow electricity$ we derive that there must be an electricity, i.e., $electricity$ is true. From this fact together with $\neg AB(bulb2)$, and the rule $\neg AB(bulb2) \wedge electricity \rightarrow glow(bulb2)$ we are able to derive $glow(bulb2)$, which obviously contradicts the observation $\neg glow(bulb2)$. Therefore, $\Delta = \emptyset$ cannot be a diagnosis. When assuming $\Delta = \{bulb2\}$, there is no contradiction anymore because the rule $\neg AB(bulb2) \wedge electricity \rightarrow glow(bulb2)$ can no longer be applied, because of $AB(bulb2)$. As a consequence $\{bulb2\}$ is a diagnosis.

Unfortunately, when using this definition we also are able to identify $\{bat\}$, or $\{bulb1, bulb2\}$ among other subsets of $COMP$ as diagnoses, which seems to be counterintuitive. The first set $\{bat\}$ is a diagnosis that is due to the fact that our model does not capture the knowledge that a bulb can only glow if there is electricity attached. If we add the following two rules to our model SD, we would not be able to derive such a diagnosis anymore:

$$glow(bulb1) \rightarrow electricity$$
$$glow(bulb2) \rightarrow electricity$$

In this case, the observation $glow(bulb1)$ allows to derive $electricity$, from which we obtain a conflict using rule $\neg AB(bulb2) \wedge electricity \rightarrow glow(bulb2)$, the assumption $\neg AB(bulb2)$, and the observation $\neg glow(bulb2)$. In literature there are a few papers describing how to improve models considering knowledge about physical necessity or impossibilities (see e.g. [16]) for eliminating counterintuitive diagnoses.

The second diagnosis $\{bulb1, bulb2\}$ cannot be handled that easily. First of all, from the definition of diagnosis (Definition 11.2) and assuming that models are only considering the behavior of healthy components (of the form $\neg AB(c) \wedge \ldots \rightarrow \ldots$), it easy to show that all supersets of diagnoses have to be diagnoses as well. Hence, we either allow also to specify the behavior in case of a fault, e.g., via stating that an abnormally behaving bulb $bulb_x$ can never glow $\neg(AB(bulb_x) \wedge glow(bulb_x))$, or to focus on parsimonious diagnoses, e.g., diagnoses that are smaller. Usually in MBD we define minimal diagnoses as follows:

Definition 11.3 (Minimal Diagnosis) Given a diagnosis problem $(SD, COMP, OBS)$. A diagnosis Δ for $(SD, COMP, OBS)$ is said to be minimal (or parsimonious) if and only if there is no other diagnosis Δ' of the same diagnosis problem that is a subset of Δ, i.e., $minimal(\Delta) \leftrightarrow \nexists \Delta' \subset \Delta \wedge diagnosis(\Delta')$.

For our example, it can be easily seen that $\{bulb1, bulb2\}$ is not a minimal diagnosis anymore because $\{bulb2\}$ itself is a diagnosis, which is also a minimal diagnosis.

Can we apply MBD directly to configure smart battery systems? Unfortunately not, because for configuration we do not have a correct behavior of a certain component but a set of possible states instead. Therefore, we have to slightly adapt the definitions of MBD for the case of configuration. For more information about the formalization of fault models used for diagnosis we refer the interested reader to [43] and [11]. For an overview of different MBD approaches readers might consult [54]. In the next subsection, we discuss how MBR can be used for configuration purposes.

11.3.2 Model-Based Configuration

The idea of using MBR and in particular MBD for configuring systems is not new. [7] and later [45] discussed the relationship between reconfiguration and diagnosis. [15] provided insights for utilizing MBD in case of configuration. In this section, we follow this previous research and adapt the definitions already discussed in order to make them applicable for solving the configuration task. There has been a lot of research in the area of automated configuration starting with [29], where rule-based expert systems were used. Later representation languages for configuration problems were introduced (see e.g. [30]) aiming at increasing expressiveness of configuration knowledge, leading to representations based on description logics [55], or general constraint satisfaction problems (CSPs) [20] including extensions like Dynamic [18] or Generative CSPs [44, 46].

In this paper, we restrict configuration to parameter configuration where we assume a fixed number of parameters for a system to be configured. Similar to the diagnosis problem, we introduce a configuration problem comprising a configuration knowledge base KB, system parameters PAR, and requirements REQ. The rules in the knowledge base KB formalize the relationships between parameters, and restrictions. The requirements REQ are for stating the expected capabilities of a system. To illustrate parameter configuration, we make again use of our simple electrical circuit depicted in Fig. 11.10. In contrast to diagnosis, configuration deals with identifying components and parameters such that the final system fulfills its requirements. We illustrate the configuration problem for the two-bulbs example in Fig. 11.11, where we depict a component library from which components are selected in order to come up with a configured system. For the two-bulb circuit, we have to identify the correct bulbs that fit to the battery. For example, we may distinguish four different bulbs two with a nominal voltage of 1.5 V, one with 12 V, and one with 220 V. Each of the bulbs has a different electrical characteristic like its power consumption, the underlying technology among other factors. We assume that we only know the required power for each bulb. In addition, we may have different batteries again with a different voltage and also with a different nominal maximum current to be provided (Table 11.1).

When putting together two bulbs in parallel with a single battery, we have to assure that the voltage levels are the same and that the sum of the current needed to light the bulbs is less than the maximum current provided by the battery. From physics we also know that electrical power is the product of voltage and current, i.e., $P = V \cdot I$. From these verbal requirements we are able to come up with the following constraints representing the requirements formally where B_1, B_2 are the selected bulbs and Bat represents the selected battery.

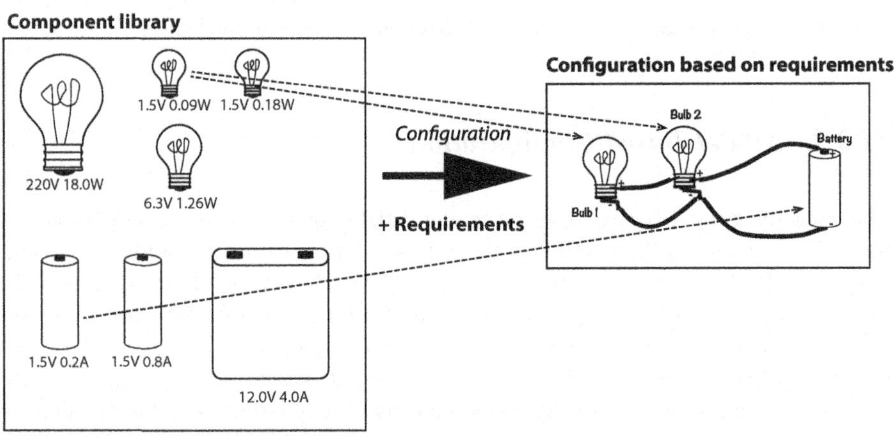

Fig. 11.11 Configuring the two-bulbs example

Table 11.1 Available bulbs and batteries to be used to design the two-bulb circuit from Fig. 11.10

Component	Parameters
$bulb1$	$v_B = 1.5\,\text{V}, p_B = 0.09\,\text{W}$
$bulb2$	$v_B = 1.5\,\text{V}, p_B = 0.18\,\text{W}$
$bulb3$	$v_B = 6.3\,\text{V}, p_B = 1.26\,\text{W}$
$bulb4$	$v_B = 220\,\text{V}, p_B = 18\,\text{W}$
$bat1$	$v_{Bat} = 1.5\,\text{V}, i_{Bat} = 0.2\,\text{A}$
$bat2$	$v_{Bat} = 1.5\,\text{V}, i_{Bat} = 0.8\,\text{A}$
$bat3$	$v_{Bat} = 12\,\text{V}, i_{Bat} = 4\,\text{A}$

$$v_{B_1} = v_{B_2} = v_{Bat}$$

$$\frac{p_{B_1}}{v_{B_1}} + \frac{p_{B_1}}{v_{B_1}} \leq i_{Bat}$$

Using the available knowledge we are able to construct the circuit using many different types of bulbs and batteries. Using both bulbs of type *bulb*1 and a battery of type *bat*1 obviously fulfills all specified requirements. The same holds for selecting both bulbs of type *bulb*2 and the battery of type *bat*2. In the latter case the higher required current is compensated by the higher current of the battery. In addition, we might select one bulb of type *bulb*1 and one from *bulb*2 together with a battery of type *bat*1. Also in this case the constraints are fulfilled. For practical purposes it is often the case to have additional constraints like selecting each bulb from the same type. We now formally define configuration problems and their solutions.

Definition 11.4 (Parameter Configuration Problem (PCP)) A tuple (KB, PAR, τ, REQ) states a parameter configuration problem where KB is a knowledge base describing general knowledge of the system to be configured and its parameters, PAR is a set of parameters to be configured, τ is a function mapping parameters to their potential values, i.e., the parameter domain, and REQ is a set of requirements the configured system has to fulfill.

Our small configuration example can be specified using a PCP as follows:

$$KB = \left\{ \begin{array}{c} type(X, bulb1) \rightarrow (v_X = 1.5\,\text{V} \wedge p_X = 0.09\,\text{W}), \\ type(X, bulb2) \rightarrow (v_X = 1.5\,\text{V} \wedge p_X = 0.18\,\text{W}), \\ type(X, bulb3) \rightarrow (v_X = 6.3\,\text{V} \wedge p_X = 1.26\,\text{W}), \\ type(X, bulb4) \rightarrow (v_X = 220\,\text{V} \wedge p_X = 18\,\text{W}), \\ type(X, bat1) \rightarrow (v_X = 1.5\,\text{V} \wedge i_X = 0.2\,\text{A}), \\ type(X, bat2) \rightarrow (v_X = 1.5\,\text{V} \wedge i_X = 0.8\,\text{A}), \\ type(X, bat3) \rightarrow (v_X = 12\,\text{V} \wedge i_X = 4\,\text{A}), \end{array} \right\}$$

$$PAR = \{B_1, B_2, Bat\}$$

$$\tau(B_1) = \tau(B_2) = \{bulb1, bulb2, bulb3, bulb4\} \text{ and } \tau(Bat) = \{bat1, bat2, bat3\}$$

$$REQ = \left\{ v_{B_1} = v_{B_2}, v_{B_2} = v_{Bat}, \frac{p_{B_1}}{v_{B_1}} + \frac{p_{B_1}}{v_{B_1}} \leq i_{Bat} \right\}$$

A solution of a PCP is an assignment of values for each parameter such that the knowledge base together with the requirements are not in contradiction.

Definition 11.5 (Configuration) Given a PCP (KB, PAR, τ, REQ). A function Δ mapping a value of $\tau(p)$ to each parameter $p \in PAR$ is a configuration, i.e., a solution of the PCP, if and only if $KB \cup REQ \cup \{type(p, \Delta(p))|p \in PAR\}$ is consistent.

Obviously the definition of diagnosis is similar to the definition of configuration. However, instead of only distinguishing correct from incorrect behavior using $\neg AB$ and AB, respectively, we now allow to have different values, i.e., the one specified in τ. In the above definition of configuration, we assume that all parameters must have an assigned value from their domains. Note that in the example used to illustrate PCP we specify system specific aspects using REQ. All other parts of the knowledge are given in KB. In KB we might also specify rules that implement further restrictions or necessities. One restriction might be that in case of specific battery type we might not be able to choose all bulbs. Or we may want to say that in case we have a certain component, we need to select the type of another component.

When using the definition of configuration, we are able to come up with the same solutions than already discussed for our two-bulb example. For example, $\Delta(B_1) = \Delta(B_2) = bulb1, \Delta(Bat) = bat1$ is a configuration according to Definition 11.5 but $\Delta(B_1) = \Delta(B_2) = bulb2, \Delta(Bat) = bat1$ is not. In the latter case, the second requirement is violated. Computing configurations can be easily done using constraint solving. In this case, a configuration problem or specifically a PCP is represented as a CSP. For more details about constraint solving have a look at Rina Dechter's book introducing the foundations behind CSPs [10].

11.3.3 Summarizing and Discussing Model-Based Reasoning

Model-based reasoning can be applied for a variety of tasks including diagnosis, i.e., the identification of root causes of a detected misbehavior, and configuration. In all these cases, the outcome of reasoning relies on the provided model. Faults in the model of course may lead to wrong diagnoses or configurations. The granularity of the model, i.e., the components or parameters represented in the model, influences the obtained results as well. If we map several physical components of a system to one component in the corresponding model, we are not able to distinguish the physical components in diagnosis. Hence, modeling is the most important part when applying model-based reasoning in practice. The outcome of model-based reasoning is only good if the model captures the essential part of system's behavior.

It is also worth noting that model-based diagnosis as introduced in Definition 11.2 basically is not restricted to a certain type of fault. In the introduced definition only the correct behavior of a component is used for determining root causes, i.e., diagnoses. Hence, model-based diagnosis can handle faults occurring in operation that have not been seen before in a smart way. Diagnosis approaches that rely on coding faults and their consequences can hardly deal with such situations. However, using fault models may reduce the number of diagnoses and provides more information, i.e., the faulty components and as well the reasons behind the faulty behavior. Let us illustrate the differences between model-based diagnosis without and with fault models using an example. Consider a digital circuit comprising basic logical gates like and-gates or inverters. When dealing with fault models we may define what happens in case of a stuck at 1 fault. If we implement diagnosis relying on stuck at 1 faults only, we may not be able to handle a case well where we need stuck at 0 faults. This is not the case when relying on modeling the correct behavior alone, where we can deal with stuck at 1 and stuck at 0 faults.

In addition, it might be not feasible to really explicitly consider all fault cases. For example, let us assume to have a program that given all observations computes potential root causes. For our two-bulbs example from Fig. 11.10 we would have to come up with a program that looks as follows:

Explicitly Computing Diagnoses for the Two-Bulbs Example

```
if (glow(bulb1)) {
   if (glow(bulb2)) {
      return "all components correct";
   } else {
      return "bulb2 is broken";
   }
} else {
   if (glow(bulb2)) {
      return "bulb1 is broken";
   } else {
      return "bat is empty or bulb1 and bulb2
       are broken";
   }
}
```

Using such programs for smaller systems might be appropriate. However, when considering larger systems like 100 bulbs in parallel, handling all cases explicitly is not feasible and someone might not handle more unlikely cases. When using model-based reasoning all the information about which components might contribute to a certain behavior is in the model. Even if there is a large number of diagnosis candidates, all of them can be enumerated using model-based diagnosis algorithms.

Model-based reasoning can also be applied in *predictive maintenance* where someone is interested in identifying situations where the system's behavior starts deviating from expectations for predicting when to carry out maintenance activities. In such situations the system might work mostly in between its boundaries but also exhibits unwanted behavior like vibrations. Model-based reasoning makes use of a model for predicting values that can be compared with observations. We can use these models also during operation for comparing the predicted outcome with the current one using monitoring. In such a way, we are able to identify discrepancies and are also able to compute diagnosis candidates. During maintenance someone might want to distinguish these candidates in order to find the right component to be replaced. This idea is very much similar to the concept of digital twins, which are virtual models of a processes, products, or services used during monitoring for identifying discrepancies.

Because of the need for having the right models available for model-based reasoning and the difficulties in providing such models, it would be very much appealing to use *machine learning* for obtaining models. Surprisingly in the context of model-based reasoning there is no work on extracting models from data directly. Most closely there is work on extracting models to be used for diagnosis from simulation models. Peischl et al. [33] introduced a method for extracting logic rules from Modelica models using simulation. The approach makes use of fault models where the corresponding physical behavior is modeled in the language Modelica. However, obtaining models from data might rely on existing machine learning approaches like decision tree induction or automata learning. What is required is knowledge about the correct and faulty case as well as a more detailed description of underlying root causes. Alternatively, someone might only want to learn the correct behavior for each component using available data. The latter approach would directly lead to a model that can be used for diagnosis as described in this chapter.

11.4 Smart Battery System Case Study

In this section, we apply MBR for computing diagnoses and configurations for a smart battery system. We make use of the power module comprising six batteries, and one wire component for a stack size k of 2 that is depicted in Fig. 11.2. We start with providing a solution for the configuration problem mentioned in Sect. 11.2. Afterwards, we illustrate how MBR can be applied to diagnose the resulting configured system in case of a detected misbehavior.

Before discussing the case study in detail it is worth discussing differences between solutions for diagnosis and configuration relying on specific programs that explicitly compute diagnoses and configurations, and the model-based reasoning approach. The latter makes use of models from which we obtain the results. Changes in the system cause changes in the model, which can be usually introduced easily. In case of the former approach, larger portions of the program might be changed causing additional effort. In addition, we may also either miss certain cases to be handled in a program or are even not able to describe all the different cases explicitly. When relying on models, and assuming that the models appropriately reflect the system, we always are able to return all solutions that can be correctly derived from the models. Finally, in model-based diagnosis as discussed in this chapter, we consider modeling the correct behavior only. Every behavior obtained from the model that deviates from the observations lead to diagnoses not considering specific faults or their corresponding fault models. Hence, we are able to provide diagnoses for all faults even when we do not know the specific fault model making the approach smart, i.e., allowing to derive all correct results using available knowledge.

11.4.1 The Configuration Case

For configuring any system we need (1) a knowledge base describing the parts we may use for combining together and the components' provided functionality, and (2) requirements the system has to fulfill. The latter usually refers to functionality we want to obtain. In case of the smart battery system, we have to specify what batteries are delivering when they are in a particular state and how this contributes to the overall required electrical characteristics. In the following we discuss the parameters needed according to Definition 11.4 for constituting a configuration problem. We start introducing the involved parameters, which are in our case the six batteries and the one wire:

$$PAR = \{b_1, b_2, b_3, b_4, b_5, b_6, w\}$$

For simplicity of formalizing the knowledge base we further introduce the subset of parameters only comprising batteries:

$$PAR_B = PAR \setminus \{w\}$$

For configuring the system, we need the possible types, which are the possible state for the given battery system:

$$\tau(b_1) = \ldots = \tau(b_6) = \{off, load, s1, s2\} \text{ and } \tau(w) = \{off, s1, s2\}$$

Finally, we specify the knowledge base. For this purpose, we make use of the constraints discussed in the previous section. In particular, we introduce different rules. We start defining rules for mapping parameter types to states. Afterwards, we introduce constraints for the parameter m that is used to specify the voltage level required. We distinguish the two cases $m = 1$ and $m = 2$. For $m = 1$ we need a battery and a wire. In the following formulation we specify all potential combinations. We do the same for the second parameter p stating the required current to be delivered. Again we consider different cases. In the rule we follow the already discussed constraints. Finally, we introduce rules that have to be fulfilled always, i.e., it is not allowed to concurrently use a wire and a battery at the same battery stack level, and there are also restrictions for the parameters.

$$KB = \left\{\begin{array}{l}
\forall X \in PAR : type(X, off) \rightarrow state(X, off), \\
\forall X \in PAR_B : type(X, load) \rightarrow state(X, load), \\
\forall X \in PAR : type(X, s1) \rightarrow state(X, s1), \\
\forall X \in PAR : type(X, s2) \rightarrow state(X, s2), \\
m = 1 \leftrightarrow \exists X \in PAR_B : \left(\begin{array}{l} (state(X, s1) \wedge state(w, s2)) \vee \\ (state(w, s1) \wedge state(X, s2)) \end{array}\right), \\
m = 2 \leftrightarrow \exists X \in PAR_B : \exists Y \in PAR_B : state(X, s1) \wedge state(Y, s2), \\
p = 1 \leftrightarrow \forall S \in \{s1, s2\} : \left(\begin{array}{l} \exists X \in PAR_B : state(X, S) \wedge \\ (\forall Y \in PAR_B : (Y \neq X) \rightarrow \neg state(Y, S)) \end{array}\right), \\
p = 2 \leftrightarrow \forall S \in \{s1, s2\} : \left(\begin{array}{l} \exists X \in PAR_B : \exists Y \in PAR_B : X \neq Y \\ \quad \wedge state(X, S) \wedge state(Y, S) \wedge \\ \quad (\forall Z \in PAR_B : (Z \neq X \wedge Z \neq Y) \\ \qquad \rightarrow \neg state(Z, S)) \end{array}\right), \\
p = 3 \leftrightarrow \forall S \in \{s1, s2\} : \left(\begin{array}{l} \exists X \in PAR_B : \exists Y \in PAR_B : \exists Z \in PAR_B : \\ \quad X \neq Y \neq Z \wedge state(X, S) \wedge \\ \quad state(Y, S) \wedge state(Z, S) \wedge \\ (\forall W \in PAR_B : (W \neq X \wedge W \neq Y \wedge W \neq Z) \\ \qquad \rightarrow \neg state(W, S)) \end{array}\right), \\
\forall X \in PAR_B : \neg(state(X, s1) \wedge X \neq w \wedge state(w, s1)), \\
\forall X \in PAR_B : \neg(state(X, s2) \wedge X \neq w \wedge state(w, s2)), \\
m \leq 2 \wedge p \leq 3
\end{array}\right\}$$

What remains to be specified are the requirements. In this case, we search for configurations where we have twice the voltage of a single battery cell, but we require only the current provided by one cell:

$$REQ = \{m = 2, p = 1\}$$

Let us now have a look whether the following mappings Δ_1 and Δ_2 are configurations as specified in Definition 11.5:

	Δ_1	Δ_2
b_1	off	off
b_2	$load$	$s1$
b_3	$s1$	off
b_4	$s2$	off
b_5	$load$	$s2$
b_6	$load$	off
w	off	$s2$

Let us have a look at Δ_1 first. Due to the first 4 rules in KB, we know that $state(b_1, off), state(b_2, load), state(b_3, s1), state(b_4, s2), \ldots$ must hold. Because of the requirement $m = 2$, we have to check the rule $m = 2 \leftrightarrow \exists X \in PAR_B : \exists Y \in PAR_B : state(X, s1) \wedge state(Y, s2)$ for satisfiability. When assigning b_3 to X and b_4 to Y, we see that $state(X, s1) \wedge state(Y, s2)$ must hold. We do the same for requirement $p = 1$ and have a look at its corresponding rule $p = 1 \leftrightarrow \forall S \in \{s1, s2\} : \begin{pmatrix} \exists X \in PAR_B : state(X, S) \wedge \\ (\forall Y \in PAR_B : (Y \neq X) \rightarrow \neg state(Y, S)) \end{pmatrix}$. Because there is only one battery assigned to $s1$ and $s2$, respectively, the rule is not violated. Finally, we have to check the last three rules in KB. All of them are obviously fulfilled, because the wire w has an assigned state of off and the requirements specify values within their boundaries. Consequently, Δ_1 is a configuration for the smart battery case study.

Obviously, Δ_2 is not a configuration because $state(b_2, s_2), state(w, s_2)$, which can be derived from the first rules of KB using the provided information in Δ_2, contradicts the rule $\forall X \in PAR_B : \neg(state(X, s2) \wedge X \neq w \wedge state(w, s2))$. When substituting X with b_5, we obtain a logical rule $state(b_5, s2) \wedge b_5 \neq w \wedge state(w, s2)$, which is true, and therefore its negation is false leading to the contradiction and the final conclusion that Δ_2 cannot be a configuration according to Definition 11.5.

What we have discussed so far is that we are able to compute configurations for smart battery systems (and other systems) only using models describing functionalities, necessities, and restrictions together with given requirements. The computation of configurations can be automated. We only need to take a mapping of parameters to values and check whether this mapping is not in contradiction with the model (i.e., the configuration knowledge base KB). In a smart battery system we can compute configurations over time and not only when installing the system. Hence, we are able to react to changes in the requirements over time and re-configure the system accordingly. Moreover, we can also re-configure the system in case of known faults. For example, if we know that a battery b is not working anymore, we only need to set its state to off. This can be easily achieved by searching for configurations Δ with $\Delta(b) = off$. A similar approach can be used when b has to be loaded and is not available anymore. Hence, the configurator is flexible allowing to react to different faults or shortcomings over time. Note that in practice of course, reconfiguration might be more complicated because of, for example, physical requirements. We may need to add a battery first before removing another one. However, with the proposed MBR solution, we are able to compute different configurations under various circumstances and can leave the necessary steps for configuration changes to the real implementation of the battery system.

11.4.2 The Diagnosis Case

In the following, we outline how to apply model-based diagnosis for locating faults in a configured smart battery system. For this purpose, we have to formulate the diagnosis problem stated in Definition 11.1. A diagnosis problem comprises a model of components and their interconnections, i.e., the system description SD, the set of components $COMP$ that might be faulty, and a set of observations OBS we want to explain. Modeling components can be done is many different ways and at many different levels of abstractions. For example, the simplest model of a battery only states that a battery delivers a certain amount of voltage between its poles. A more detailed model adds an internal resistor considering that withdrawing a larger current influences also the provided voltage. In addition, we might consider the uncharging characteristics, which varies in case of different underlying technologies. Depending on the purpose of a model we might switch models and consider different aspects of a component or system.

In diagnosis, it is often sufficient dealing with abstract models, which only describe the behavior of components in a qualitative way, e.g., a battery provides voltage, ignoring details like the exact amount of voltage. We are following this principle and state that a non-empty battery delivers a voltage v_B and a current i_B. If a battery is empty, no voltage is provided anymore. In the following formal description of the model, we do not use the term *empty* but use the predicate AB instead meaning that the battery is empty. In case $\neg AB$ holds, we know that the battery provides voltage. Furthermore, we only consider the case where the battery is working as expected:

$$\forall B \in COMP : bat(B) \rightarrow (\neg AB(B) \rightarrow (volt(B, v_B) \wedge cur(B, i_B)))$$

For the wire component used in a smart battery, we know that there is no voltage drop in case it is working, i.e., we can formalize this as follows:

$$\forall W \in COMP : wire(W) \rightarrow (\neg AB(W) \rightarrow volt(W, 0))$$

In addition, to this behavior, we have to formalize the structure of the system. For the smart battery system example we have six batteries and one wire that might be connected at position $s1$ or $s2$. When using the same predicate $state$ as for configuration, we are able to formulate whether a component is in a particular state, e.g., battery b_3 is at position $s1$. In this case, we write $state(b_3, s1)$ and add this to our model. The structure of the smart battery system determines the behavior. Depending on the number of batteries that are in parallel we get information about the maximum current, and depending on the number of batteries in serial the voltage we obtain. Because of stack size 2 we can only put two batteries in series at the maximum. Hence, we have to state how this influences the output voltage level. For the output voltage level we introduce a predicate $vout$. The serial behavior can be expressed as follows:

$$
\begin{aligned}
&\forall X, Y \in COMP: \\
&\qquad state(X, s1) \wedge state(Y, s2) \wedge volt(X, 0) \wedge volt(Y, 0) \rightarrow vout(0) \\
&\forall X, Y \in COMP: \\
&\qquad state(X, s1) \wedge state(Y, s2) \wedge volt(X, 0) \wedge volt(Y, v_B) \rightarrow vout(v_B) \\
&\forall X, Y \in COMP: \\
&\qquad state(X, s1) \wedge state(Y, s2) \wedge volt(X, v_B) \wedge volt(Y, 0) \rightarrow vout(v_B) \\
&\forall X, Y \in COMP: \\
&\qquad state(X, s1) \wedge state(Y, s2) \wedge volt(X, v_B) \wedge volt(Y, v_B) \rightarrow vout(2v_B) \\
&\forall X, Y \in COMP: \\
&\qquad state(X, s1) \wedge state(Y, s2) \wedge vout(0) \rightarrow (volt(X, 0) \wedge volt(Y, 0)) \\
&\forall X, Y \in COMP: \\
&\qquad state(X, s1) \wedge state(Y, s2) \wedge vout(2v_B) \rightarrow (volt(X, v_B) \wedge volt(Y, v_B))
\end{aligned}
$$

$$
\neg(vout(0) \wedge vout(v_B)) \wedge \neg(vout(0) \wedge vout(2v_B)) \wedge \neg(vout(v_B) \wedge vout(2v_B))
$$

For the parallel composition of batteries for increasing the current, we may come up with a very much similar model. We have to look at how many batteries are put together in parallel. This determines the maximum current. There must be the same number of batteries in parallel in order to deliver the combined current except in case of a wire. The following first-order logic rules formalize this behavior, where the first three rules are for obtaining the maximum current for each state and the other rules are combining this information to finally obtain the maximum current that is handled using the *iout* predicate:

$$
\begin{aligned}
&\forall S \in \{s1, s2\} : state(w, S) \rightarrow iout(0, S) \\
&\forall S \in \{s1, s2\} : \neg state(w, S) \wedge \forall X \in COMP : (bat(X) \wedge state(X, S) \wedge \\
&\quad (\nexists Y \in COMP : X \neq Y \wedge state(Y, S))) \rightarrow iout(i_B, S) \\
&\forall S \in \{s1, s2\} : \neg state(w, S) \wedge \forall X, Y \in COMP : (bat(X) \wedge state(X, S) \wedge \\
&\quad X \neq Y \wedge bat(Y) \wedge state(Y, S) \\
&\quad (\nexists Z \in COMP : X \neq Z \wedge Y \neq Z \wedge state(Z, S))) \rightarrow iout(2i_B, S) \\
&\forall S \in \{s1, s2\} : \neg state(w, S) \wedge \forall X, Y, Z \in COMP : (bat(X) \wedge state(X, S) \wedge \\
&\quad X \neq Y \wedge bat(Y) \wedge state(Y, S) \wedge X \neq Z \wedge Y \neq Z \wedge bat(Z) \wedge state(Z, S) \\
&\quad (\nexists W \in COMP : X \neq W \wedge Y \neq W \wedge Z \neq W \wedge state(W, S))) \rightarrow iout(3i_B, S)
\end{aligned}
$$

$$
\begin{aligned}
&\forall X \in \{0, i_B, 2i_B, 3i_B\} : \forall S \in \{s1, s2\} : iout(X, S) \rightarrow iout(X) \\
&\forall X \in \{0, i_B, 2i_B, 3i_B\} : (iout(0, s1) \wedge iout(X, s2) \rightarrow iout(X) \\
&\forall X \in \{i_B, 2i_B, 3i_B\} : (iout(i_B, s1) \wedge iout(X, s2) \rightarrow iout(i_B) \\
&\forall X \in \{2i_B, 3i_B\} : (iout(2i_B, s1) \wedge iout(X, s2) \rightarrow iout(2i_B) \\
&\forall X \in \{0, i_B, 2i_B, 3i_B\} : (iout(X, s1) \wedge iout(0, s2) \rightarrow iout(X) \\
&\forall X \in \{i_B, 2i_B, 3i_B\} : (iout(X, s1) \wedge iout(i_B, s2) \rightarrow iout(i_B) \\
&\forall X \in \{2i_B, 3i_B\} : (iout(X, s1) \wedge iout(2i_B, s2) \rightarrow iout(2i_B)
\end{aligned}
$$

$$
\begin{aligned}
&\neg(iout(0) \wedge iout(i_B)) \wedge \neg(iout(0) \wedge iout(2i_B)) \wedge \neg(iout(0) \wedge iout(3i_B)) \\
&\neg(iout(i_B) \wedge iout(2i_B)) \wedge \neg(iout(i_B) \wedge iout(3i_B)) \wedge \neg(iout(2i_B) \wedge iout(3i_B))
\end{aligned}
$$

It is worth noting that depending on the underlying formalism models might become easier readable. For example, when relying on constraint satisfaction [10] the models would be much similar to mathematical equations.

In addition to the discussed equations, we have to add structural information like the state of each component. Let us assume that battery b_3 and b_4 are in state $s1$ and $s2$, respectively, and the other components of the six battery and one wire system are either in the off or $load$ state. Therefore, we have to add $state(b_3, s1)$, $state(b_4, s2)$, and in addition $state(b_1, off)$, $state(b_2, load)$, $state(b_5, load)$, $state(b_6, load)$, $state(w, off)$ to the model. Let us further assume that we observe an output voltage of v_B and an output current of i_B, i.e.

$$OBS = \{vout(v_B), iout(i_B)\}$$

For this example, when assuming all batteries and the wire to work correctly, we obtain a conflict with the model SD. We know that $\neg AB(b_3)$ and $\neg AB(b_4)$ from which we obtain $volt(b_3, v_B)$, $cur(b_3, i_B)$, $volt(b_4, v_B)$, $cur(b_4, i_B)$. From this information, knowing that $state(b_3, s1)$ and $state(b_4, s2)$, and the equation $\forall X, Y \in COMP : state(X, s1) \wedge state(Y, s2) \wedge volt(X, v_B) \wedge volt(Y, v_B) \rightarrow vout(2v_B)$, we can derive that $vout(2v_B)$ has to hold, which contradicts our observation $vout(v_B)$. Hence, there is a fault in the system, and we have to find the reason behind it. Let us assume $AB(b_3)$ and $\neg AB(b_4)$. In this case, we cannot derive $volt(b_3, v_B)$ anymore. Therefore, we are also not able to derive $vout(2v_B)$ and there is no contradiction anymore. Hence, $\{b_3\}$ is a diagnosis. A further analysis would reveal that $\{b_4\}$ is also a diagnosis. To distinguish one diagnosis from the other we need further information. Either we add new measurements like the voltage drop at each battery, or we simply assume one battery to fail, i.e., assume one diagnosis to be valid, and re-configure the system using our configuration model. If the re-configured system meets its specification, we know that the assumption is right. Otherwise, we would need another reconfiguration step but this time assuming the other diagnosis to be valid. The latter procedure of distinguishing diagnoses is also called active diagnosis, because it requires additional steps like the replacement of components.

Obviously, this active diagnosis can only be applied if a change of the system does not influence the health state of a component. For example, consider a fuse connected with an electronic device. The fuse might be broken due to a fault in the electronic device. Replacing the fuse alone, would lead to a broken new fuse. Replacing the electronic device would also not be a valid repair. Hence, in such cases we need to consider dependent faults in the overall diagnostic procedure. Weber and Wotawa [51] introduced a solution to this problem based on the foundations behind MBR as described in this chapter.

Let us consider that we have additional sensors for measuring the voltage like specified in Fig. 11.8. In this case, diagnosis becomes easy because we would obtain not only a value for v_{out} but also for the different voltages at the different stages of the battery stack. For example, if assuming that we have the same serial connection of batteries than before but this time considering the observations to be:

$$OBS = \{vout(v_B), volt(b_3, v_B), volt(b_4, 0), iout(i_B)\}$$

We obtain only the one diagnosis $\{b_4\}$ that is needed to eliminate the inconsistencies.

11.5 Related Research

Pell and colleagues [34] and Rajan and colleagues [37] were the first describing a system based on model-based reasoning that was able to react on faults during operation based on models. Their work was integrated into the control software of NASA's deep space one spacecraft for improving autonomy and also successfully tested in space. In [21] Hofbauer et al. introduced a diagnostic system that is able to adapt the kinematics model in order to compensate faults in motors of a robotics drive. Steinbauer and Wotawa [42] discussed several approaches for implementing self-adaptive systems utilizing model-based reasoning including repairing software systems in case of faults. Most recently, Wotawa [54] showed how to use model-based reasoning in the context of autonomous systems. Readers interested in more details behind model-based reasoning including different diagnosis approaches like abductive diagnosis shall consult [54].

Configuration has been in focus of research in Artificial Intelligence for a long time. McDermott [29] introduced a rule-based system for configuration. Later on configuration based on knowledge has been more in the focus of research including Klein [24] or Klein et al. [25]. The use of configuration for technical system based on reasoning has also been reported, e.g., in Biswas and colleagues [3] or Yost and Rothenfluh [56]. Because of the switch from rule-based configuration to the use of models and reasoning a theory behind modeling became more and more important, e.g., see [47, 52], or [14]. Interestingly model-based reasoning has also been used to debug such knowledge bases. Felfernig et al. [13] used model-based diagnosis as introduced in this chapter for finding faults in configuration models. Most recently, Uta and Felfernig [48] reported on the use of configuration in the area of energy transmission networks.

Regarding model-based reasoning there is a lot of work dealing with algorithms including Reiter's hitting set algorithm [19, 38] that has been further improved since then, e.g., see [53] and most recently [36]. Further improvements in computation can be achieved combining the hitting set algorithm with algorithms that allow to search for minimal conflicts effectively like QUICKXPLAIN [23]. There has also been several algorithms introduced that directly return diagnosis not requiring to compute conflicting assumptions, i.e., conflicts, see e.g. [31]. For a more recent comparison of the runtime of several diagnosis algorithms we refer to Nica et al. [32]. For a comparison of QUICKXPLAIN with MaxSAT have a look at Walter et al. [50]. In addition to algorithms it is important to provide models. There are a lot of papers dealing with modeling real systems including but not limited to [2, 4, 27, 28, 35, 40, 49]. What is important to come up with the right model for diagnosis or configuration is to find the right level of abstraction. Sachenbacher and Struss [39] provided the foundations behind automated domain abstraction to find such models that fit a particular task. In addition, the use of models implemented in different programming languages might also be of interest, see e.g. [26].

Diagnosis and health prediction for battery systems have been considered as well. Andoni et al. [1] discussed the use of data analysis for prognostics and health management. Similarly Chen and Pecht [6] tackled the problem of prognostics using both model-based and data-driven methods. In [17] Gao et al. shows how to use charging data to predict the remaining useful lifetime of Lithium-Ion batteries. For a discussion on battery models we refer to Jongerden and Hverkort's paper [22].

11.6 Conclusions

In this chapter, we discussed the use of model-based reasoning in the context of smart batteries that can be adapted and diagnosed during runtime for achieving their requirements and specifications. In particular, we introduced the application of models describing potential functionality for coming up with parameter configuration, and models capturing the system's behavior for implementing diagnosis. Both configuration and diagnosis can be fully automated using reasoning engines either based on logic, constraint satisfaction, or any other means for deriving conflicts based on models and observations or requirements. It is worth noting that we often use a qualitative representation of values in models. In this case, we have to provide a mapping between the real values and their corresponding representation. Noise and uncertainty occurring in reality can be handled in this case if they do not impact the outcome of the mapping. In addition, we illustrate the application of model-based reasoning using a smart battery system focusing on basic principles, and discuss related research where we mainly reference to literature dealing with applications.

The suggested model-based reasoning approach to be used for configuration and diagnosis brings in the following advantages:

- The whole reasoning is based on available models only. Hence, when having a model, we can fully automated both configuration and diagnosis.
- The presented approach is flexible and can be easily adapted for configuring and diagnosing different systems. We only need to change the models.
- Providing that the models are correct, we know that all configurations and diagnosis must be correct. Furthermore, we are able to compute all configuration and diagnoses making the approach complete.
- There exists various algorithms for diagnosis and configuration. In the paper, we introduced the most appropriate related scientific papers.
- Although computing configuration and diagnosis is computationally demanding, current computers are able to deal with reasonably sized systems with up to

10,000 components within a reasonable time from less to 1 s to minutes at the maximum. Hence, the presented approach is feasible considering today's available computational resources.

Hence, in summary model-based reasoning offers a flexible, correct, complete and feasible approach for solving tasks like diagnosis and configuration. Especially, for systems that require increased flexibility and autonomy model-based reasoning can be effectively used. We outline the application of model-based reasoning using a smart battery system where we focussed on configuration and a more high-level diagnosis. However, it would have also been possible to use more sophisticated physical models of the battery and the electronics for diagnostic purposes. It is important to recall that the requirements of models to reflect the physical world accurately depend on the application domain. For the purpose of high-level diagnostics the introduced model seems to be sufficient.

Acknowledgements The financial support by the Austrian Federal Ministry for Digital and Economic Affairs and the National Foundation for Research, Technology and Development is gratefully acknowledged.

References

1. M. Andoni, W. Tang, V. Robu, D. Flynn: Data analysis of battery storage systems. CIRED - Open Access Proc. J. **2017**(1), 96–99 (2017). https://doi.org/10.1049/oap-cired.2017.0657
2. A. Beschta, O. Dressler, H. Freitag, M. Montag, P. Struss, A model-based approach to fault localization in power transmission networks. Intell. Syst. Eng. **2**, 3–14 (1992)
3. G. Biswas, K. Krishnamurthy, P. Basu, Applying qualitative reasoning techniques for analysis and evaluation in structural design, in *Proceedings of the Seventh IEEE Conference on Artificial Intelligence Application* (1991), pp. 265–268
4. F. Cascio, L. Console, M. Guagliumi, M. Osella, A. Panati, S. Sottano, D.T. Dupré, Generating on-board diagnostics of dynamic automotive systems based on qualitative models. AI Commun. **12**(1/2), 33–43 (1999)
5. C.L. Chang, R.C.T. Lee, *Symbolic Logic and Mechanical Theorem Proving* (Academic Press, Cambridge, 1973)
6. C. Chen, M. Pecht, Prognostics of lithium-ion batteries using model-based and data-driven methods, in *Proceedings of the IEEE 2012 Prognostics and System Health Management Conference (PHM-2012 Beijing)* (2012), pp. 1–6. https://doi.org/10.1109/PHM.2012.6228850
7. J. Crow, J. Rushby, Model-based reconfiguration: toward an integration with diagnosis. In: *Proceedings of the National Conference on Artificial Intelligence (AAAI)* (Morgan Kaufmann, Los Angeles, 1991), pp. 836–841
8. R. Davis, Diagnostic reasoning based on structure and behavior. Artif. Intell. **24**, 347–410 (1984)
9. R. Davis, H. Shrobe, W. Hamscher, K. Wieckert, M. Shirley, S. Polit, Diagnosis based on structure and function, in *Proceedings of the National Conference on Artificial Intelligence (AAAI), Pittsburgh* (1982), pp. 137–142
10. R. Dechter, *Constraint Processing* (Morgan Kaufmann, Los Angeles, 2003)
11. J. de Kleer, A.K. Mackworth, R. Reiter, Characterizing diagnosis and systems. Artif. Intell. **56**, 197–222 (1992)
12. J. de Kleer, B.C. Williams, Diagnosing multiple faults. Artif. Intell. **32**(1), 97–130 (1987)

13. A. Felfernig, G. Friedrich, D. Jannach, M. Stumptner, Consistency based diagnosis of configuration knowledge bases, in *Proceedings of the European Conference on Artificial Intelligence (ECAI), Berlin* (2000)
14. A. Felfernig, G. Friedrich, D. Jannach, M. Stumptner, An integrated development environment for the design and maintenance of large configuration knowledge bases, in *Proceedings Artificial Intelligence in Design* (Kluwer Academic Publishers, Worcester, 2000)
15. G. Friedrich, M. Stumptner, Consistency-based configuration. Tech. Rep., Technische Universität Wien (1992)
16. G. Friedrich, G. Gottlob, W. Nejdl, Physical impossibility instead of fault models, in *Proceedings of the National Conference on Artificial Intelligence (AAAI), Boston* (1990), pp. 331–336. Also appears in Readings in Model-Based Diagnosis (Morgan Kaufmann, 1992)
17. D. Gao, M. Huang, J. Xie, A novel indirect health indicator extraction based on charging data for lithium-ion batteries remaining useful life prognostics. SAE Int. J. Alternative Powertrains **6**(2), 183–193 (2017). https://www.jstor.org/stable/26169183
18. E. Gelle, R. Weigel, Interactive configuration with constraint satisfaction, in *Proceedings of the 2nd International Conference on Practical Applications of Constraint Technology (PACT)* (1996)
19. R. Greiner, B.A. Smith, R.W. Wilkerson, A correction to the algorithm in Reiter's theory of diagnosis. Artif. Intell. **41**(1), 79–88 (1989)
20. M. Heinrich, E. Jüngst, A resource-based paradigm for the configuring of technical systems from modular components, in Proceedings of the IEEE Conference on Artificial Intelligence Applications (CAIA), pp. 257–264 (1991)
21. M.W. Hofbaur, J. Köb, G. Steinbauer, F. Wotawa, Improving robustness of mobile robots using model-based reasoning. J. Intell. Robot. Syst. **48**(1), 37–54 (2007)
22. M.R. Jongerden, B.R. Haverkort, Which battery model to use? IET Softw. **3**(6), 445–457 (2009). https://doi.org/10.1049/iet-sen.2009.0001
23. U. Junker, QUICKXPLAIN: preferred explanations and relaxations for over-constrained problems, in *Proceedings of the 19th national conference on Artifical intelligence (AAAI)* (AAAI Press/The MIT Press, Menlo Park/Cambridge, 2004), pp. 167–172
24. R. Klein, Model representation and taxonomic reasoning in configuration problem solving, in *Proceedings of the German Workshop on Artificial Intelligence (GWAI), Informatik-Fachberichte*, vol. 285 (Springer, Berlin, 1991), pp. 182–194
25. R. Klein, M. Buchheit, W. Nutt, Configuration as model construction: the constructive problem solving approach, in *Proceedings of Artificial Intelligence in Design '94* (Kluwer, Dordecht, 1994), pp. 201–218
26. M.E. Klenk, J. de Kleer, D.G. Bobrow, B. Janssen, Qualitative reasoning with Modelica models, in *Proceedings of the Twenty-Eighth AAAI Conference on Artificial Intelligence (AAAI)*. (AAAI Press, Menlo Park, 2014), pp. 1084–1090
27. A. Malik, P. Struss, M. Sachenbacher, Case studies in model-based diagnosis and fault analysis of car-subsystems, in *Proceedings of the European Conference on Artificial Intelligence (ECAI)* (1996)
28. I. Matei, A. Ganguli, T. Honda, J. de Kleer, The case for a hybrid approach to diagnosis: a railway switch, in DX@Safeprocess, *CEUR Workshop Proceedings*, vol. 1507, (2015), pp. 225–234. CEUR-WS.org
29. J.McDermott, R1: a rule-based configurer of computer systems. Artif. Intell. **19**, 39–88 (1982)
30. S. Mittal, F. Frayman, Towards a generic model of configuration tasks, in *Proceedings of the 11th International Joint Conference on Artificial Intelligence (IJCAI)* (1989), pp. 1395–1401
31. I. Nica, F. Wotawa, Condiag – computing minimal diagnoses using a constraint solver, in *Proceedings of the 23rd International Workshop on Principles of Diagnosis* (2012)
32. I. Nica, I. Pill, T. Quaritsch, F. Wotawa, The route to success - a performance comparison of diagnosis algorithms, in *International Joint Conference on Artificial Intelligence (IJCAI), Beijing* (2013)

33. B. Peischl, I. Pill, F. Wotawa, Using Modelica programs for deriving propositional horn clause abduction problems, in *Proceedings of the 39th German Conference on Artificial Intelligence (KI)*, Lecture Notes in Artificial Intelligence, vol. 9904. Springer, Klagenfurt (2016)
34. B. Pell, D. Bernard, S. Chien, E. Gat, N. Muscettola, P. Nayak, M. Wagner, B. Williams, A remote-agent prototype for spacecraft autonomy, in *Proceedings of the SPIE Conference on Optical Science, Engineering, and Instrumentation*. Volume on Space Sciencecraft Control and Tracking in the New Millennium, Bellingham, Washington (Society of Professional Image Engineers, Albany, 1996)
35. C. Picardi, R. Bray, F. Cascio, L. Console, P. Dague, O. Dressler, D. Millet, B. Rehfus, P. Struss, C. Vallée, Idd: integrating diagnosis in the design of automotive systems, in *Proceedings of the European Conference on Artificial Intelligence (ECAI)* (IOS Press, Lyon, 2002), pp. 628–632
36. I. Pill, T. Quaritsch, Rc-tree: a variant avoiding all the redundancy in Reiter's minimal hitting set algorithm, in *ISSRE Workshops*. IEEE Computer Society, Washington, 2015), pp. 78–84
37. K. Rajan, D. Bernard, G. Dorais, E. Gamble, B. Kanefsky, J. Kurien, W. Millar, N. Muscettola, P. Nayak, N. Rouquette, B. Smith, W. Taylor, Y. Tung, Remote agent: an autonomous control system for the new millennium, in *Proceedings of the 14th European Conference on Artificial Intelligence (ECAI), Berlin* (2000)
38. R. Reiter, A theory of diagnosis from first principles. Artif. Intell. **32**(1), 57–95 (1987)
39. M. Sachenbacher, P. Struss, Task-dependent qualitative domain abstraction. Artif. Intell. **162**(1–2), 121–143 (2005)
40. M. Sachenbacher, P. Struss, C.M. Carlén, A prototype for model-based on-board diagnosis of automotive systems. AI Commun. **13** (2000). Special Issue on Industrial Applications of Model-Based Reasoning, pp. 83–97
41. R.M. Smullyan, *First Order Logic* (Springer, Berlin, 1968)
42. G. Steinbauer, F. Wotawa, Model-based reasoning for self-adaptive systems - theory and practice, in *Assurances for Self-Adaptive Systems*. Lecture Notes in Computer Science, vol. 7740 (Springer, Berlin, 2013), pp. 187–213
43. P. Struss, O. Dressler, Physical negation — integrating fault models into the general diagnostic engine, in *Proceedings 11th International Joint Conference on Artificial Intelligence, Detroit* (1989), pp. 1318–1323
44. M. Stumptner, A. Haselböck, A generative constraint formalism for configuration problems, in *3rd Congress Italian Association for Artificial Intelligence*. Lecture Notes in Artificial Intelligence, vol. 729 (Springer, Torino, 1993)
45. M. Stumptner, F. Wotawa, Model-based reconfiguration, in *Proceedings Artificial Intelligence in Design, Lisbon* (1998)
46. M. Stumptner, A. Haselböck, G. Friedrich, COCOS - a tool for constraint-based, dynamic configuration, in *Proceedings of the 10th IEEE Conference on AI Applications (CAIA), San Antonio* (1994)
47. T. Tomiyama, From general design theory to knowledge-intensive engineering. Artif. Intell. Eng. Des. Anal. Manuf. **8**(4), 319–333 (1994)
48. M. Uta, A. Felfernig, Towards knowledge infrastructure for highly variant voltage transmission systems, in *ConfWS, CEUR Workshop Proceedings*, vol. 2220, (2018), pp. 109–118. CEUR-WS.org
49. T. Voigt, S. Flad, P. Struss, Model-based fault localization in bottling plants. Adv. Eng. Inform. **29**(1), 101–114 (2015)
50. R. Walter, A. Felfernig, W. Küchlin, Inverse QuickXplain vs. MaxSAT - a comparison in theory and practice, in *Configuration Workshop CEUR,Workshop Proceedings*, vol. 1453, CEUR-WS.org (2015), pp. 97–104
51. J. Weber, F. Wotawa, Diagnosis and repair of dependent failures in the control system of a mobile autonomous robot. Appl. Intell. **36**(4), 511–528 (2012)
52. B.J. Wielinga, A.T. Schreiber, KADS: a modeling approach to knowledge engineering. Knowl. Acquis. **4**(1), 5–53 (1992)
53. F. Wotawa, A variant of Reiter's hitting-set algorithm. Inf. Process. Lett. **79**(1), 45–51 (2001)

54. F. Wotawa, Reasoning from first principles for self-adaptive and autonomous systems, in *Predictive Maintenance in Dynamic Systems – Advanced Methods, Decision Support Tools and Real-World Applications*, ed. by E. Lughofer, M. Sayed-Mouchaweh (Springer, Berlin, 2019). https://doi.org/10.1007/978-3-030-05645-2
55. J.R. Wright, E.S. Weixelbaum, K. Brown, G.T. Vesonder, S.R. Palmer, J.I. Berman, H.G. Moore, A knowledge-based configurator that supports sales, engineering, and manufacturing at AT&T network systems, in *Proceedings of the 5th Conference on Innovative Applications of AI* (AAAI Press, Menlo Park, 1993)
56. G. Yost, T. Rothenfluh, Configuring elevator systems. Int. J. Hum.-Comput. Stud. **44**, 521–568 (1996)

Chapter 12
Data-Driven Predictive Flexibility Modeling of Distributed Energy Resources

Indrasis Chakraborty, Sai Pushpak Nandanoori, Soumya Kundu, and Karanjit Kalsi

12.1 Introduction

The last couple of decades have seen a significant increase in the deployment of renewable energy resources into the electricity grid. Renewable generation is, however, mostly significantly more intermittent and uncertain compared to traditional fossil-fueled generation, e.g., wind power generation is heavily reliant on the variability in the wind speed [33]. Traditional techniques used by grid operators to nullify any imbalance between the generation and load include the use of spinning reserves, backup generators, as well as load curtailment methods, depending on the nature of the power fluctuations [35]. However, these traditional

This manuscript has been authored work was produced by Battelle Memorial Institute under Contract No. DE-AC05-76RL01830 with the U.S. Department of Energy. The US Government retains and the publisher, by accepting the article work for publication, acknowledges that the US Government retains a non-exclusive, paid-up, irrevocable, world-wide license to publish or reproduce the published form of this manuscript work, or allow others to do so, for US Government purposes. The Department of Energy will provide public access to these results of federally sponsored research in accordance with the DOE Public Access Plan (http://energy.gov/downloads/doe-public-access-plan).

I. Chakraborty (✉) · S. P. Nandanoori · S. Kundu · K. Kalsi
Lawrence Livermore National Laboratory, Livermore, CA, USA
e-mail: chakraborty3@llnl.gov; saipushpak.n@pnnl.gov; soumya.kundu@pnnl.gov; karanjit.kalsi@pnnl.gov

M. Sayed-Mouchaweh (ed.), *Artificial Intelligence Techniques for a Scalable Energy Transition*, https://doi.org/10.1007/978-3-030-42726-9_12

methods are not suitable, nor sufficient, when there is a very high penetration of renewable generation. For example, building large spinning reserves to compensate for the variability in renewable generation is prohibitively expensive. Moreover, it is counter-productive to deploy high carbon-footprint diesel generators as backup while increasing renewable penetration. Finally, curtailment of loads to mitigate variability in generation is not a desirable solution, either.

Concurrently over the last decade, use of flexible end-use loads and other devices for grid services have gained increased attention both in the academia as well as the industry. In the context of power grid operations, any energy (consuming and/or generating) resource that is capable of offering some flexibility in its net electricity demand (over a certain duration) is generally referred to as a *distributed energy resource* (or, DER). Advanced sensing, controls and communications infrastructure, especially at the medium-to-low voltage power distribution networks, have enabled the proliferation of connected, *smart* DERs—such as heating, ventilating, and air conditioning (HVAC) units, electric water heaters (EWHs), energy storage units, lighting, washer/dryer units, refrigerators, etc.—which are able to communicate with each other and/or a resource coordinator. In particular, Internet-of-Things (IoT) devices, such as smart thermostats and sensors, are capable of operating interactively and autonomously while remaining connected with other devices and/or the building automation system, and are often enabled by low cost cloud and computing platforms for local in-device data analytics and controls implementation [34, 51]. It is becoming increasingly feasible for these distribution side end-use resources to assist conventional generators in providing grid ancillary services [9, 49]. Due to the increased communication capabilities and fast-acting nature of these devices, it is imperative to use these existing resources for providing grid services [2, 9, 10, 14, 17, 26, 28, 41, 49, 56], as relatively much faster, cleaner, and cost-effective alternatives to more traditional measures.

Several works including [2–4, 9, 12, 15, 27–29, 32, 36–38, 40, 43, 48, 53, 55, 56] have proposed that residential/commercial HVAC systems, EWHs, electric vehicles (EVs), refrigerators, and heat pumps can be aggregated to provide certain grid services by leveraging their capability to store thermal energy, thereby achieving flexibility in power consumption. Furthermore, devices such as washer/dryer units and EVs come under the category of deferrable loads and they can be scheduled to provide grid services as well. The major challenge in deploying these DERs lies in quantifying the amount of load flexibility available in order to provide grid services. As there will be millions of these devices, it is not possible for the control authority to communicate with each device individually. Hence, an aggregator is required to establish faster communication and implement various control strategies. The aggregator acts as a mediator between the grid and the individual DERs. It is the task of an aggregator to characterize the available flexibility for the ensemble of DERs to provide grid services.

12.1.1 Virtual Battery Characterization of Flexibility

A straightforward method to achieve aggregation is by taking a Minkowski sum of these loads to quantify the load flexibility [29]. However, this is a very computationally expensive method when there are thousands of various DERs in the ensemble. In this context, recent works [11, 16, 17, 22–25, 42] proposed the notion of a virtual battery (VB) to characterize the aggregate flexibility similar to a physical battery. It is important to mention that the use of virtual storage to represent the flexibility of certain loads to retain/store thermal energy, not necessarily electrochemical storage, is acquiring momentum in the community. This VB has characteristics such as self-dissipation, capacity limits, power limits, and charging/discharging rate limits similar to a real battery. Further to emphasize on the difference between a real and VB, the energy in a real battery is stored in the form of chemical energy, whereas in the VB it is stored in the form of thermal energy. The mathematical representation of the VB model is not fixed and could depend on the characteristics of the underlying devices in the ensemble. Most of the recent works considering homogeneous/heterogeneous devices assume a first-order linear model representation for VBs.

Most recent works on VB model identification can be seen in [17, 18, 24, 39]. These works include characterizing a VB model for a wide range of systems from small residential thermostatically controllable loads (TCLs) to complex systems such as commercial building HVAC loads. Similar to a real battery, the VB also has self-dissipation, energy capacity, and power (charge/discharge) limits as parameters. In this work, we propose an alternative, data-driven, deep neural network framework of characterizing the aggregate flexibility of ensembles of air conditioners (ACs) and water heaters (EWHs) using the available end-use measurements. Since the training of the deep networks is an offline process requiring high computational effort, it is not desirable to retrain the network if the number of TCLs in the ensemble change over time due to changing availability of end-use appliances. Henceforth, we propose a transfer learning based approach to identify the VB parameters of the new ensemble with minimal retraining.

The VB model, in its simple form, is a linear continuous-time first-order system with control input. The state of the VB indicates the energy of the underlying ensemble. The VB parameters such as the battery capacity indicate the size of the ensemble, self-dissipation indicates any leakage in the state of charge of the battery, the charge/discharge limits (also referred to as ramp rates) indicate how fast or slow can the battery be charged/discharged. The regulation signal that acts as an input to the VB must remain within the power limits. Furthermore, depending on the type of service, there exists lower level controllers that assign the frequency or power thresholds to the individual devices as shown in [41]. Another important objective is to maintain the end-user comfort constraints while simultaneously providing the grid services. It is advantageous to represent the ensemble with an equivalent resource representation such as VB as this helps to avoid unnecessary control and communication complexity.

12.1.2 Advancements in Deep Learning

In recent years, deep learning has been gaining popularity among researchers due to its inherent nature imitating the human brain learning process. Specifically, several research works show the applicability of deep learning in capturing representative information from a raw dataset using multiple nonlinear transformations as shown in [19]. Deep learning based methods can model high level abstractions in data utilizing multiple processing layers, compared to shallow learning methods. In [6], deep learning methods are used for simplifying a learning task from input examples. Based on scientific knowledge in the area of biology, cognitive humanoid autonomous methods with deep learning architecture have been proposed and applied over the years [5, 20, 31, 52, 54]. Deep learning replaces handcrafted feature extraction by learning unsupervised features as shown in [50]. Although deep learning methods are very popular in applications like feature extraction, data visualization, and forecasting, they are rarely visited in the context of power systems applications. Our current application of using deep learning methods to identify the VB state for an ensemble of devices has not been explored in existing literature, although this application falls broadly under the scope of a dimensional reduction problem.

12.2 Modeling

The VB model is presented in this section and the identification of VB parameters such as power limits, initial state of charge is discussed for an ensemble of AC and EWH devices. The temporal evolution of temperature for AC or EWH devices is governed by a first-order differential equation and they are shown in [41, 44].

12.2.1 Virtual Battery

The evolution of the VB state is governed by the following dynamics:

$$\dot{x}(t) = -ax(t) - u(t)\,, \quad x(0) = x_0 \tag{12.1a}$$

$$C_1 \leq x(t) \leq C_2, \tag{12.1b}$$

$$P^-(t) \leq u(t) \leq P^+(t), \tag{12.1c}$$

where $x(t) \in \mathbb{R}$ denotes the state of charge (*soc*) of the VB at time t with the initial *soc* x_0, a denoting the self-dissipation rate, and the lower and upper energy limits of the VB are denoted by C_1 and C_2, respectively. The regulation signal $u(t)$ acts as an input to the VB and must always lie within the time-varying power limits $P^-(t)$ and $P^+(t)$. Finally, the VB parameters are denoted by vector $\phi = [a, C_1, C_2, x_0, P^-(t), P^+(t)]$. The first-order VB model is applied to characterize the aggregated flexibility of DERs and many building loads [17, 18, 24].

Homogeneous or heterogeneous ensemble of DERs are represented using this first-order VB model. For example, if an EV or a real battery is considered along with ACs or EWHs, then the VB of ACs and EWHs appear in parallel to other batteries. However, there will be additional challenges to dispatch EVs to provide grid services due to their charging constraints and this needs further investigation. The inclusion of solar photovoltaic (PV) results only in additional power limits as this is a free, instantaneous source of energy, and hence does not result in an increased battery capacity. This scenario changes if solar PV is connected to battery storage. The scope of this work extends to ACs, EWHs, and a heterogeneous ensemble of ACs and EWHs. In what follows, we discuss how the VB parameters ϕ are computed.

12.2.1.1 VB Parameters: Power Limits

The time-varying power limits of the VB are identified by extending the binary search algorithm proposed in [23] to all time points. The algorithm to compute $P^+(t)$ is shown below.

Algorithm 1 Algorithm for finding upper power limit, $P^+(t)$ at each t

Input: Initialize $\alpha = 0$, *Lower bound*
  Initialize $\beta = 1$, *Upper bound*
Output: $P^+(t)$
  while the ensemble tracks $\hat{u}(t) = \beta + P_{base}(t)$ **do** *No instant violation*
    $\alpha = \beta, \beta = 2 \times \beta$ *Update α and β*
  end while
  while $(\beta - \alpha) > \epsilon$ **do** *Testing a new bound*
    $\gamma = (\alpha + \beta)/2$
    if ensemble tracks $\hat{u}(t) = \gamma + P_{base}(t)$ **then**
      $\alpha = \gamma$
    else
      $\beta = \gamma$
    end if
  end while
  $P^+(t) = \alpha$
  return $P^+(t)$

Similarly, the algorithm to find $P^-(t)$ can be formulated by changing β to $-\beta$ and γ to $-\gamma$ in the above algorithm.

The power limits corresponding to an ensemble of 100 AC devices and 120 EWH devices are shown in Fig. 12.1. In Fig. 12.1, the time-varying lower and upper power

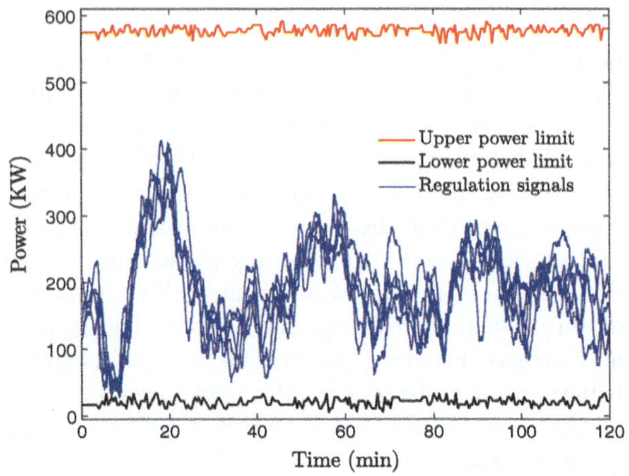

Fig. 12.1 Time-varying power limits for 100 AC devices and (scaled) PJM regulation signals

limits are shown in black and red. The regulation signals from PJM are considered and scaled appropriately to the size of the ensemble. A few regulation signals that satisfy the power limits are shown in blue in Fig. 12.1. If a regulation signal satisfies the power limits, then it can be applied to the VB to track that regulation signal. The charge/discharge limits corresponding to the VB are computed by taking the time derivative of the power limits.

12.2.1.2 VB Parameters: Initial Condition of VB State

The initial condition of the VB system corresponding to an AC ensemble is denoted by $x_0 = x(t)|_{t=0}$, where

$$x(t) = \sum_i \frac{T_i(t) - (T_i^{set} - \delta T/2)}{\eta_i/C_i}. \tag{12.2}$$

Similarly, the initial condition to the first-order VB model corresponding to an EWH ensemble is given by $x_{w0} = x_w(t)|_{t=0}$, where

$$x_w(t) = \sum_i \frac{T_{w_i}^{set} + \delta T_w/2 - T_{w_i}(t)}{1/C_{w_i}}. \tag{12.3}$$

Finally in the case of a VB for a heterogeneous ensemble with ACs and EWHs, the initial condition for the VB can be obtained as a sum of initial conditions defined in Eqs. (12.2) and (12.3).

After the time-varying power limits and initial condition of the VB corresponding to the ensemble are computed, the self-dissipation and capacity limits of the VB need to be computed. In VB parameter computation, it is important to ensure that if the ensemble fails to track a regulation signal, then the (linear first-order) VB also fails to track the regulation signal. In order to ensure this, the time-series data for the devices are generated after applying the regulation signals to the ensemble. Any feasible regulation signal corresponding to an ensemble is computed by adding the normalized regulation signals (from PJM) to the baseline power consumption of the ensemble. The baseline power consumption is the actual power consumption of the ensemble without any regulation signal being applied. Any regulation signal sent to the ensemble for tracking is met by appropriately switching ON/OFF the devices, which changes the aggregate power consumption of the ensemble to maintain it close to the regulation signal. The following section discusses one such control problem formulated as an optimization problem.

12.2.1.3 Tracking Regulation Signals

For a given ensemble of ACs/EWHs, the solution to the following optimization problem gives a control law indicating which devices have to be in an ON/OFF state such that the regulation signal is tracked at all times and user defined set points are satisfied.

$$\text{minimize}_s \; ||T(t+\Delta T) - T^{set}||_2^2 + ||T_w(t+\Delta T) - T_w^{set}||_2^2 \tag{12.4}$$

$$\text{subject to } T^{set} - \delta T/2 \leq T(t+\Delta T) \leq T^{set} + \delta T/2 \tag{12.5}$$

$$T_w^{set} - \delta T_w/2 \leq T_w(t+\Delta T) \leq T_w^{set} + \delta T_w/2 \tag{12.6}$$

$$|u_i(t) - s^\top P| \leq \varepsilon \tag{12.7}$$

$$s \in \{0, 1\}, \tag{12.8}$$

where

$$T_i(t+\Delta T) = e^{\frac{-1}{R_i C_i}\Delta T} T_i(t) + \left(T_{a_i} - s_i \eta P_i R_i\right)\left(1 - e^{\frac{-1}{R_i C_i}\Delta T}\right),$$

$$T_{w_i}(t+\Delta T) = e^{-a_i(t)\Delta T}\left(T_{w_i}(t) + 1\right) + \left(e^{-a_i(t)\Delta T} - 1\right)\frac{b_i(s_i, t)}{a_i(t)},$$

$\varepsilon > 0$, $u_i(t)$ is the ith regulation signal at time t and $T(t)$, $T_w(t)$ denote the vector of temperatures corresponding to AC and EWH devices at time t, respectively. The temperature set points and dead band temperature limits for AC (EWH) devices are given by T^{set} (T_w^{set}) and δT (δT_w).

The constraints, Eqs. (12.5) and (12.6), refer to the temperature bounds on AC and EWH devices. Notice that this optimization problem is nonconvex due to the binary constraints on the optimization variable s_i (ON/OFF status of each device) given in Eq. (12.8). Equation (12.7) guarantees that the ensemble power is maintained close to the regulation signal.

The tracking optimization problem is a mixed integer quadratic program and it can be solved by applying solvers such as Gurobi or the binary constraints on switching status and can be relaxed and solved using IPOPT as shown in [42]. For each regulation signal, the tracking regulation signals optimization problem is solved to identify whether the ensemble can be able to track the regulation signal at every time step. If at any time-point, the solution does not exist, then it implies that the ensemble failed to track the regulation signal.

An ensemble of 100 AC devices and 120 EWH ensemble devices are considered here. The temperature evolution of AC and EWH devices is given in [42]. The parameters such as set-point temperature, outside air temperature, thermal resistance, and thermal capacitance for the 100 AC devices are chosen such that no two devices have the same parameters. Similarly, the device parameters for all the EWHs are chosen. A medium water flow profile is considered for all the EWHs. The baseline power corresponding to 100 AC devices and a regulation signal are shown in Fig. 12.2.

Next, it remains to compute the VB parameters, capacity limits, and self-dissipation. The time-series data for training the autoencoder for transfer learning or for training the variational autoencoder is generated by applying several regulation signals to the ensemble and solving the optimization problem given in Eqs. (12.4)–(12.8) in order to make sure the regulation signal is tracked. If the ensemble of ACs and EWHs fails to track the regulation signal, then the time-series data is considered until the time-point at which the tracking fails. Sections 12.4 and 12.5 consider this time-series data for the deep neural network training to identify the self-dissipation and capacity limits of the VB.

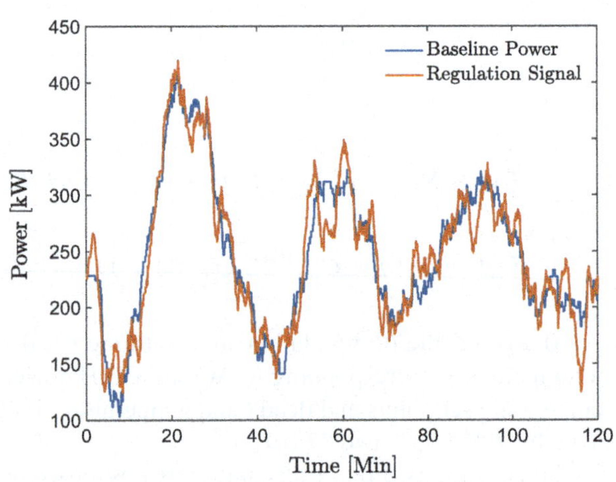

Fig. 12.2 A regulation signal and the baseline power of the ensemble of 100 AC devices and 120 EWHs

12.2.2 Validation of VB Model

This section contains a brief discussion on validating the VB models. This work proposes to use VB models as a measure of aggregate flexibility of the DERs such as ACs and EWHs. The VB parameters are identified for an ensemble with varying population as well as for an ensemble with uncertain device parameters. However, these VB parameters need to be validated. This VB parameter validation can be done in two different ways, open-loop and closed-loop. As the name suggests, the open-loop method does not involve applying a regulation signal while in the closed-loop method, a regulation signal is applied to the ensemble as well as the VB and their performance is compared. As the open-loop method to validate VB parameters seems intractable, the best way to compare these VB parameters is by applying a regulation signal. For any given regulation signal, if the ensemble fails to track that regulation signal, then an accurate fit of the VB model should capture this phenomenon in terms of VB capacity or power limit violations. It is important to mention here that the work by Hughes et al. [24] and Hughes [22] and subsequent works [42] enforces the condition that the linear first-order VB fails when the ensemble fails as a constraint in the computation of VB parameters.

12.2.3 Our Contributions

The aggregation of the DERs to form an equivalent battery can be based on several factors such as heterogeneity of devices and location, among other factors. We identified two major challenges in identifying the correct VB models for a given ensemble.

1. **Variability in the Ensemble Population** (Sect. 12.4): The number of devices in a given ensemble might vary as the end-use customers have the ability to opt-in or opt-out of providing grid services. A transfer learning based approach has been proposed to accommodate for these additional devices in the existing ensemble. The proposed framework utilizes a stacked *autoencoder* to mimic the dimensional reduction problem by finding VB representation at the encoding dimension. Furthermore, the addition of Net2DeeperNet and Net2WiderNet allows for the addition/subtraction of devices from the ensemble, which provides the required flexibility needed to account for variability in the ensemble population.
2. **Parametric Uncertainty in the Devices** (Sect. 12.5): Uncertainty in device parameters cannot be captured using the already proposed stacked autoencoder, due to its inherent deterministic architecture. This type of uncertainty can be classified into the aleatoric type. For example, one can introduce aleatoric uncertainties (stochastic processes) by running the aggregate simulations with different water flow profiles for EWHs. A machine learning algorithm utilizing a

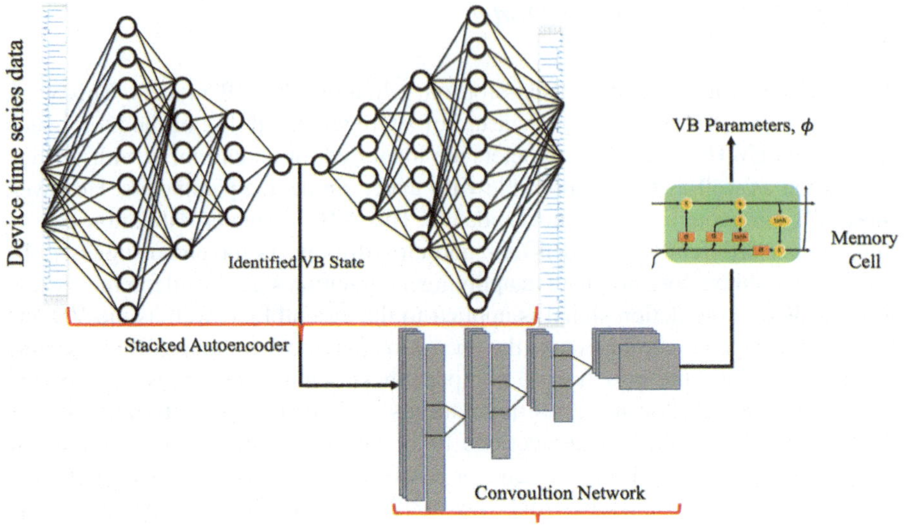

Fig. 12.3 Schematic of our proposed framework

variational autoencoder is proposed to discover uncertainty propagation patterns and thereby identifying a stochastic VB model.

Some preliminary findings of this work can be found in [11, 42]. The rest of this chapter is organized as follows. Section 12.2 discusses the mathematical models and describes how the VB parameters are identified. Section 12.3 discusses the necessary preliminaries before going to the technical details of the proposed frameworks. A transfer learning based VB computation is discussed in Sect. 12.4 to handle the variability in the ensemble population. Section 12.5 consists of stochastic VB computations applying the variational autoencoder framework to overcome the uncertainty in the device parameters. In Sect. 12.6, we discuss some numerical results using the proposed method for few example ensemble of devices. In Sect. 12.7 this chapter ends with final thoughts on this work and discusses future extensions.

Now before describing the technical details on the proposed frameworks, we draw the schematic of the whole framework and mention some preliminaries in the next section (Fig. 12.3).

12.3 Preliminaries

In this section we will describe the major components of our proposed deterministic and stochastic deep network based frameworks. The components are stacked autoencoder (SAE), long short-term memory (LSTM) network, convolution network (ConvNet), and probabilistic encoder and decoder.

12.3.1 Description of Stacked Autoencoder (SAE)

Autoencoder [1] (AE) is a type of deep neural network which is trained by restricting the output values to be equal to the input values. This also indicates both input and output spaces have the same dimensionality. The reconstruction error between the input and output of the network is used to adjust the weights of each layer. Therefore, the features learned by AE can well represent the input data space. Moreover, the training of AE is unsupervised, since it does not require label information.

We have considered a supervised learning problem with a training set of n (input,output) pairs $S_n = \{(x^{(1)}, y^{(1)}), \ldots, (x^{(n)}, y^{(n)})\}$, that is sampled from an unknown distribution $q(X, Y)$. X is a d dimensional vector in $\mathbb{R}^d$. $Z \in \mathbb{R}^{d'}$ is a lower $(d' < d)$ dimensional representation of X. Z is linked to X by a deterministic mapping f_θ, where θ is a vector of trainable parameters. We will now briefly mention the terminology associated with AE.

Encoder Encoder involves a deterministic mapping f_θ which transforms an input vector $\boldsymbol{x}$ into hidden representation $\boldsymbol{z}$. f_θ is an affine nonlinear mapping defined as

$$f_\theta(\boldsymbol{x}) = s(\boldsymbol{W}\boldsymbol{x} + \boldsymbol{b}), \tag{12.9}$$

where $\theta = \{\boldsymbol{W}, \boldsymbol{b}\}$ is a set of parameters, and $\boldsymbol{W}$ is $d' \times d$ weight matrix, $\boldsymbol{b}$ is a bias vector of dimension d', and $s(.)$ is an activation function.

Decoder The hidden dimensional representation $\boldsymbol{z}$ is mapped to dimension d, using mapping $g_{\theta'}$ and represented as $\hat{\boldsymbol{y}}$. The mapping $g_{\theta'}$ is called the decoder. Similar to f_θ, $g_{\theta'}$ is also an affine nonlinear mapping defined as

$$g_{\theta'}(\boldsymbol{z}) = s(\boldsymbol{W}'\boldsymbol{z} + \boldsymbol{b}'), \tag{12.10}$$

where $\theta' = \{\boldsymbol{W}', \boldsymbol{b}'\}$. Also, $\hat{\boldsymbol{y}}$ is not an exact reconstruction of $\boldsymbol{x}$, but rather in probabilistic terms as the parameters of a distribution $p(X|\hat{Y} = \hat{\boldsymbol{y}})$ that may generate $\boldsymbol{x}$ with high probability. We can equate the encoded and decoded outputs as $p(X|Z = z) = p(X|\hat{Y} = g_{\theta'}(\hat{\boldsymbol{y}}))$. The reconstruction error to be optimized is $L(\boldsymbol{x}, \hat{\boldsymbol{y}}) \propto -\log p(\boldsymbol{x}|\hat{\boldsymbol{y}})$. For real valued $\boldsymbol{x}$, $L(\boldsymbol{x}, \hat{\boldsymbol{y}}) = L_2(\boldsymbol{x}, \hat{\boldsymbol{y}})$, where $L_2(.,.)$ represents the Euclidean distance between two variables. In other words, we will use the squared error objective for training our autoencoder. For this current work, we will use affine and linear encoder along with affine and linear decoder with squared error loss.

Autoencoder training consists of minimizing the reconstruction error by carrying out the optimization

$$J_{AE} = \arg\min_{\theta,\theta'} \mathbb{E}_{q^0(X)}[L(X, \hat{Y}(X)],$$

where $\mathbb{E}_{(.)}[\times]$ denotes the expectation, and q^0 is the empirical distribution defined by samples in S_n. For the loss function defined before, the optimization problem can be rewritten as

$$J_{AE} = \arg\min_{\theta,\theta'} \mathbb{E}_{q^0(X)}[\log p(X|\hat{Y})] = g_{\theta'}(f_\theta(X))].$$

Intuitively, the objective of training an autoencoder is to minimize the reconstruction error amounts by maximizing the lower bound on shared information between X and hidden representation Z.

We utilize the proposed AE and stack them to initialize a deep network in a way similar to deep belief networks [19] or ordinary AEs [6, 30, 46]. Once the AEs have been properly stacked, the innermost encoding layer output is considered as a VB representation of the ensemble of TCLs. Furthermore, the number of layers of the stacked AEs are designed based on the reconstruction error for AEs. Keeping in mind a sudden change in dimension in both encoding and decoding layers can cause difficulty in minimizing the reconstruction error in J_{AE}. The parameters of all layers are fine-tuned using a stochastic-gradient descent approach [8].

12.3.2 Description of Long Short-Term Memory (LSTM) Network

We have used LSTM [21] to learn the long range temporal dependencies of the VB state of a given ensemble (encoded representation of the TCL states). For a LSTM cell with N memory units, at each time step, the evolution of its parameters is determined by

$$\begin{aligned}
i_t &= \sigma(W_{x_i} x_t + W_{h_i} h_{t-1} + W_{c_i} c_{t-1} + b_{i_1}),\\
f_t &= \sigma(W_{x_f} u_t + W_{h_f} h_{t-1} + W_{c_f} c_{t-1} + b_{i_f}),\\
z_t &= \tanh(W_{x_c} x_t + W_{h_c} h_{t-1} + b_c),\\
c_t &= f_t \odot c_{t-1} + i_t \odot z_t,\\
o_t &= \sigma(W_{u_o} u_t + W_{h_o} h_{t-1} + W_{c_o} c_{t-1} + b_{i_o}),\\
h_t &= o_t \odot \tanh(c_t),
\end{aligned}$$

where the $W_{x_{()}}$ and $W_{h_{()}}$ terms are the respective rectangular input and square recurrent weight matrices, $W_{c_{()}}$ are peephole weight vectors from the cell to each of the gates, σ denotes sigmoid activation functions (applied element-wise) and the i_t, f_t, and o_t equations denote the input, forget, and output gates, respectively; z_t is the input to the cell c_t. The output of a LSTM cell is o_t and denotes pointwise vector products. The forget gate facilitates resetting the state of the LSTM, while the peephole connections from the cell to the gates enable accurate learning of timings.

The goal of a LSTM cell training is to estimate the conditional probability $p(o_t|i_t)$ where i_t consists of a concatenated set of variables (VB state of previous time steps and control input) and o_t consists of the VB state of the current time step. The proposed LSTM calculates this conditional probability by first obtaining fixed dimensional representation v_t of the input i_t given by the hidden state h_t. Subsequently, the conditional probability of o_t is calculated by using the hidden state representation h_t. Given a training dataset with input S and output T, training of the proposed LSTM is done by maximizing the log probability of the training objective

$$J_{LSTM} = \frac{1}{|\mathcal{S}|} \sum_{(T,S)\in\mathcal{S}} \log p(T|S), \tag{12.11}$$

where $\mathcal{S}$ denotes training set. After successful training, the forecasting is done by translating the trained LSTM as

$$\hat{T} = \arg\max_{T} p(T|S), \tag{12.12}$$

where $\hat{T}$ is the LSTM based prediction of output dataset T.

12.3.3 Description of Convolution Neural Network (ConvNet)

A simple convolution neural network (ConvNet) is a sequence of layers, and every layer of a ConvNet transforms one volume of activations to another through a differentiable function. In this work, we used two types of layers to build ConvNet architectures: a convolution layer and a pooling layer. We stacked these two layers alternately to form a ConvNet. We can write the output of a one-dimensional ConvNet as follows:

- One-dimensional convolution layer
  - Accepts a volume of size $W_1 \times H_1 \times D_1$, where W_1 is the batch size of the training data set
  - Requires four hyperparameters: number of filters K_C, their spatial extent F_C, length of stride S_C, and the amount of zero padding P_C
  - Produces a volume of size $W_2 \times H_2 \times D_2$, where

$$\begin{aligned} W_2 &= W_1 \\ H_2 &= (H_1 - F_C + 2P_C)/S_C + 1 \\ D_2 &= K_C. \end{aligned} \tag{12.13}$$

  - Each filter in a convolution layer introduces $F \times F \times D_1$ weights, and in total $(F \times F \times D_1) \times K$ weights and K biases

- Pooling layer
  - Accepts a volume of size $W_2 \times H_2 \times D_2$
  - Requires two hyperparameters: spatial extent F_P and stride S_P
  - Produces a volume of size $W_3 \times H_3 \times D_3$, where

$$\begin{aligned} W_3 &= W_2 \\ H_3 &= (H_2 - F_P)/S_P + 1 \\ D_3 &= D_2. \end{aligned} \tag{12.14}$$

  - Introduces zero weights and biases, since the pooling layer computes a fixed function of the input.

Next, for our proposed ConvNet, we will outline the architecture as applied to predict VB state and simultaneously learn and estimate VB parameters.

- Input $[b_s \times l_b \times 1]$, where b_s denotes the batch size of our training process (2 h, with 1 s resolution).
- The convolutional (CONV) layer computes the output of neurons that are connected to local regions in the input, each computing a dot product between their weights and a small region they are connected to in the input volume. We have multiple CONV layers in our proposed ConvNet, each with a different filter size K_C. For an input layer of size $[b_s \times l_b \times 1]$, the output of a CONV layer will be $[b_s \times l_b \times K_C]$.
- The rectified linear unit (Relu) layer will apply an element-wise activation function, such as $\max(0, x)$ thresholding at zero. This leaves the size of the output volume unchanged to $[b_s \times l_b \times K_C]$.
- The pooling (POOL) layer will perform a down-sampling operation along the spatial dimension (width, height), resulting in an output volume such as $[b_s \times \frac{(l_b - F_P)}{S_P} + 1 \times K_C]$, with a filter size of $F_P \times S_P$.

12.3.4 *Overview of Probabilistic Encoder and Decoder*

Before we can say that our model is representative of our dataset, we need to make sure that for every data point X in the dataset, there is one (or many) setting of the latent variables which causes the model to generate something very similar to X. Formally, say we have a vector of latent variables z in a single dimensional (VB) space Z which we can easily sample according to some probability density function (PDF) $P(z)$ defined over Z. Then, say we have a family of deterministic functions $f(z; \theta)$, parameterized by a vector θ in some space Θ,where $f : Z \times \Theta \rightarrow X$. f is deterministic, but if z is random and θ is fixed, then $f(z; \theta)$ is a random variable in the space X. We wish to optimize θ such that we can sample z from $P(z)$ and with high probability $f(z; \theta)$ will be similar to X in our original dataset.

Now we define the previous description mathematically, by aiming to maximize the probability of each X in the original dataset under the entire generative process, according to

$$P(X) = \int P(X|z;\theta)P(z)dz. \tag{12.15}$$

In our later proposed variational autoencoder, the choice of output distribution is considered to be Gaussian, i.e., $P(X|z;\theta) = \mathcal{N}(X|f(z;\theta), \sigma^2 * I)$. In other words, it has mean $f(z;\theta)$ and covariance of the product of the identity matrix and a hyperparameter σ.

Now we describe the proposed framework for handling the variation in number of devices in the ensemble.

12.4 Variability in DER Population

Before going into the details of describing the method of defining VB state (x), along with the unknown VB modeling parameters (a, C_1, and C_2), we will demonstrate two important parameters of the DER ensemble and their associated variability and uncertainty. First, the variability is associated with the total number of DERs in the ensemble. For example, we can compute a VB model for an ensemble of 100 AC and 120 EWH devices. However, any addition/subtraction of AC/EWH devices in this existing ensemble results in recalculation of VB parameters. All the existing methods in literature fail to cope with the retraining in any other way than the standard "vanilla" way of retraining from scratch, which is not practically feasible and computationally burdensome. This motivates the necessity of developing a transfer learning based method to answer the variability in total number of devices and we will describe this method in detail in this current section.

12.4.1 Dataset Description and Dataset Splitting

We propose VB state x and subsequently VB parameter ϕ mentioned in Sect. 12.2.1 for an ensemble of homogeneous (AC or EWH devices) and heterogeneous devices (mixture of AC and EWH devices). The regulation signals from PJM [45] are considered and scaled appropriately to match the ensemble of ACs and EWHs. The devices in each ensemble have to change their state in order to follow a regulation signal. In doing so, while keeping the aggregate power of the ensemble close to the regulation signal, the switching action of the ensemble should not violate the temperature constraints of individual devices. The switching strategy is determined by the solution of an optimization problem as mentioned in Sect. 12.2.1.3.

For an ensemble of AC devices, a combination of 100 ACs is considered and the ON and OFF devices at every time instance are identified by solving an optimization problem. This generates the temperature state of each device, $T(t)$, at each time iteration, for 200 distinct regulation signals. If the ensemble fails to track a regulation signal, then the time-series data is considered up to the point where tracking fails. The outside air temperature and user set-point for each device are assumed to be same in this analysis.

The power limits of the ensemble are computed through a one-sided binary search algorithm as described in the Algorithm in Sect. 12.2.1. The 100 AC ensemble is simulated for 2 h with 1 s time resolution, for each regulation signal. For some regulation signals the ensemble violates the power limits P^- and P^+ before the 2 h running time and only the temperature of each AC is considered, until the time when the ensemble satisfies the power limit. Finally, for making a suitable dataset for applying stacked autoencoder (SAE), we stack the temperature of each AC device, followed by temperature set points for each device, and load efficiency and thermal capacity of each AC device, by column, and then stack the data points for each regulation signal by row. For the selected ensemble, this stacking results in a dataset of dimension $\mathbb{R}^{1440199 \times 203}$. To obtain the input stack to SAE for an ensemble of EWHs, a similar approach as described above for AC devices is followed. While generating this data, it is assumed that water flow into the WHs is at a *medium* rate.

We have used a tenfold cross validation for training and validation of our proposed SAE. We have kept the testing set separated as an indicator of generalized performance. For the given dataset (Sect. 12.4.1) 30% of the dataset is separated and kept as a test dataset. The remaining 70% of the dataset has been used in the random cross validation, for both training and validation of the proposed SAE.

12.4.1.1 Transfer Learning via Net2Net for SAE

The structure (input node numbers) for the proposed SAE depends on the number of TCLs in the ensemble. This requires retraining of the SAE if we change the number of TCL in the ensemble. We are proposing to use the developed Net2Net strategy [13] where there is a change in number or type of TCLs in the ensemble. In order to explain this idea in the context of VB state modeling, we define "source system" (S) as an ensemble of N devices (where N is a defined integer).[1] We will further define "target system" (T) as an ensemble of M devices (where $M \neq N$). The goal is that S will provide a good internal representation of the given task for T to copy and begin refining. This idea was initially presented as FitNets [47] and subsequently modified by Chen et al. in [13] as Net2Net.

We are proposing to combine two Net2Net strategies, namely Net2WiderNet and Net2DeeperNet.[2] Both of them are based on initializing the "target" network to

[1]For clarity we discuss Net2Net in the context of homogeneous device ensemble.

[2]Net2WiderNet and Net2DeeperNet were first introduced by Chen et al. in [13].

represent the same function as the "source" network. As an example, let the SAE representing S be represented by a function $\hat{\boldsymbol{y}} = f(\boldsymbol{x}, \theta)$, where $\boldsymbol{x}$ is input to the SAE, $\hat{\boldsymbol{y}}$ is the output from the SAE (which for SAE is the reconstruction of input x), and θ is the trainable parameters of the SAE. We propose to choose a new set of parameters θ' for the SAE representing T such that

$$\forall \boldsymbol{x}, f(\boldsymbol{x}, \theta) = g(\boldsymbol{x}, \theta').$$

Net2DeeperNet As the name suggests, Net2DeeperNet allows us to transform a network into a deeper one. Mathematically, Net2DeeperNet replaces one layer with two layers, i.e., $h^{(i)} = \phi(h^{(i-1)T} W_1^{(i)})$ gets replaced by $h^{(i)} = \phi(W_2^{(i)T} \phi(W_1^{(i)T} h^{(i-1)}))$. The new weight matrix W_2 is initialized as identity matrix and get updated in the training process. Moreover, we need to ensure that $\phi(I\phi(v)) = \phi(v)$ for all v, in order to ensure Net2DeeperNet can successfully replace the original network with deeper ones.

Net2WiderNet Net2WiderNet allows a layer to be replaced with a wider layer, meaning a layer that has more neurons (it can be also narrower if needed). Suppose that layer i and layer $i + 1$ are both fully connected layers, and layer i uses an element-wise non-linearity. To widen layer i, we replace $W^{(i)}$ with $W^{(i+1)}$. If layer i has m inputs and n outputs, and layer $i + 1$ has p outputs, then $W^{(i)} \in \mathbb{R}^{m \times n}$ and $W^{(i+1)} \in \mathbb{R}^{n \times p}$. Net2WiderNet allows us to replace layer i with a layer that has q outputs, with $q > n$. We will introduce a random mapping function $g : \{1, 2, \ldots, q\} \to \{1, 2, \ldots, n\}$, that satisfies

$$g(j) = j, j \leq n$$
$$g(j) = \text{randomsamplefrom} 1, 2, \ldots, n, j > n.$$

The new weights in the network for target T is given by

$$U_{k,j}^{(i)} = W_{k,g(j)}^{(i)},$$
$$U_{j,h}^{(i+1)} = \frac{1}{|\{x|g(x) = g(j)\}|} W_{g(j),h}^{(i+1)}.$$

Here, the first n columns of $W^{(i)}$ are copied directly into $U^{(i)}$. Columns $n + 1$ through q of $U^{(i)}$ are created by choosing a random sample as defined in g. The random selection is performed with replacement, so each column of $W^{(i)}$ is potentially copied many times. For weights in $U^{(i+1)}$, we must account for the replication by dividing the weight by replication factor given by $\frac{1}{|\{x|g(x)=g(j)\}|}$, so all the units have exactly the same value as the unit in the network in source S.

The addition of devices can be handled by the Net2Net architecture of the proposed framework. Table 12.1 shows two examples of device addition for an

Table 12.1 Performance comparison between with and without transfer learning for the proposed deterministic framework in Fig. 12.4

	Number of devices		Untrained parameters		Pretrained parameters		Epoch		Reconstruction error	
Method	AC	EWH	AC	EWH	AC	EWH	AC	EWH	AC	EWH
w/o transfer learning	100	0	205,880	NA	0	NA	2000	NA	0.0028	NA
w/o transfer learning	112	0	220,280	NA	0	NA	2000	NA	0.0032	NA
with transfer learning	112	0	14,400	NA	205,880	NA	215	NA	0.0028	NA
w/o transfer learning	0	120	NA	222,980	NA	0	NA	2000	NA	0.0069
w/o transfer learning	0	135	NA	241,890	NA	0	NA	2000	NA	0.072
with transfer learning	0	135	NA	18,910	NA	222,980	NA	718	NA	0.0061

ensemble of AC and EWH devices, along with the computational benefit of using Net2Net based method. Device subtraction from the ensemble cannot be handled in similar manner, as this results in empty (untrainable) neurons in the proposed framework. Our future work will try to address this aspect.

12.4.2 Proposed Method Description–Example Problem 100 AC Device Ensemble

The SAE introduced in Sect. 12.4.1 is trained on the dataset described in Sect. 12.4.1 for an ensemble of 100 AC devices. The objective of the training of SAE is to represent the given 203 dimensional dataset into a 1 dimensional encoded space, and subsequently transforms the 1 dimensional encoded representation back into the 203 dimensional original data space, with tolerable loss. The selected layer dimension of the proposed SAE is 203-150-100-50-20-1-20-50-100-150-203, where all the activation functions are linear. Moreover, the variables in 203 input dimensions are not normalized, to represent the VB state dependency on the input variables. That also motivates the necessity of having unbounded linear activation functions, throughout the proposed SAE.

Next, when more AC devices are added to the given 100 AC devices ensemble, we leverage the proposed Net2Net framework introduced in Sect. 12.4.1.1, for the retraining and subsequent representation of the VB state for the dataset representing the new ensemble. Obviously the robust way is to retrain the proposed SAE architecture from scratch for the new dataset, but that includes higher computation cost and time. We can utilize the red network already trained on 100 AC devices ensemble, for the new ensemble dataset, which results in significant savings of computation cost and time.

Finally, we introduced a convolution based LSTM network for forecasting the VB state evolution, given any regulation signal. Given the SAE is only able to represent the VB state for the given time the state of TCL is available, we must utilize a deep network for predicting time evolution of VB state for the ensemble of TCLs. Simultaneously, this proposed convolution based LSTM network can be used to estimate the remaining unknown parameter a in the vector ϕ, which represents all the parameters related to VB.

12.4.3 Long Term VB State Prediction–Two-Stepped Training Process

Figure 12.4 shows the proposed deep network architecture addresses the variability of total number of DERs in the ensemble. We have to keep in mind the SAE is only active during the time when we have measurements from devices in the

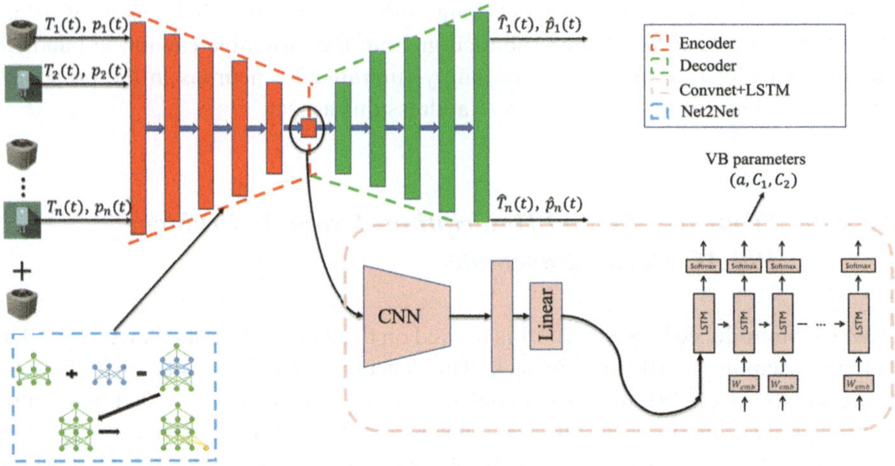

Fig. 12.4 Proposed transfer learning based deterministic framework

ensemble. Outside of that time window, we depend on the prediction of VB state from the proposed ConvNet+LSTM architecture (as shown in Fig. 12.4). For the dataset described in Sect. 12.4.1, the VB state is predicted using a novel two-stepped training process. During the first step, given X and Y, the network described in Fig. 12.4 is trained to find a nonlinear mapping F_1, which satisfies the tolerance criteria (ϵ) associated with prediction accuracy. After successful first step training, F_1 is applied on the input testing matrix, $X \triangleq [X_0^T, X_1^T, \ldots, X_k^T]^T, k = n_T + 1, n_T + 2, \ldots, N$, and the mapping output $F_1(X)$ can be denoted as $\hat{Y}$, for the testing data. Before going to the next training step, the previous VB state (identified by the SAE) is replaced by $\hat{Y}$. This will ensure that while predicting VB state in long term scale, we are using a previous prediction using the ConvNet+LSTM network in Fig. 12.4 instead of identified VB state from the SAE.

Before explaining the two-stepped training algorithm, we want to point out the motivation behind our definition of this training. Intuitively, the proposed network will learn the unsupervised prediction problem during the first training step. Successful satisfaction of stopping criteria defined by (ϵ) does not guarantee the network's performance in the testing phase (as the network is never trained on the testing period). This results in an accumulation of prediction error, while applying the trained network on the testing data, after the first step of training. We are proposing a second step training process, where the network will learn a way to mitigate this prediction error accumulation by looking at the desired output when SAE is in operation. In Algorithm 2 we showed the necessary steps for the second step training process. Output $\hat{X}$ from Algorithm 2, replaces the input data X for the proposed deep network for the second stepped training, while output Y of the deep network remains unchanged. Traditional one-step training method cannot rectify the prediction error accumulation problem over time, unlike the proposed two-stepped training.

Algorithm 2 Algorithm for 2nd step of the training process for finding long term "accurate" prediction of VB state

Input: Trained Network after first step (F_1), window size d, **X, Y**
Output: $\hat{X}$
  for i in range(0,N) **do**
    count=-d
    if $i < d$ **then**
      for j in range(0,5d) **do**
        $\alpha[0, j, 0] = \mathbf{X}(j + 5d * i)$
        $\gamma[i, j, 0] = \alpha[0, j, 0]$
      end for
      $\beta[i] = \mathbf{Y}(i)$
    else
      for j in range(1,5d) **do**
        $\alpha[0, j, 0] = \mathbf{X}(j + 5d * i)$
        $\gamma[i, j, 0] = \alpha[0, j, 0]$
      end for
      for k in range(0,5d,5) **do**
        $\alpha[0, k, 0] = \beta[i + count]$
        $\gamma[i, j, 0] = \alpha[0, j, 0]$
        $count = count + 1$
      end for
      $\beta[i] = F_1(\alpha)$
    end if
  end for
  $\hat{X} = \gamma$
  return $\hat{X}$

12.4.4 Probabilistic Moments

At this point, it is important to find the analytic expression of respective (mean,standard deviation) of the VB state (single dimensional) representation, given the (mean,standard deviation) of the input space, as we have normalized input data using its mean and standard deviation. Let us consider $\mathbf{X} \sim \mathcal{N}(\mu_X, \Sigma_X)$ is the distribution of the input data **X**. Input data **X**, passes through a network of the form (Affine, Affine, Affine, Relu) as shown in Fig. 12.5, before transforming to VB state (z) in the encoding space. For calculation simplicity, we break down this series of transformation into the following two-stepped structure $\mathbf{X} \xrightarrow{\text{(Affine,Affine)}} \mathbf{Y} \xrightarrow{\text{(Affine,Relu)}} z$, then we can write $\mathbf{Y} = \mathbf{W}_2\Big(\mathbf{W}_1\mathbf{X} + \mathbf{B}_1\Big) + \mathbf{B}_2$, where $\mathbf{Y} \in \mathbb{R}^{150}$, $\mathbf{W}_2 \in \mathbb{R}^{150\times 200}$, $\mathbf{B}_2 \in \mathbb{R}^{150\times 1}$, $\mathbf{W}_1 \in \mathbb{R}^{200\times 295}$, and $\mathbf{B}_1 \in \mathbb{R}^{200\times 1}$. After some algebraic manipulations, equivalent (mean-standard deviation) for **Y**, corresponding to (μ_X, Σ_X), can be written as:

$$\Sigma_Y = \Sigma_X, \tag{12.16}$$

$$\mu_Y = \mathbf{W}_2\mathbf{W}_1\mu_X + (1 - \Sigma_X)\Big(\mathbf{W}_2\mathbf{B}_1 + \mathbf{B}_2\Big). \tag{12.17}$$

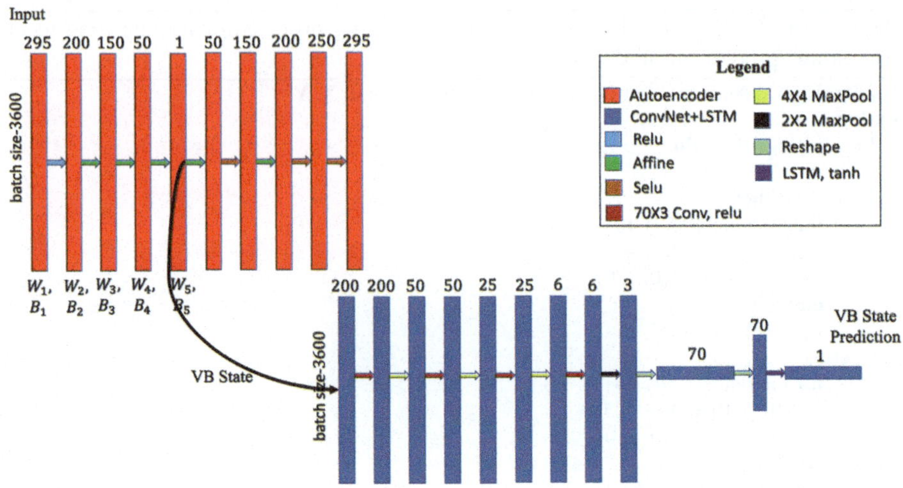

Fig. 12.5 Deep network architecture of the proposed SAE based ConvNet+LSTM, for VB state identification and prediction

Now using (12.17), we have designed $\mathbf{B}_2$ to enforce a zero mean distribution at $\mathbf{Y}$, i.e., $\mathbf{Y} \sim \mathcal{N}(0, \Sigma_Y)$. $\mathbf{B}_2$ is designed as:

$$\mathbf{B}_2 = -\frac{\mathbf{W}_2 \mathbf{W}_1 \mu_X}{(1 - \Sigma_X)} - \mathbf{W}_2 \mathbf{B}_1, \tag{12.18}$$

to achieve zero mean distribution at $\mathbf{Y}$.

Let us consider $\mathbf{Y} \sim \mathcal{N}(\mu_Y = 0, \Sigma_Y)$ is the output in the $\mathbf{Y}$ space, as discussed before where $\mu_Y = 0$ and Σ_Y is the known mean and standard deviation of vector $\mathbf{Y}$. After passing $\mathbf{Y}$, through a network of the form (Affine, ReLU, Affine), the functional form of the network output is $z = q(\mathbf{Y}) = \mathbf{W}_4 \max(\mathbf{W}_3 \mathbf{Y} + \mathbf{B}_3, \mathbf{0}_{50}) + \mathbf{B}_4$, where max(.) is an element-wise operator. For our application $q : \mathbb{R}^{150} \rightarrow \mathbb{R}$, $\mathbf{W}_3 \in \mathbb{R}^{50 \times 150}$, $\mathbf{W}_4 \in \mathbb{R}^{1 \times 50}$, $\mathbf{B}_3 \in \mathbb{R}^{50 \times 1}$, $\mathbf{B}_4 \in \mathbb{R}$, and $\mathbf{0}_{50}$ is a 50-dimensional vector of zeros. Now we will state two theorems to give analytic expression of first and second statistical moments of $q(\mathbf{Y})$, where $\mathbf{Y} \sim \mathcal{N}(\mu_Y = 0, \Sigma_Y)$ (see [7] for proof of these theorems).

Theorem *First Moment Theorem: For any function in the form of* $q(\mathbf{Y})$, *we get*

$$\mathbb{E}[q_i(\mathbf{Y})] = \sum_{j=1}^{50} \mathbf{W}_4(i, j) \Big(\frac{1}{2} \mu_j - \frac{1}{2} \mu_j erf\Big(\frac{-\mu_j}{\sqrt(2) \sigma_j} \Big)$$

$$+ \frac{1}{\sqrt{2\pi}} \sigma_j \exp\left(\frac{-\mu_j^2}{2\sigma_j^2} \right) \Big) + \mathbf{B}_4(i), \tag{12.19}$$

where $\mu_j \triangleq \mathbf{W}_3\mu_Y + \mathbf{B}_3$, $\sigma_j \triangleq \bar{\Sigma}(j, j)$, $\bar{\Sigma} \triangleq \mathbf{W}_3\Sigma_Y\mathbf{W}_3^T$ *and* $erf(x) \triangleq \frac{2}{\sqrt{\pi}}\int_0^x e^{-t^2}dt$.

Theorem *Second Moment Theorem: For any function in the form of* $q(\mathbf{Y})$ *where* $\mathbf{Y}\sim\mathcal{N}(0, \Sigma_Y)$ *and* $\mathbf{B}_3 = \mathbf{0}_{50}$ *we get*

$$\mathbb{E}\left[q_i^2(\mathbf{Y})\right] = 2\sum_{j_1=1}^{50}\sum_{j_2=1}^{j_1-1}\mathbf{W}_4(i, j_1)\mathbf{W}_4(i, j_2)\left(\frac{\sigma_{j_1,j_2}}{2\pi}\sin^{-1}\left(\frac{\sigma_{j_1,j_2}}{\sigma_{j_1}\sigma_{j_2}}\right)\right.$$
$$\left.+\frac{\sigma_{j_1}\sigma_{j_2}}{2\pi}\sqrt{1-\frac{\sigma_{j_1,j_2}^2}{\sigma_{j_1}^2\sigma_{j_2}^2}+\frac{\sigma_{j_1,j_2}}{4}}\right)+\frac{1}{2}\sum_{r=1}^{50}\mathbf{W}_4(i, r)^2\sigma_r^2 + \mathbf{B}_4(i). \tag{12.20}$$

Now mean μ_z and standard deviation Σ_z of single dimensional VB state Z can be calculated using (12.19) and (12.20) as:

$$\mu_z \triangleq \mathbb{E}[q_i(\mathbf{Y})],\ \Sigma_z \triangleq \sqrt{\mathbb{E}[q_i^2(\mathbf{Y})] - \left(\mathbb{E}[q_i(\mathbf{Y})]\right)^2}. \tag{12.21}$$

Now we describe the framework for handling the parametric uncertainties in the devices, in the next section.

12.5 Parametric Uncertainties in the Devices

In this section we consider uncertainty in one of the device parameters, namely the water draw profile of the EWH in the ensemble. Furthermore, we have considered an ensemble of 100 AC devices (no uncertainties of device parameters) and 150 EWH devices (uncertainty in the water draw profile) for VB state calculation. As described in Sect. 12.4.3, the long term prediction algorithm is borrowed in this section as well. In Fig. 12.6, we have plotted the stochastic (on the right) and deterministic (on the left) water draw profile, used in the EWH for generating device level data. We propose a variational autoencoder (VAE) based framework to address these types of uncertainties in the ensemble, and consequently SAE is replaced by this designed VAE in Fig. 12.7.

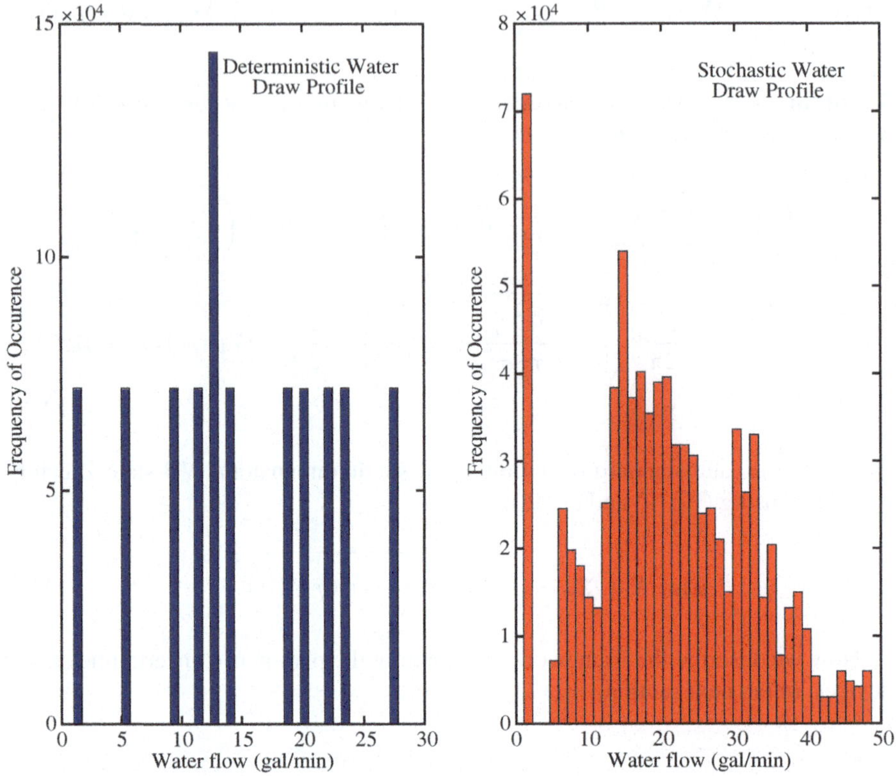

Fig. 12.6 Deterministic and stochastic water draw profile used for EWH, used for training the SAE and VAE in Figs. 12.5 and 12.7, respectively

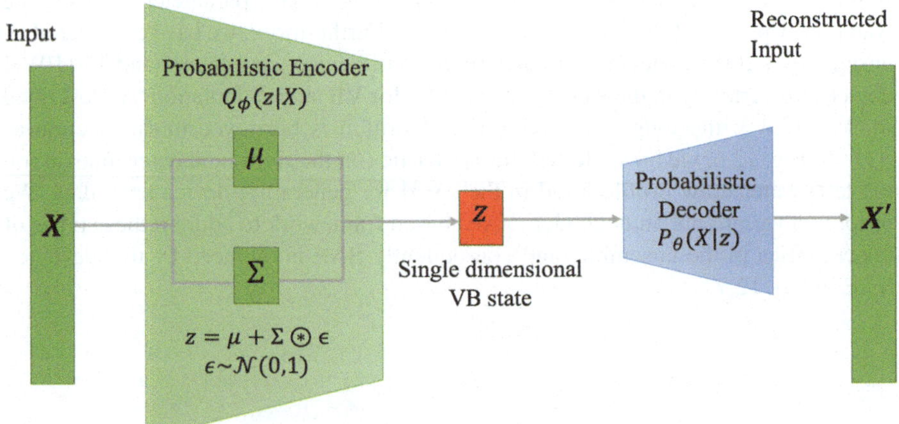

Fig. 12.7 Schematic architecture of proposed VAE, **X** represents the device level data from the ensemble as described in Sect. 12.4.1, **z** represents the single dimensional VB state representation and finally **X**′ represents the reconstruction of the original device level dataset. It is important to note that this framework can also be used to create realistic synthetic device level dataset

12.5.1 Description of the Proposed Variational Autoencoder (VAE)

The key idea behind training our proposed VAE is to attempt to sample z, which is likely to produce a data point similar to X, and simultaneously compute $P(X)$, just for the sample z. We now define a function $Q(z|X)$ which can take a value of X and give us a distribution over z values that are likely to produce similar data points as in X.

First we define Kullback–Leibler divergence between $P(z|X)$ and $Q(z)$, for some arbitrary Q, i.e.,

$$\mathcal{D}[Q(z)||P(z|X) = E_{z\sim Q}[\log Q(z) - \log P(z|X)]. \tag{12.22}$$

We next apply the Bayes rule to $P(z|X)$ in Eq. (12.22), and rewrite Eq. (12.22) as

$$\mathcal{D}[Q(z)||P(z|X) = E_{z\sim Q}[\log Q(z) - \log P(X|z) - \log P(z)] + \log P(X). \tag{12.23}$$

Now after negating both sides, and contracting part of $E_{z\sim Q}$ into a KL-divergence the terms yield

$$\log P(X) - \mathcal{D}[Q(z)||P(z|X)] = E_{z\sim Q}[\log P(X|z)] - \mathcal{D}[Q(z)||P(z)]. \tag{12.24}$$

Since we want to infer $P(X)$, we construct Q which does not depend on X, and in particular makes $\mathcal{D}[Q(z)||P(z|X)]$ small, i.e.,

$$\log P(X) - \mathcal{D}[Q(z)||P(z|X)] = E_{z\sim Q}[\log P(X|z)] - \mathcal{D}[Q(z|X)||P(z)]. \tag{12.25}$$

We want to maximize the left-hand side of Eq. (12.25), while we also want to optimize the right-hand side of Eq. (12.25) using stochastic-gradient descent, given right choice of Q. Moreover, the right-hand side of Eq. (12.25) behaves similar to an autoencoder, where Q encodes X into z and P decodes it to reconstruct X.

In order to perform stochastic-gradient descent on the right side of Eq. (12.25), we choose $Q(z|X) = \mathcal{N}(z|\mu(X;\theta), \Sigma(X;\theta))$, where μ and Σ are deterministic functions with parameters θ, that can be learned from data. In our framework Σ is constrained to be a diagonal matrix. Because of the choice of Q, the last term of the right-hand side becomes

$$\mathcal{D}[Q(z|X)||P(z)] = \frac{1}{2}(\mathrm{tr}(\Sigma(X)) + (\mu(X))^T(\mu(X) - k - \log(\det(\Sigma(X))))), \tag{12.26}$$

where k is the dimensionality of the distribution. Finally our objective function for optimization associated with training of VAE (shown in Fig. 12.8) can be written as

$$E_{X\sim D}[\log P(X) - \mathcal{D}[Q(z|X)||P(z|X)]] = E_{X\sim D}[E_{z\sim Q}[\log P(X|z)] - \mathcal{D}[Q(z|X)||P(z)]]. \tag{12.27}$$

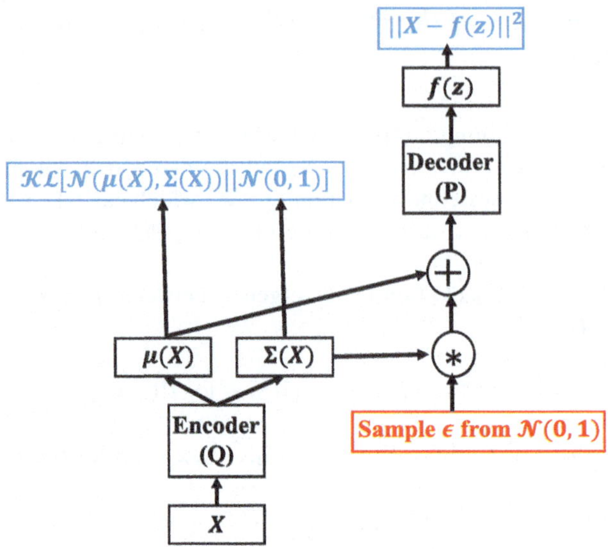

Fig. 12.8 Training algorithm of the proposed VAE in Fig. 12.7 is shown, where red color shows sampling operations and blue shows loss functions. ϵ can be tuned to generate various realistic device state information once the VAE is trained (especially using the Probabilistic Decoder block). This adds as a byproduct of the proposed VAE architecture

In Fig. 12.7 we describe the proposed VAE architecture, with z denoting the single dimensional VB state. Similar as before (as in Sect. 12.4.1), the dataset for this VAE training consists of the individual device state and parameter for all the ACs and WHs. Furthermore, in Fig. 12.8 we show a schematic of the proposed training algorithm of the VAE. We have used stochastic gradient for this training.

Now we evaluate performance of proposed SAE and VAE and consequently identify VB parameter (ϕ, as introduced in Sect. 12.2.1) for few example ensemble of ACs and EWHs.

12.6 Numerical Results

We simulate few ensembles of a combination of ACs and EWHs ensembles, for evaluating the performance of our proposed deterministic and stochastic VB framework. The deterministic VB framework helps to address variability in the total number of devices in the ensemble via the principle of transfer learning. Table 12.1 shows a comparison study between our proposed transfer learning based framework and the "vanilla" way of retraining from scratch (without transfer learning), for two different homogeneous ensembles, namely an ensemble of AC and an ensemble of EWH devices. In this context, Table 12.1 shows an average computation time[3] savings of 77%, when identifying VB states for different ensembles.[4]

Figure 12.7 describes the proposed VAE architecture for addressing the parametric uncertainty, such as the uncertain water draw profile of EWHs in the ensemble. We have selected an ensemble of 100 ACs and 150 EWHs for numerical demonstration of this framework. Figures 12.9 and 12.10 show the identified VB parameters along with their associated uncertainties (rest of the VB parameter, namely P^+ and P^- are identified in Sect. 12.2.1.1). We have also calculated the 95% confidence interval values of both energy capacities and self-dissipation rate of the proposed VB model, for the selected heterogeneous ensemble with parametric uncertainty in water draw profile as in Fig. 12.6.

Similarly, identified VB state using the deterministic framework in Fig. 12.8 is shown in Fig. 12.11, for a 2-h window of operation. We have also plotted temporal evolution of a few AC and EWH devices subjected to a regulation signal, during the same time period.

To sum up, we have showed performance of our proposed SAE and VAE based frameworks for several homogeneous and heterogeneous ensembles. The SAE based framework shows significant computational time savings without any performance compromise. VAE based framework successfully identifies VB model parameter in case of parametric uncertainty in the device level. We want to point out that no information regarding the device level uncertainty is used while training the proposed VAE based framework, which shows promise of utilization of this type of framework on real device data.

12.7 Concluding Remarks and Future Directions

In this chapter we have proposed two deep learning based frameworks for identifying VB state, along with the parameters in the first-order VB model. By virtue of these two frameworks, we have addressed the variability of total number of devices in the ensemble and the parametric uncertainty in the devices. The first framework is a transfer learning based SAE, which identifies the VB state when there is variability in the total number of devices in the ensemble, in addition we have proposed a ConvNet+LSTm based extension of this proposed framework which identifies the unknown parameter in the first-order VB model. We have to point out that this framework identifies VB parameters in a deterministic fashion. However, the second framework identifies VB parameters in case of presence uncertainty

[3]We used Epoch (number of training iterations) from Table 12.1 as a measure of computation time.

[4]There is no direct correlation between the reconstruction error and the flexibility of device ensemble. However, we have used reconstruction error, a way to validate our claim that virtual battery model is an equivalent model of the device ensemble. For future validation we have planned to use tracking performance of device ensemble and the virtual battery representation.

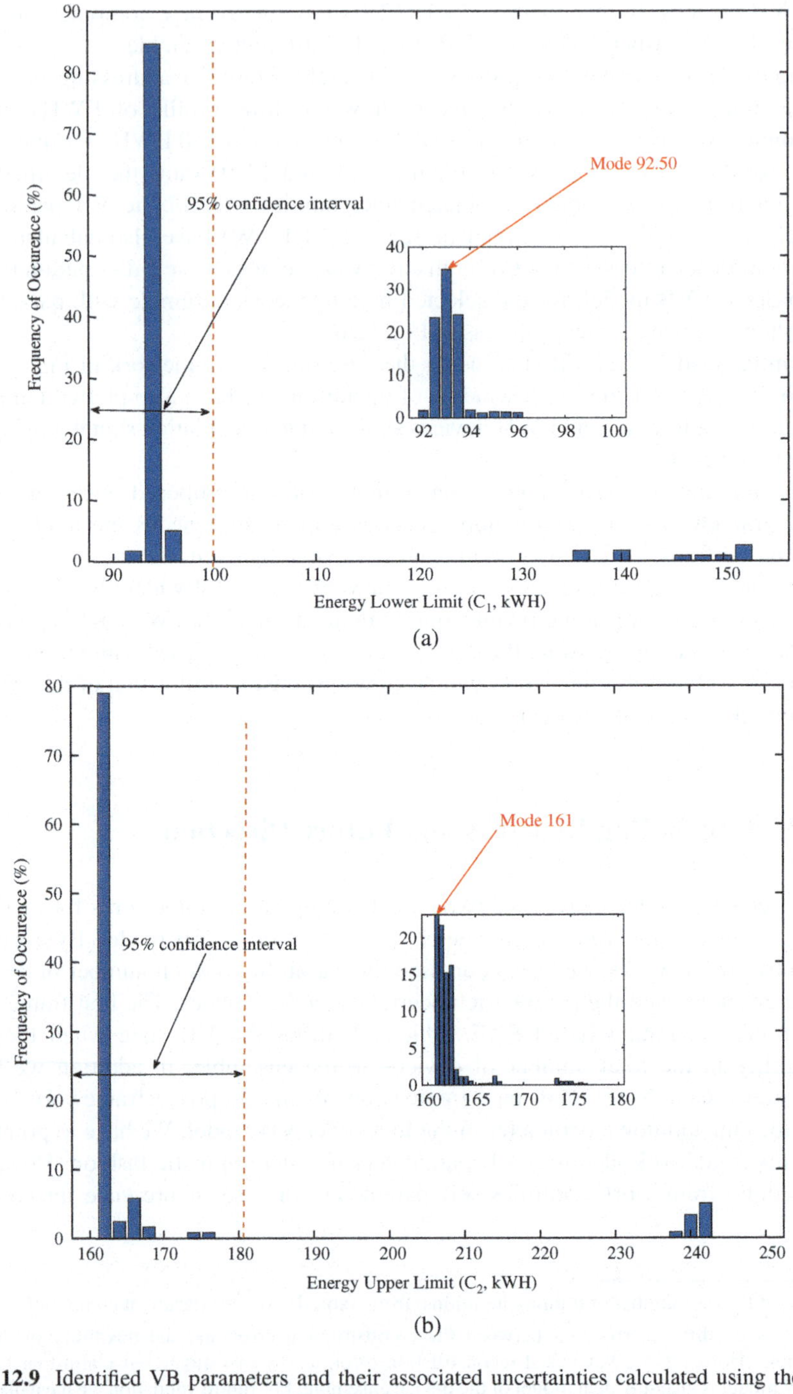

Fig. 12.9 Identified VB parameters and their associated uncertainties calculated using the proposed VAE in Fig. 12.7. (**a**) Lower bound of VB energy capacity, C_1. (**b**) Upper bound of VB energy capacity, C_2

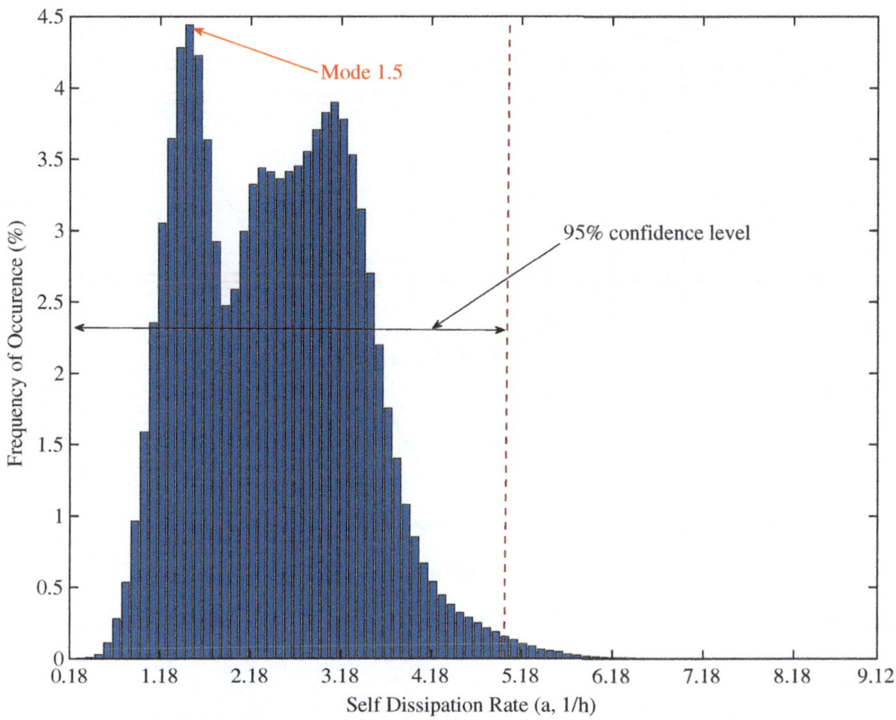

Fig. 12.10 Identified VB parameters (self-dissipation rate) and its associated uncertainties calculated using the proposed VAE in Fig. 12.7

in the device parameters. This framework utilized VAE based architecture to identify first-order VB model parameters and their possible variations, in case of presence uncertainty. Finally, we have shown several numerical demonstrations of our proposed frameworks in the context of homogeneous and heterogeneous ensembles of ACs and WHs.

Existing state of the art methods for identification of VB parameters for an ensemble of thermostatic loads can be either via closed-form expressions or via optimization-based techniques. Both of these require the knowledge of the individual load models and parameters. In real-world applications, however, it is expected that very little about the individual load models and parameters are known. End-use measurements such as power consumption, room temperature, and device on/off status are the only sources of information available. This motivates the necessity of using our proposed frameworks for real in-field applications. Furthermore, our proposed frameworks require the dataset to be generated by using a series of regulation signals. These regulation signals might not be available in a physical application. We want to explore our proposed frameworks in this context and evaluate the effectiveness of our proposed frameworks when regulation signals are not available.

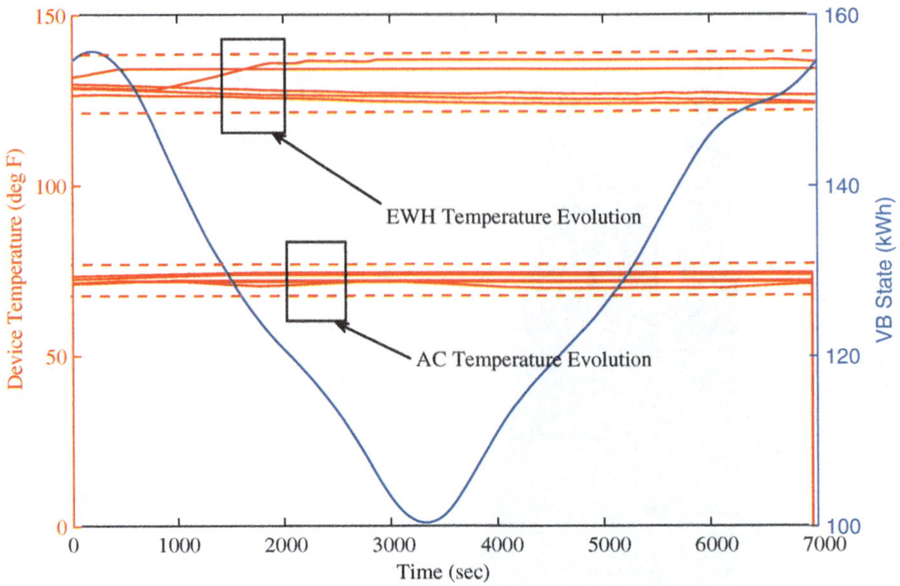

Fig. 12.11 Device states for both ACs and EWHs along with the time evolution of the deterministic VB state during a 2-h period

Furthermore, this framework requires sufficient device level data available for sufficiently long time to handle the temporal transience. For our application we have chosen to run the device level simulation for 2 h period of time, with 1 s resolution, as previously mentioned. As we have worked with simulated dataset, data quality is not a concern but for practical implementation of our framework, data quality can be a concern while using stacked autoencoder. Our future work will try to address this aspect.

Acknowledgements The technical work was funded by U.S. Department of Energy under the ENERGISE program (contract DE-AC02-76RL01830). The writing of this book chapter was supported by Pacific Northwest National Laboratory Directed Research and Development (LDRD) Funds. The authors would like to thank Dr. Slaven Peles and Dr. Vaibhav Donde for their helpful comments and suggestions. The authors would also like to thank Colleen Winters for her detailed technical edits.

References

1. P. Baldi, Autoencoders, unsupervised learning, and deep architectures, in *Proceedings of ICML Workshop on Unsupervised and Transfer Learning* (2012), pp. 37–49
2. S. Bashash, H.K. Fathy, Modeling and control insights into demand-side energy management through setpoint control of thermostatic loads, in *American Control Conference (ACC'11)* (IEEE, Piscataway, 2011), pp. 4546–4553

3. S. Bashash, H.K. Fathy, Robust demand-side plug-in electric vehicle load control for renewable energy management, in *Proceedings of the 2011 American Control Conference* (IEEE, Piscataway, 2011), pp. 929–934
4. S. Bashash, H.K. Fathy, Modeling and control of aggregate air conditioning loads for robust renewable power management. IEEE Trans. Control Syst. Technol. **21**(4), 1318–1327 (2013)
5. Y. Bengio, H. Lee, Editorial introduction to the neural networks special issue on deep learning of representations. Neural Netw. **64**(C), 1–3 (2015)
6. Y. Bengio, P. Lamblin, D. Popovici, H. Larochelle, Greedy layer-wise training of deep networks, in *Advances in Neural Information Processing Systems* (2007), pp. 153–160
7. A. Bibi, M. Alfadly, B. Ghanem, Analytic expressions for probabilistic moments of PL-DNN with gaussian input, in *Proceedings of the IEEE Conference on Computer Vision and Pattern Recognition* (2018), pp. 9099–9107
8. L. Bottou, Large-scale machine learning with stochastic gradient descent, in *Proceedings of COMPSTAT'2010*, pp. 177–186. Springer, Berlin (2010)
9. D.S. Callaway, Tapping the energy storage potential in electric loads to deliver load following and regulation, with application to wind energy. Energ. Conver. Manage. **50**(5), 1389–1400 (2009)
10. D.S. Callaway, I.A. Hiskens, Achieving controllability of electric loads. Proc. IEEE **99**(1), 184–199 (2011)
11. I. Chakraborty, S.P. Nandanoori, S. Kundu, Virtual battery parameter identification using transfer learning based stacked autoencoder, in *2018 17th IEEE International Conference on Machine Learning and Applications (ICMLA)*, (IEEE, Piscataway, 2018), pp. 1269–1274
12. C.Y. Chang, W. Zhang, J. Lian, K. Kalsi, Modeling and control of aggregated air conditioning loads under realistic conditions, in *2013 IEEE PES Innovative Smart Grid Technologies Conference (ISGT)* (IEEE, Piscataway, 2013), pp. 1–6
13. T. Chen, I. Goodfellow, J. Shlens, Net2net: accelerating learning via knowledge transfer. (2015, preprint). arXiv:1511.05641
14. C. Guille, G. Gross, A conceptual framework for the vehicle-to-grid (v2g) implementation. Energy Policy **37**(11), 4379–4390 (2009)
15. H. Hao, A. Kowli, Y. Lin, P. Barooah, S. Meyn, Ancillary service for the grid via control of commercial building HVAC systems, in *2013 American Control Conference* (IEEE, Piscataway, 2013), pp. 467–472
16. H. Hao, B.M. Sanandaji, K. Poolla, T.L. Vincent, A generalized battery model of a collection of thermostatically controlled loads for providing ancillary service, in 2013 51st Annual Allerton Conference on Communication, Control, and Computing (Allerton) (IEEE, Piscataway, 2013), pp. 551–558
17. H. Hao, B.M. Sanandaji, K. Poolla, T.L. Vincent, Aggregate flexibility of thermostatically controlled loads. IEEE Trans. Power Syst. **30**(1), 189–198 (2015)
18. H. Hao, D. Wu, J. Lian, T. Yang, Optimal coordination of building loads and energy storage for power grid and end user services. IEEE Trans. Smart Grid **9**, 4335–4345 (2017)
19. G.E. Hinton, R.R. Salakhutdinov, Reducing the dimensionality of data with neural networks. Science **313**(5786), 504–507 (2006)
20. G.E. Hinton, S. Osindero, Y.W. Teh, A fast learning algorithm for deep belief nets. Neural Comput. **18**(7), 1527–1554 (2006)
21. S. Hochreiter, J. Schmidhuber, Long short-term memory. Neural Comput. **9**(8), 1735–1780 (1997)
22. J.T. Hughes, A framework for enabling the utilization of flexible loads to provide frequency regulation. Ph.D. Thesis, University of Illinois at Urbana-Champaign (2016)
23. J.T. Huges, A.D. Domínguez-García, K. Poolla, Virtual battery models for load flexibility from commercial buildings, in *2015 48th Hawaii International Conference on System Sciences (HICSS)* (IEEE, Piscataway, 2015), pp. 2627–2635
24. J.T. Hughes, A.D. Domínguez-García, K. Poolla, Identification of virtual battery models for flexible loads. IEEE Trans. Power Syst. **31**(6), 4660–4669 (2016)

25. G. Kats, A. Seal, Buildings as batteries: the rise of 'virtual storage'. Electr. J. **25**(10), 59–70 (2012)
26. S. Koch, J.L. Mathieu, D.S. Callaway, Modeling and control of aggregated heterogeneous thermostatically controlled loads for ancillary services, in *Proceedings of the Power Systems Computation Conference (PSCC)* (2011), pp. 1–7
27. S. Koch, M. Zima, G. Andersson, Active coordination of thermal household appliances for load management purposes. IFAC Proc. Vol. **42**(9), 149–154 (2009)
28. S. Kundu, N. Sinitsyn, S. Backhaus, I. Hiskens, Modeling and control of thermostatically controlled loads. (2011, preprint). arXiv:1101.2157
29. S. Kundu, K. Kalsi, S. Backhaus, Approximating flexibility in distributed energy resources: a geometric approach, in *2018 Power Systems Computation Conference (PSCC)* (IEEE, Piscataway, 2018), pp. 1–7
30. H. Larochelle, Y. Bengio, J. Louradour, P. Lamblin, Exploring strategies for training deep neural networks. J. Mach. Learn. Res. **10**, 1–40 (2009)
31. Y. LeCun, Y. Bengio, G. Hinton, Deep learning. Nature **521**(7553), 436 (2015)
32. M. Maasoumy, B.M. Sanandaji, A. Sangiovanni-Vincentelli, K. Poolla, Model predictive control of regulation services from commercial buildings to the smart grid, in *2014 American Control Conference* (IEEE, Piscataway, 2014), pp. 2226–2233
33. Y.V. Makarov, C. Loutan, J. Ma, P. De Mello, Operational impacts of wind generation on California power systems. IEEE Trans. Power Syst. **24**(2), 1039–1050 (2009)
34. M. Manic, K. Amarasinghe, J.J. Rodriguez-Andina, C. Rieger, Intelligent buildings of the future: cyberaware, deep learning powered, and human interacting. IEEE Ind. Electron. Mag. **10**(4), 32–49 (2016)
35. PJM Manual, Energy & Ancillary Services Market Operations, Revision 87, Effective Date: March 23, 2017, Prepared by Forward Market Operations
36. J.L. Mathieu, D.S. Callaway, State estimation and control of heterogeneous thermostatically controlled loads for load following, in *2012 45th Hawaii International Conference on System Sciences* (IEEE, Piscataway, 2012), pp. 2002–2011
37. J.L. Mathieu, M. Kamgarpour, J. Lygeros, D.S. Callaway, Energy arbitrage with thermostatically controlled loads, in *2013 European Control Conference (ECC)* (IEEE, Piscataway, 2013), pp. 2519–2526
38. J.L. Mathieu, S. Koch, D.S. Callaway, State estimation and control of electric loads to manage real-time energy imbalance. IEEE Trans. Power Syst. **28**(1), 430–440 (2013)
39. J.L. Mathieu, M. Kamgarpour, J. Lygeros, G. Andersson, D.S. Callaway, Arbitraging intraday wholesale energy market prices with aggregations of thermostatic loads. IEEE Trans. Power Syst. **30**(2), 763–772 (2015)
40. S. Meyn, P. Barooah, A. Bušić, J. Ehren, Ancillary service to the grid from deferrable loads: the case for intelligent pool pumps in Florida, in *52nd IEEE Conference on Decision and Control* (IEEE, Piscataway, 2013), pp. 6946–6953
41. S.P. Nandanoori, S. Kundu, D. Vrabie, K. Kalsi, J. Lian, Prioritized threshold allocation for distributed frequency response, in *2018 IEEE Conference on Control Technology and Applications (CCTA)* (IEEE, Piscataway, 2018), pp. 237–244
42. S.P. Nandanoori, I. Chakraborty, T. Ramachandran, S. Kundu, Identification and validation of virtual battery model for heterogeneous devices (2019, preprint). arXiv:1903.01370
43. A. Nayyar, J. Taylor, A. Subramanian, K. Poolla, P. Varaiya, Aggregate flexibility of a collection of loadsπ, in *52nd IEEE Conference on Decision and Control* (IEEE, Piscataway, 2013), pp. 5600–5607
44. C. Perfumo, E. Kofman, J.H. Braslavsky, J.K. Ward, Load management: model-based control of aggregate power for populations of thermostatically controlled loads. Energy Convers. Manag. **55**, 36–48 (2012)
45. PJM, http://www.pjm.com
46. M. Ranzato, P. Taylor, J. House, R. Flagan, Y. LeCun, P. Perona, Automatic recognition of biological particles in microscopic images. Pattern Recogn. Lett. **28**(1), 31–39 (2007)

47. A. Romero, N. Ballas, S.E. Kahou, A. Chassang, C. Gatta, Y. Bengio, FitNets: hints for thin deep nets. (2014, preprint). arXiv:1412.6550
48. B.M. Sanandaji, H. Hao, K. Poolla, Fast regulation service provision via aggregation of thermostatically controlled loads, in *2014 47th Hawaii International Conference on System Sciences* (IEEE, Piscataway, 2014), pp. 2388–2397
49. F.C. Schweppe, R.D. Tabors, J.L. Kirtley, H.R. Outhred, F.H. Pickel, A.J. Cox: Homeostatic utility control. IEEE Trans. Power Apparatus Syst. **PAS-99**(3), 1151–1163 (1980)
50. H.A. Song, S.Y. Lee, Hierarchical representation using NMF, in *International Conference on Neural Information Processing* (Springer, Berlin, 2013), pp. 466–473
51. E. Upton, G. Halfacree, *Raspberry Pi User Guide* (Wiley, Hoboken, 2014)
52. G. Varol, A.A. Salah, Efficient large-scale action recognition in videos using extreme learning machines. Exp. Syst. Appl. **42**(21), 8274–8282 (2015)
53. E. Vrettos, G. Andersson, Combined load frequency control and active distribution network management with thermostatically controlled loads, in *2013 IEEE International Conference on Smart Grid Communications (SmartGridComm)* (IEEE, Piscataway, 2013), pp. 247–252
54. X.L. Zhang, J. Wu, Denoising deep neural networks based voice activity detection, in *2013 IEEE International Conference on Acoustics, Speech and Signal Processing (ICASSP)* (IEEE, Piscataway, 2013), pp. 853–857
55. W. Zhang, K. Kalsi, J. Fuller, M. Elizondo, D. Chassin, Aggregate model for heterogeneous thermostatically controlled loads with demand response, in *2012 IEEE Power and Energy Society General Meeting* (IEEE, Piscataway, 2012), pp. 1–8
56. W. Zhang, J. Lian, C.Y. Chang, K. Kalsi, Aggregated modeling and control of air conditioning loads for demand response. IEEE Trans. Power Syst. **28**(4), 4655–4664 (2013)

Chapter 13
Applications of Artificial Neural Networks in the Context of Power Systems

Jan-Hendrik Menke, Marcel Dipp, Zheng Liu, Chenjie Ma, Florian Schäfer, and Martin Braun

13.1 Introduction

Grid operators face challenges due to an increasing number of volatile assets in their grids. On the generation side, distributed energy resource (DER) such as photovoltaic generators, wind energy converters or combined heat and power plants feed in according to weather conditions or individual plans. These conditions could result in unstable grid states. On the demand side, volatile loads such as charging stations for electric cars draw considerable power, sometimes tens to hundreds of kW, while being subject to customer preference. Storage systems may put additional stress on future power systems when they are optimized for maximum profits and not operated in favor of the distribution system operator.

There are different key performance indicators and security measures a grid operator has to uphold. For example, in Germany, grid loss is a performance indicator of how efficiently a grid is operated and therefore, an indicator that regulators use to incentivize and penalize different grid operators. Real-time monitoring of power grids is often a prerequisite for security measures, as a violation of operational limits can be detected, allowing the operator to act accordingly. The single contingency

J.-H. Menke (✉) · M. Dipp · Z. Liu · F. Schäfer
University of Kassel, Kassel, Germany
e-mail: jan-hendrik.menke@uni-kassel.de; marcel.dipp@uni-kassel.de; zheng.liu@uni-kassel.de

C. Ma
Fraunhofer Institute for Energy Economics and Energy System Technology, Kassel, Germany
e-mail: chenjie.ma@iee.fraunhofer.de

M. Braun
University of Kassel, Kassel, Germany

Fraunhofer Institute for Energy Economics and Energy System Technology, Kassel, Germany
e-mail: martin.braun@uni-kassel.de

M. Sayed-Mouchaweh (ed.), *Artificial Intelligence Techniques for a Scalable Energy Transition*, https://doi.org/10.1007/978-3-030-42726-9_13

policy is used to identify if the grid is inside operational limits even when one grid asset fails. All these areas can benefit from machine learning, which can make use of existing data to train a model in advance for specific tasks.

Another area of research, which complements grid operation, is grid planning. The aforementioned changing conditions in today's power grids often necessitate reinforcement or extension of power grids. In grids with a high number of substations and lines, the optimal reinforcement or extension path is not trivial to calculate. To find the minimum necessary improvements, possible measures are validated with a large number of scenarios. In this case, machine learning models can speed up the validation process considerably while still retaining a high accuracy for the validation.

Especially the use of artificial neural networks (ANN) [6] has gained popularity in recent years due to breakthroughs in research and computing power catching up to fulfill more complex tasks. Therefore, this chapter presents selected works which apply machine learning techniques, especially using ANN, in the field of power system analysis. These exemplified applications demonstrate the potential of machine learning for different technical applications. The research is mainly based on supervised learning, which is introduced in the next section. Another popular field of study is reinforcement learning, which can be applied to a variety of problems related to power systems. Possible applications of reinforcement learning will be discussed at the end of the chapter.

13.2 Methodological Background

13.2.1 Supervised Learning

In general, machine learning algorithms fulfill specific tasks [6]. Tasks depend on inputs, also called features and generate an output, called label. Tasks are often too complex or challenging to be solved by analytic algorithms or rule-based approaches or the calculation time with these methods is not within desired limits. To fulfill a task, a model is usually selected and trained by an algorithm. Examples of popular models are ANN, gradient boosted trees or support vector machines.

Two types of tasks are often used: classification and regression. Classification tasks seek to assign a discrete label to a set of inputs, i.e., the N (numerical) features are assigned to one of k specific categories or classes: $\mathbb{R}^N \mapsto \{1, \ldots, k\}$. Famous examples are found in computer vision, where neural networks can identify individual animals in pictures. The types of animals, which can be identified, are the classes of the classification task. On the other hand, there are regression tasks. Here, the labels are continuous numerical values, so that the task is to transform N features to a label: $\mathbb{R}^N \mapsto \mathbb{R}$. Regression problems can describe technical processes, e.g., transform an array of sensor data into a relevant process variable. These problems are often found in the field of power systems, as many load flow-related variables

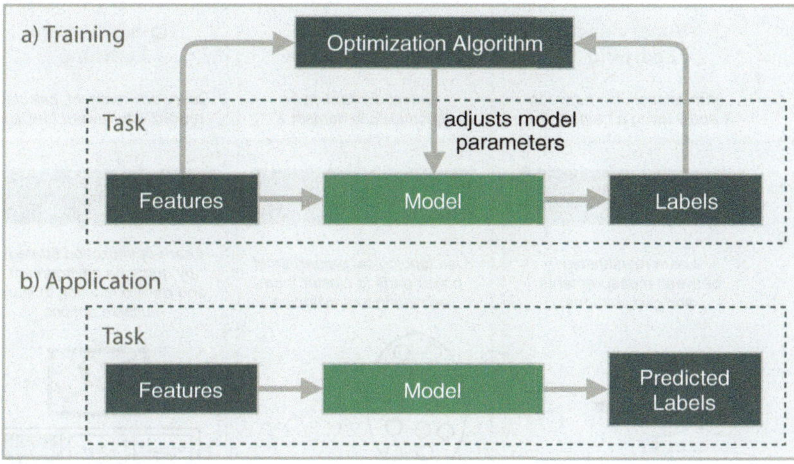

Fig. 13.1 Schema of a machine learning task. (**a**) During the training phase of a supervised learning problem, an optimization algorithm uses the training dataset's features and labels to adjust the model parameters. At the end of the optimization, the model parameters are set in such a way, that a given set of features is transformed into the corresponding labels. (**b**) Prediction/Inference stage of the task [6]. After the model has been sufficiently trained, unseen features can be fed into the model, which in turn predicts labels based on its learned parameters. The inference process is usually very fast and, with suitable training, accurate

are numerical and continuous. Many problems can be defined as a regression task for machine learning models to solve.

Supervised learning algorithms make use of a dataset which consists both of features and corresponding labels. Thus, the algorithm which trains a model can fit the model in such a way that the model output for the given features matches those of the corresponding labels. The training dataset can be made up of historical data, manual labeling by humans, or generated by performing expensive analytical calculations. A model is trained from the dataset, which can, if correctly trained, generalize for unknown data and predict labels features not seen or trained before. Figure 13.1 depicts this process. The use cases presented in the following sections generally make use of supervised learning to solve their particular problems.

13.2.2 Other Areas of Machine Learning

Before presenting applications of supervised learning and ANN, we want to take a look at other disciplines of machine learning and their potentials in future applications, namely unsupervised learning and reinforcement learning.

Figure 13.2 compares the different methods of machine learning using illustrative examples. As introduced in the previous subsections, supervised learning is relying on labeled data to be available, i.e., there exists a list of inputs for which the

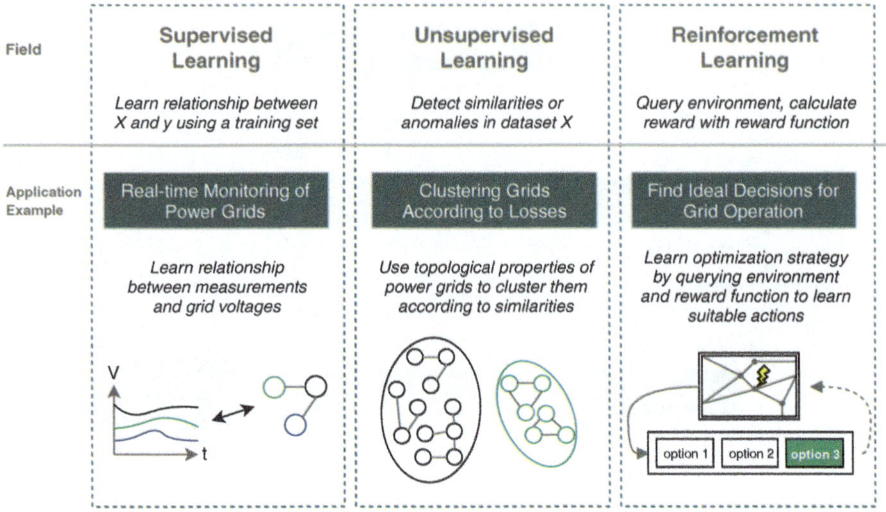

Fig. 13.2 Comparison of supervised learning, unsupervised learning, and reinforcement learning with a short description and an application example

outputs are known in advance. With this knowledge, a model can be trained and is then able to infer outputs for new inputs. An example, which is introduced in detail in Sect. 13.4, is the monitoring of grid voltages from using a low number of live measurements. A model is trained using a dataset consisting of a number of measurements X and corresponding grid voltages y. Afterwards, the trained model converts live measurements to grid voltages.

In unsupervised learning tasks, only the features of the dataset are known; the data is not labeled [6], i.e., it dataset contains only features X without any known outputs. Therefore, the task of unsupervised learning algorithms is to identify useful structures hidden in the features, i.e., the underlying probability distribution that generated a dataset. The result could be used to eliminate noise in the features or to assign new samples into discrete clusters subsequently. A suitable method of unsupervised learning, e.g., k-means clustering, can categorize the power systems into a predefined number of clusters. The clusters can then be used for further analyses. This example is used in Fig. 13.2.

Reinforcement learning means learning how to map situations to actions to maximize a numerical reward signal [23]. An agent finds itself in a specific state S_T and is allowed to take an action $a_T \in A$. It learns which action to take by receiving reward signals from the actions taken in many subsequent episodes. The agent is not told which actions to take, e.g., by a set of rules, but only by discovering reward signals and maximizing them. In complex environments, the situations following an action may depend on the taken action a_T. Therefore, the reward can be delayed if only a particular sequence of actions yields a high reward. Contrary to supervised learning, where a labeled set of examples is available to learn and use for generalizing for unknown inputs, reinforcement learning is also suited for interactive problems and problems for which an adequate training dataset cannot be built beforehand. The agent learns from its own choices, which in turn are made by using experience. It continually improves itself. The difference to unsupervised learning is that reinforcement learning seeks to discover a reward signal from the environment instead of finding a fixed structure or distribution in static features.

Reinforcement learning is used to tackle many different research questions related to power system operation and planning. A deep policy gradient and Q-learning as methods of reinforcement learning are used in [14] to optimize energy flows for a building. Using online training, the reward increases over many episodes, i.e., the system learns good behavior without training before use. Further applications are presented in the review paper [28]: reinforcement learning is employed to improve the decision making to maximize monetary value for flexibilities, such as energy storage, in [9]. Cost reduction for a cluster of residential loads is also achieved in [3] by employing convolutional neural networks and Q-learning. The electricity cost for a cluster of water heaters, constrained by daily time-of-use price profiles, is minimized through RL in [17]. Simulation results show that Q-learning can be successfully applied to reduce costs for the operation of the cluster of water heaters.

While many publications focus on one asset class, e.g., water heaters or electrical energy storage, little research is available from the point of view of a grid operator. Most relevant is microgrid operation, for which some research has been published (e.g., [8, 25]). Yan et al. perform load frequency control of a stochastic power system using deep reinforcement learning in [27]. Generator turbines are controlled by an optimized policy, which is continuously improved. Compared to an optimized PID-controller, the reinforcement learning approach yields fewer frequency deviations. The researched grid, however, is also of limited scale.

To the authors' knowledge, (deep) reinforcement learning is not yet applied to whole interconnected distribution or transmission grids from the point of view of the power system operator. Since reinforcement learning is a highly active field of research, applying this research to power system operation could be promising. The application example we give in Fig. 13.2 is the task of operating a grid within limits even for unexpected issues. The environment could create several events after which the grid's operating limits are breached. The ideal strategy to counteract this problem is found by reinforcement learning over the course of many runs of an algorithm.

13.2.3 Artificial Neural Networks

ANNs are a type of model used in supervised learning. An ANN consists of multiple neurons. Mathematically, neurons perform a summing operation of different weighted inputs and "activate" if the summed signal is large enough to trigger the neuron's activation function. The resulting signal is often a scaled value, e.g., between 0 and 1. Multiple neurons connect in layers. The signal undergoes several nonlinear transformations as it travels through the layers, depending on the individual connection weights between the neurons in different layers. Parallel neurons in a single layer increase the capacity of a model since they can generate different signals for the same input signal. An ANN architecture is described by the number of layers that are connected serially and the individual layer sizes (the number of neurons in the layer). The activation function also considerably changes the behavior of an ANN.

During training, an optimization algorithm adjusts the connection weights between the individual neurons, so that a set of features yields the target labels as outputs. Training-related parameters are the optimization algorithm, its learning rate (for gradient-based algorithms) and other parameters like learning rate decrease or weight decay. The following hyperparameters generally define an ANN model. Specific values are given for the specific tasks in the later sections.

- *Number of layers:* The number of sequential layers used in the ANN
- *Layer size:* The number of parallel neurons inside a specific layer
- *Activation function:* Function used to activate a specific neuron
- *Epochs:* (Maximum) number of epochs the model is trained
- *Batch size:* The number of training samples processed per batch
- *Learning rate:* The learning rate used in the optimizer
- *Learning rate decrease:* If an adaptive learning rate is used, it describes the factor with which the learning rate is multiplied at every decrease step

A more in-depth explanation of ANN and the relevant hyperparameters can be found in the book *Deep Learning* by Ian Goodfellow et al. [6]. We choose ANNs as the model to be used in the upcoming sections. ANNs provide state-of-the-art performance on many data science tasks. They also have a key property required for our research: support for multiple outputs without computational overhead. Many machine learning models, e.g., linear regression or decision trees, support only a single output and thus require an individual model to be fitted for every output. For ANN, only the number of neurons and layers would be tuned to accommodate the number of outputs, while a single model is fitted regardless of the number of outputs.

While ANNs are often used as the model for supervised learning problems, reinforcement learning can be performed without the use of ANN. However, ANNs are often used in state-of-the-art research of reinforcement learning, e.g., in the popular AlphaGo Zero program used to master the game of Go [21]. Here, ANNs help accelerate the reinforcement learning process by providing estimates on which action is most advantageous to take given the current observations. While performing the reinforcement learning process, the ANN is continuously trained with the current results. Over the course of many episodes, it can learn to estimate beneficial actions depending on the current state and overall increase the performance of reinforcement learning.

13.3 Speeding Up Time Series Calculations

13.3.1 Introduction

Traditional power system planning approaches are based on the "fit-and-forget" method, which is the analysis of few worst case scenarios. For this method, a small number of power flow (PF) calculations are sufficient to evaluate compliance with voltage limits or thermal transformer/line loading limits. Additional PF calculations are necessary if the single contingency policy (SCP) is taken into account. The SCP or "N-1 criterion" ensures that if a transmission line or transformer fails, the system as a whole is still in an operational state. Testing for compliance with the SCP requires to put every asset out of operation in a simulation for a given loading situation of the grid. The computational effort is still manageable for grids with less than a few hundred lines, so that full alternating current (AC) PF calculations can be used. For larger grids, linearization methods such as the line outage distribution factor (LODF) method exist.

Transmission planning or modern distribution planning methods are based on time series. These sophisticated and computationally expensive simulations are needed to model the time-depended characteristics of distributed energy resources, storage systems or flexible loads. In time series-based planning, typically one or multiple years are analyzed with quasi-static PF or optimal power flow (OPF) simulations. To simulate one year in a 15 min resolution, $T = 35{,}040$ PF calculations are required. If additionally, compliance with the SCP is taken into account, an additional PF per outage is required for each time step. For a total of N assets ($N - 1$ cases), $N \cdot T$ PF evaluations are necessary. For a grid with $N = 100$ lines and an average calculation time of 25 ms per AC power flow calculation, the whole evaluation would take more than 24 h if not performed parallelly. Therefore, a method is needed which checks if a system state complies with voltage limits, line loading limits, and the SCP.

In this section, we describe a method, first mentioned in [19], which rapidly evaluates system states based on an approximation of the PF calculation results. The method is based on open-source machine learning libraries and tools [16] in combination with pandapower [24], an open-source power system analysis tool. With this combination a fast and automated regression calculation in Python is possible. We show how an ANN must be trained to deliver fast and accurate results for the PF approximation. Results are shown for the SimBench dataset [12]. To benchmark accuracy and timings, results of the full AC PF calculations are taken into account as reference values. We show the limitations of the ANN approach and evaluate calculation times as well as prediction errors.

13.3.2 Implementation

13.3.2.1 Overview: Regression Method

We train a multilayer perceptron (MLP), a specific architecture of ANNs, to predict PF results for an annual simulation including the SCP analysis. The line current flows I_n for each branch n, as well as the bus voltage magnitudes V_m for each bus k, are predicted by a MLP-based regression method. Inputs X to the prediction are the known variables of a power flow calculation (namely the voltage magnitude V_m and voltage angle V_a for slack buses, P, V_m for generator buses and P, Q for PQ buses). The computationally expensive PF calculations are then substituted with ANN predictions to speed up the time series simulation considerably.

The required inputs for training the ANN are the grid data, e.g., line and transformer impedances, the distribution of loads and generators, and the P and Q injections per bus. For the following analysis, we use time series of 1 year in 15-min resolution ($T = 35{,}136$ time steps) as provided in the SimBench dataset [12].

Figure 13.3 shows an overview of the analysis process. Based on the input data—the grid data and bus power injections for each time step T—AC PF results are calculated. Results for the base case, with no line out of service, and optionally results for the contingency cases, are calculated and stored. Only a certain percentage of these previously calculated PF results are used to train the MLP. For this, we randomly select T_{train} time steps and the corresponding results from the previous simulation. The training data then consists of the input variables, X, and output variables y. y contains the reference values of line loadings I_n and bus voltage magnitudes V_m. For the remaining time steps $T_{predict} = T - T_{train}$,

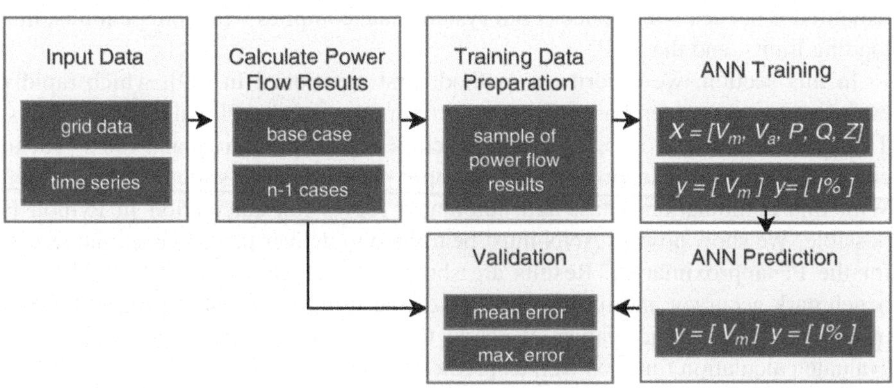

Fig. 13.3 Overview of the analysis process. Based on the input data, AC power flow results are calculated for the base case and optionally for the $N - 1$ cases. A percentage of these results are used for training of the ANN. The remaining results are then predicted and compared to the correct values

the PF results are predicted by the previously trained MLP. As a validation of the ANN's accuracy, the predictions are then compared with the calculated PF results to evaluate the prediction error.

13.3.2.2 ANN Architecture

We use a MLP architecture for the ANN with a hidden layer size of 100 perceptrons with the rectified linear unit (ReLU) activation function [6]. The ANN is fully connected with bias. Training lasts for 200 epochs; the training process is performed by the *Adam* optimizer [7]. Similar architectures have shown robust results for multiple regression problems, e.g., in [19]. The hyperparameters are manually optimized for the specific problem as is general practice. A normalizer scales the inputs with the maximum norm and a principal component analysis (PCA) is performed with a randomized singular value decomposition. A MinMaxScaler scales the outputs. All implementations are taken from the scikit-learn library [16].

We generate the input matrix X_{base} by stacking the values known before a power flow calculation, as shown in Eq. (13.1). These values include the voltage magnitudes and angles of the slack buses, the voltage magnitudes and real power injections of the generators buses as well as real power injections P and reactive power injections Q of PQ buses. Additionally, the impedances of all lines are defined as inputs z_l. If a line is out of service, the impedance is tripled, which has shown to be sufficient so that the model can learn the influence of the line outage on the power flow result.

$$X_{base} = \left[\left(v_{m,s}\right)\left(v_{a,s}\right)\left(v_{m,g}\right)\left(p_i\right)\left(q_i\right)\left(z_l\right)\right] \tag{13.1}$$

$(v_{m,s}), (v_{a,s}) \in \mathbb{R}^{T_{train} \times N_{slack}}$ are the voltage magnitude and angle matrices for each time step $t \in T_{train}$ and slack bus $s \in N_{slack}$, with N_{slack} the number of slack buses. $(v_{m,g}) \in \mathbb{R}^{T_{train} \times N_{gen}}$ are the voltage magnitude matrices for each time step $t \in T_{train}$ and generator bus $g \in N_{gen}$, with N_{gen} the number of generator buses. $(p_i), (q_i) \in \mathbb{R}^{T_{train} \times N_{bus}}$ are the bus power injection matrices for each time step $t \in T_{train}$ and bus $i \in N_{bus}$. $(z_i) \in \mathbb{R}^{T_{train} \times N_{line}}$ are the absolute magnitudes of the line impedance values for each time step $t \in T_{train}$ and line $l \in N_{line}$. An ANN architecture trained with this data allows to predict results, without disconnected lines, for the remaining time steps $T - T_{train}$. If results for line outages are to be predicted, training data must be generated by taking the particular line out of service and calculating power flow results. The impedance z_o of the disconnected line o in the input vector X_{n-1} is then multiplied by a constant c to have an unusual high impedance value. We chose $c = 3$, i.e., a tripling of the impedance. X_{n-1} is then stacked to X_{base} so that the ANN model can learn the change in power flow results, when a line is out of service. In total, X has $T_{train} \cdot (N_{line} + 1)$ rows (all outages + without outages) and $2 \cdot N_{slack} + N_{gen} + 2 \cdot N_{bus} + N_{line}$ columns.

The output matrix y contains the desired targets. We train separate ANNs to predict either the relative line loadings or voltage magnitudes as outputs. For the prediction of line loadings, the output matrix $y_{loadings}$ has a maximum of $\mathbb{R}^{(T_{train} \cdot N_{line}) \times (N_{line}+1)}$ entries if all contingency cases are calculated. Similarly, predicting voltages yields a total of $\mathbb{R}^{(T_{train} \cdot N_{bus}) \times (N_{line}+1)}$ entries in the bus voltage prediction matrix $y_{voltages}$.

$$y_{loadings} = \left[(I_l)\right] y_{voltages} = \left[(V_i)\right] \tag{13.2}$$

13.3.3 Results

Results are shown for the SimBench high voltage (HV) "mixed" dataset [12], which includes an annual time series in 15-min resolution. The topology of the grid is depicted in Fig. 13.4. The grid, which is operated on the 110 kV level, consists of 306 buses, 95 lines, and three slack buses which are the connections to the 380 kV, voltage level. Loads are connected to 58 of the buses, and 103 distributed generators are installed in the grid—all other buses model busbars or bushings.

The method is evaluated on a modern consumer laptop (Intel i7-8550U CPU, 16 GB RAM). A single power flow calculation for the SimBench HV grid takes about 17 ms, the whole year with 35,136 time steps takes 600 s, respectively. To calculate each contingency case for $N_{line} = 95$ lines and for every time step takes $95 \cdot 600\,\text{s} = 16\,\text{h}$ to complete. The time to generate training data for the ANNs is 1.6 h, since only 10% of the time steps are calculated directly. On top, the ANN training, which includes training for the $N-1$-cases, takes an additional 27 s. The prediction for the remaining results only needs 20 ms. All in all, there is a large difference in calculation time for the purely PF-based calculations compared to the ANN-based approach.

The absolute values of the prediction results are shown for the base case prediction in Fig. 13.5. The results are sorted in ascending order so that the lines/buses with the smallest absolute error are shown on the left side. The mean absolute errors of the line loading predictions are 0.1% and $6.3 \cdot 10^{-5}$ p.u. for the voltage magnitudes. Maximum errors are 7.7% for the line loading and $2 \cdot 10^{-3}$ p.u.

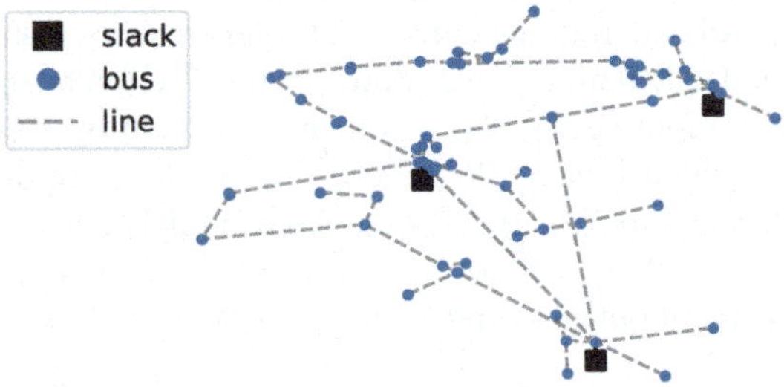

Fig. 13.4 SimBench 110 kV "mixed" grid with $N_{slack} = 3$, $N_{bus} = 306$ and $N_{line} = 95$

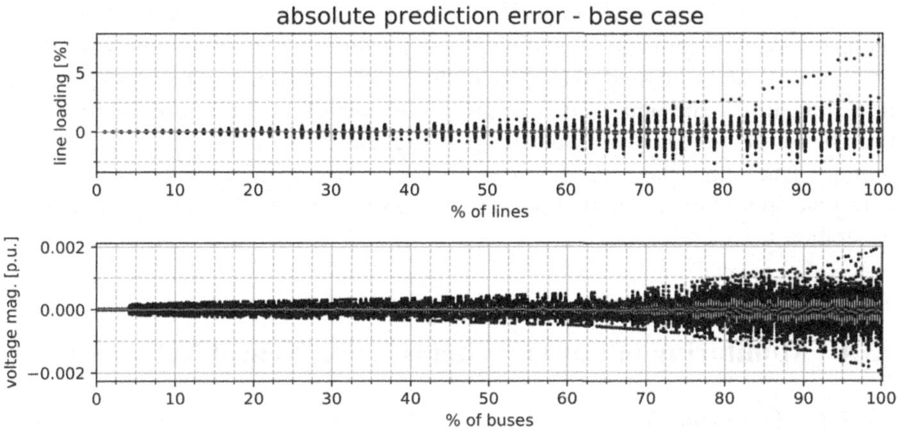

Fig. 13.5 Absolute prediction errors for line loadings (top) and voltage magnitudes (bottom). Only results for the base case was trained and predicted

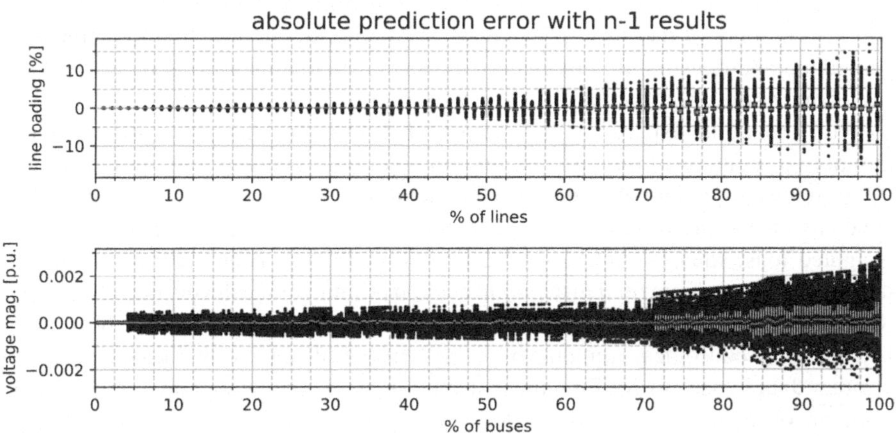

Fig. 13.6 Absolute prediction errors for line loadings (top) and voltage magnitudes (bottom) with N-1 cases included

for the voltage magnitude predictions. In 99.9% of all predictions the line loading error is less than 2.19% and the voltage magnitude error less than $7.8 \cdot 10^{-4}$ p.u.

Absolute prediction errors including $N - 1$-cases are shown in Fig. 13.6. When taking the $N - 1$-cases into account, the mean errors and max. errors increase compared to the base case prediction. For the line loadings, the absolute error is twice as high (mean 0.35%, max. 16.8%). The voltage magnitude prediction error, however, does not increase much (mean $8.85 \cdot 10^{-5}$ p.u., max. $3 \cdot 10^{-3}$ p.u.). In 99.9% of all predictions the line loading error is less than 6.18% and the voltage magnitude error less than $1.6 \cdot 10^{-3}$ p.u.

The proposed regression method to predict power flow results can accurately predict voltage magnitudes and line loading values for a given time period. In practice this can be applied to analyze multiple scenarios in power system planning in a short time. The approach is promising to identify critical loading situations fast. In future research we want to compare the required minimum amount of training data within the range of an acceptable approximation error. Also we want to test the approximation with different control strategies, such as the curtailment of generation.

13.4 Monitoring of Power Systems Using Neural Networks

13.4.1 Introduction

The state of an electrical grid is commonly completely measured and determined on the high- and extra-high voltage levels. On the medium and low-voltage level, only a limited number of measuring devices are available. In the past, due to predictable generation and consumption of electrical energy, the low measuring densities were sufficient for the operation of distribution systems.

With the rapid expansion of volatile DER and the changing load behavior, monitoring methods at the distribution grid level have become relevant to estimate safety-relevant parameters such as line loadings and bus voltage magnitudes. Whereas on the high- and extra-high voltage levels the state estimation based on weighted least squares (WLS) state estimation (SE) has been state of the art for decades, the measurement density on low and medium voltage levels is not sufficient to identify the complete state of the power system. A method to solve this problem is to create pseudo-measurements by using load profiles, historical time series, and weather data to use WLS SE on the distribution level. Such data may not always be available or in the case of load profiles, is not sufficient to achieve a high estimation accuracy.

In this context, we have developed a monitoring method based on ANNs that functions with only the available measurements and provides an accurate estimation of electrical variables in distribution grids with a high percentage of DER. The techniques covered in this section are based on detailed preliminary studies and simulations, which are described in [13]. Figure 13.7 shows the flow chart of the methodology.

A scenario generator is used to generate suitable training and test sets for the ANN using the open-source tool pandapower. The creation and training of the ANNs are accomplished with the deep learning library PyTorch [15].

This section describes our monitoring method based on ANNs. First, the scenario generator and the training process are outlined. Then the ANN architectures and the validation cases used for the simulations are illustrated. Finally, the simulation results for estimating the security-relevant parameters line loading and voltage magnitude are presented.

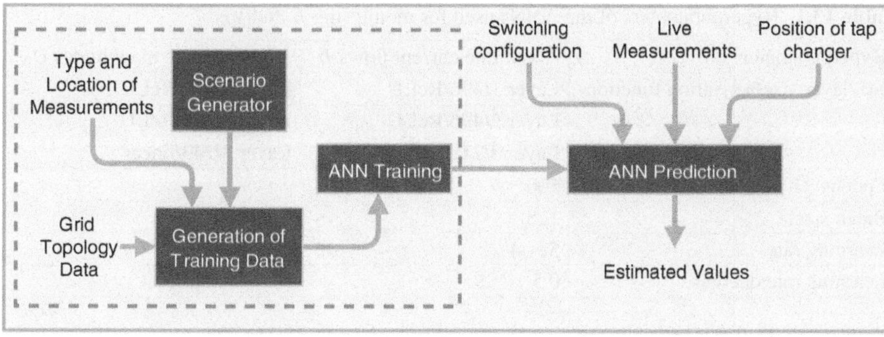

Fig. 13.7 Flow chart of the monitoring scheme: the left side represents the preparation phase, while the right box shows the operating phase, in which live measurements, the switching configuration, and the position of the transformers tap changer for the trained ANN are used to predict the target values

13.4.2 Scenario Generator

To achieve a high estimation accuracy, the ANN must be properly trained. Therefore, a high number of potential grid states must be generated. We have developed a scenario generator, which generates various grid states that are subsequently used for the ANN training process. The trained ANN will then be capable of estimating the output parameters for the corresponding input parameters. The scenario generator provides the option of scaling individual feed-ins or load components as well as using time series to generate a suitable training and test set. Each component in the SimBench dataset is scaled with its time series in a 30-min resolution over a test year, which results in $T_{year} = 17{,}568$ time steps. Next, a load flow calculation is performed for every 30-min interval. The results of the power flow calculations, the corresponding measurements, the switch configuration, and the current position of the transformer tap are the input values. In this way, a training and test set is generated to train and test the ANN. Predicted values are the voltage magnitude V_k for each bus k and the line current flows I_l for each line l.

13.4.3 Case Study and Results

Table 13.1 summarizes the applied hyperparameters. The number of neurons in the hidden layers is set to two times the maximum size of the output data for each ANN. Thus, the ANN has a hidden layer size of $N_{I,hidden} = 2 \cdot (N_{ln,lv} \cdot N_{ln,mv}) = 2 \cdot (127 + 121) = 496$ for estimating the line current flows I_l and a size of $N_{V,hidden} = 2 \cdot (N_{bus,lv} \cdot N_{bus,mv}) = 2 \cdot (115 + 128) = 486$ for estimating the voltage magnitudes V_k. The training of the ANNs is performed with the Adam optimizer [7].

Table 13.1 Hyperparameters of the ANNs used for monitoring I_l and V_K

Hyperparameter	ANN: line current flows I_l	ANN: voltage magnitudes U_K
No./layer size/activation function	Layer 1/496/ReLU	Layer 1/486/ReLU
	Layer 2/496/ReLU	Layer 2/486/ReLU
	Layer 3/248/linear	Layer 3/243/linear
Epochs	500	
Batch sizel	128	
Learning rate	5e−4	
Learning rate decrease	0.5	

The SimBench grid (semi-urban variant) [12] used for the simulations includes the low and medium voltage level and is connected to the high voltage level via two parallel HV/MV transformers. The grid has $N_{ln,mv} = 121$ cables in the medium voltage level connecting a total of $N_{bus,mv} = 115$ nodes. In addition, a low-voltage grid is connected by an MV/LV transformer containing $N_{ln,lv} = 127$ cables and $N_{bus,lv} = 128$ additional nodes. The remaining low-voltage grids attached to the medium voltage level are represented as aggregated equivalents.

All measurement devices in the LV/MV are shown in Fig. 13.8. In the grid, measuring devices are located at nine buses. They measure the active power P, reactive power Q, and voltage magnitude V. Also, in the first line of each feeder outgoing from the substation, P, Q, and I are measured. Measurement device tolerance is given by accuracy classes, which describe the accuracy that can be expected for a measured value. The accuracy class 0.5 is applied for voltage measurements. The associated standard deviation is $\frac{0.5}{3} \approx 0.167\%$. Thus 99.7% of all measurements have a maximum error of 0.5% within the $3\,\sigma$ confidence interval. For measurements of active and reactive power, a standard deviation of 0.67% is assumed. Figure 13.8 shows one of three tested switching states.

We train two ANNs, one to estimate line loading and the other to estimate voltage magnitude. The hyperparameters of the ANNs are selected according to the specifications in Table 13.1. The training and test set is generated using the scenario generator based on the time series specified in the SimBench dataset in 15-min intervals over 1 year. The training set includes the first 6 months (January to June) and the test set contains the last 6 months (July to December) of the respective year. Three switching configurations are considered which combine different available open ring structures. A total of seven different positions for the tap changers of the HV/MV transformers are covered. This results in a total of $3 \cdot 7 \cdot 17{,}568 = 368{,}928$ different scenarios for the respective training and test set.

Figures 13.9 and 13.10 show the mean and maximum estimation error for the line loadings I_l and the voltage magnitudes U_k for each line l and bus k over the complete test period. The mean absolute error for estimating the line loading I_l is 0.57% (max. error: 12.45%) at the medium voltage level and 0.32% (max. error: 7.42%) at the low-voltage level. The mean absolute error for the estimation of the

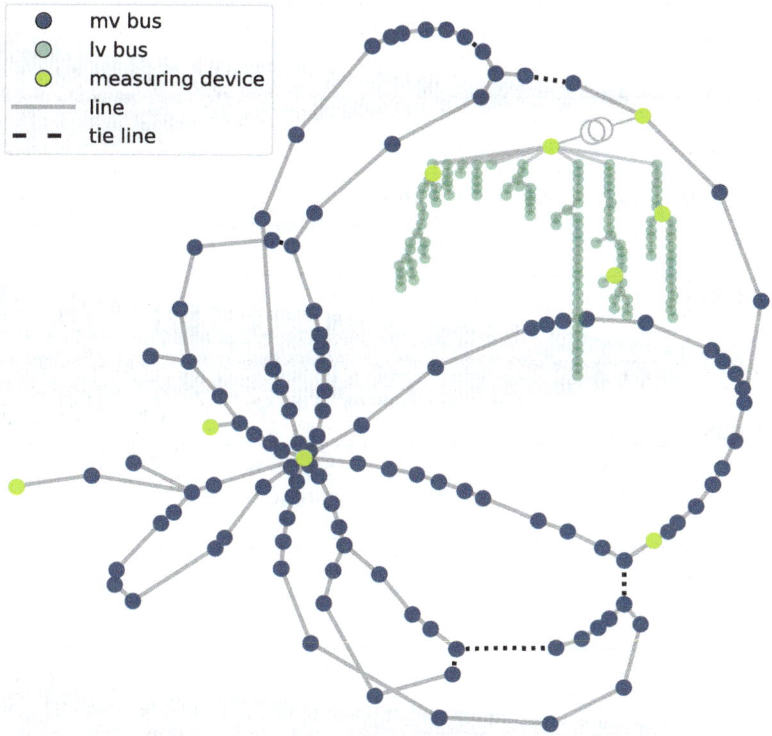

Fig. 13.8 SimBench LV/MV semi-urban grid with the default switching configuration

voltage magnitude U_k at the medium voltage level is $3.5 \cdot 10^{-4}$ p.u. (max. error: $8.3 \cdot 10^{-3}$ p.u.) and $4.2 \cdot 10^{-4}$ p.u. (max. error: $6 \cdot 10^{-3}$ p.u.) at the low voltage the level.

13.4.4 Conclusion

Monitoring methods capable of identifying safety-relevant parameters, such as I_l and U_k of a grid with a low measurement density, help to increase the reliability of distribution grids economically. The presented approach of estimating grid parameters with ANNs can consider the changed load behavior as well as the volatile feed-in of renewable energies. Other components, such as storages, can additionally be included in the methodology. Voltage magnitudes and line loadings are estimated with high accuracy. These variables are important for the grid operator to determine if the system's operational constraints are met. In addition to the live measurements as input values, various switching configurations and the transformer tap changer are also considered. With this type of monitoring method, the grid operator can identify critical grid states with high accuracy for most tested cases.

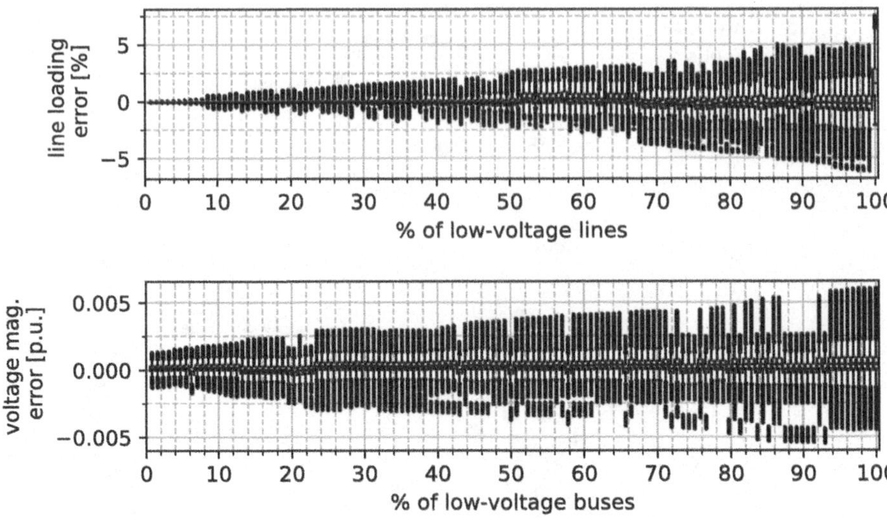

Fig. 13.9 Monitoring error of line loadings and voltage magnitudes in the SimBench grid at the LV level

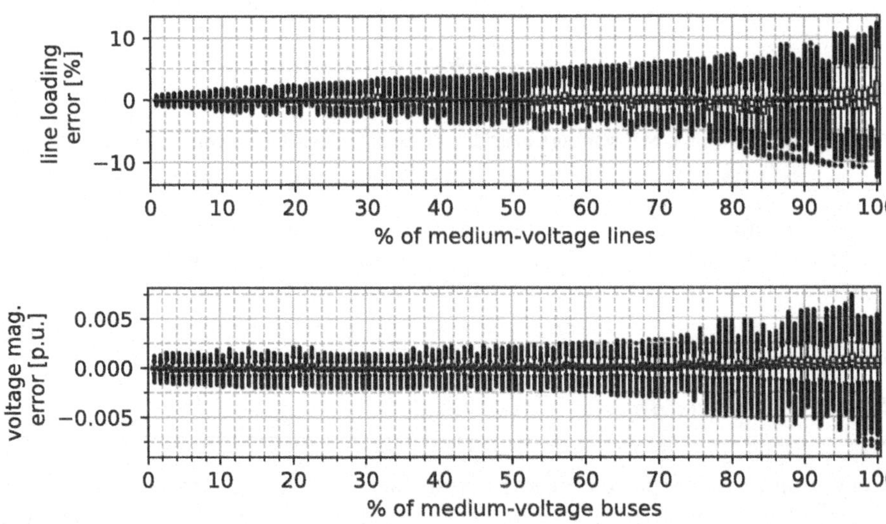

Fig. 13.10 Monitoring error of line loadings and voltage magnitudes in the SimBench grid at the MV level

However, there are also limitations to this method. The results only include the estimations of the test set. Furthermore, the computation time of the scenario generator and the training time of the ANNs increase significantly with higher time resolutions, increased switching states, or a higher number of transformers with tap changers. Additional information on this methodology can be found in [13].

13.5 Using Neural Networks as Grid Equivalents

13.5.1 Introduction

Modern power systems often consist of areas, which are operated by different grid operators but connected by tie lines. To perform grid analyses in one of the grid areas, the other connected areas are typically represented by reduced models that can approximate the behavior of the actual system. The reasons for using reduced models can be as follows [1]:

- limitation of computing power for large-scale power systems;
- usually, the interconnected areas are operated by different grid operators, each of which is often unwilling to share the complete system information to keep details confidential.

Conventional grid equivalence techniques such as Ward equivalents and radial equivalent independent (REI) equivalents or some variations of these two basic methods are mostly used to deal with these issues. The accuracy of these approaches is, however, limited for significant changes in the grid states, e.g., large power fluctuations of some assets or topology changes [5, 20]. In this section, ANN is used to approximate the interaction at the interconnection with increased accuracy.

13.5.2 Grid Equivalent Techniques

According to the representation of the model and its application, grid equivalent techniques can be classified as static and dynamic [5]. In this section, the term "grid equivalent" refers only to the static equivalent methods which are used for static analysis only, such as power flow calculations and system operation. To create an equivalent grid, the user has to determine the boundary which will divide a solved load flow grid model into an internal and external area. The inner area is typically modeled in detail because a regional utility is usually interested in this area. The external area is represented by a reduced model through the grid equivalent methods, Fig. 13.11.

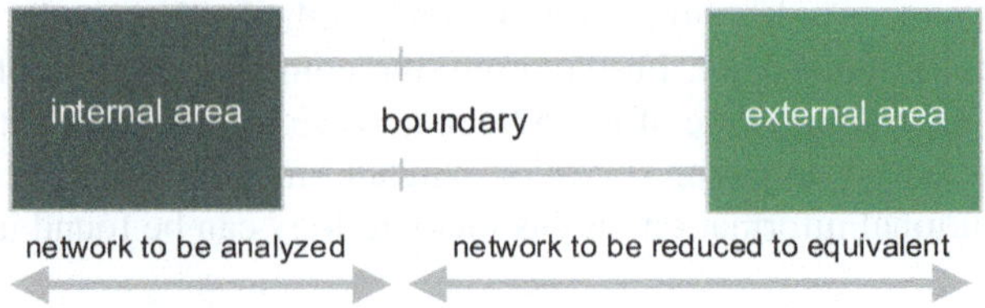

Fig. 13.11 Interconnected power system

13.5.2.1 Conventional Grid Equivalents

Assuming that the grid is at a stable operating condition, the basic Ward equivalent [26] disregards all of the external buses and presents the external system by a set of equivalent lines, shunts, and power injections attached at boundary buses. In this process, the Gaussian elimination plays a crucial role in reducing the bus admittance matrix of the original grid. To approximate the reactive power response of the external grid, an extended Ward equivalent method is developed, i.e., an additional PV bus is added without power injection at each boundary bus. Its voltage is equal to the original boundary bus voltage. However, the accuracy of Ward class equivalents is limited if either load or generator injections in the external network change. Over the last decades, attempts have been made to minimize load flow errors; one of them is called REI.

The idea behind REI equivalent is the aggregation of the power and current injection in external areas to a fictitious REI bus [4, 18]. Depending on the number of desired consumer and generator types, any number of REI buses can be created. In principle, it is advantageous to separate at least consumer and generator types [22]. Then, the grid resulting from the power aggregation can be reduced by Gaussian elimination. The aggregated consumers (loads) or generators at REI buses are still existent and configurable. An adaptation to other operating scenarios is possible without repeating the equivalent process, with the help of a simple scaling of the operating point of the aggregated devices. REI equivalents are only sufficiently accurate if the power changes in the external areas are not significant.

13.5.2.2 Grid Equivalents Using Artificial Neural Networks

The interaction between the connected areas is reflected by the power exchange at the interconnection lines, which is essentially affected by continuously changing load and DER feed-in. The relationship between the power exchange and the grid state changes corresponds to the power flow calculation, which requires further grid topology data. An ANN is an imitation of a biological neural network and can be used to model any input–output relationship without the exact knowledge of the system. Assuming that the historical time series of loads and DER, as well as the related power exchange at the boundary lines, are available, and the other influential

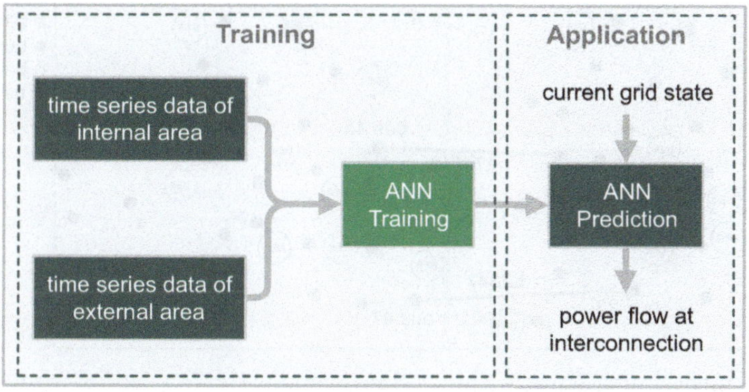

Fig. 13.12 Application of ANN for grid equivalents

factors such as grid topology, etc., remain constant, a training dataset can be built for a supervised learning task. This dataset enables a suitable model such as an ANN to learn the relationship between power exchange and grid state changes without knowledge of other grid data. The variables related to the grid state are the features while the power exchange at the boundary lines is the target. In this application, the grid operators exchange the current grid state with each other so that the ANN predicts power exchange at the boundary, Fig. 13.12.

The interconnected grid operators are sometimes reluctant to share their information with others. Because of this situation, another kind of ANN for unsupervised learning, autoencoder, could be used, whose primary purpose is to learn the important properties in the dataset. Hence the dataset is compressed or rather encrypted.

13.5.3 Case Study and Modeling

13.5.3.1 Test Case

DER are technically capable of providing ancillary services, which could be used in various system services such as voltage control. Therefore, our test scenario focuses on the performance of the equivalent grid by any changes in reactive power of DER and variable load. Using pandapower [24] as the tool for grid simulation, an interconnected 60 bus system, consisting of two IEEE 30-bus systems [2], is used as the test case. It is shown in Fig. 13.13. The 30 buses in the left area are treated as the internal grid. The remaining 30 buses on the right belong to the external grid. Both grids are connected by the tie lines L2745 and L2347, and then 14 DER units are added. We assume that the external grid will be reduced for this study. For the conventional grid equivalent, all the required data are available, e.g., grid topology

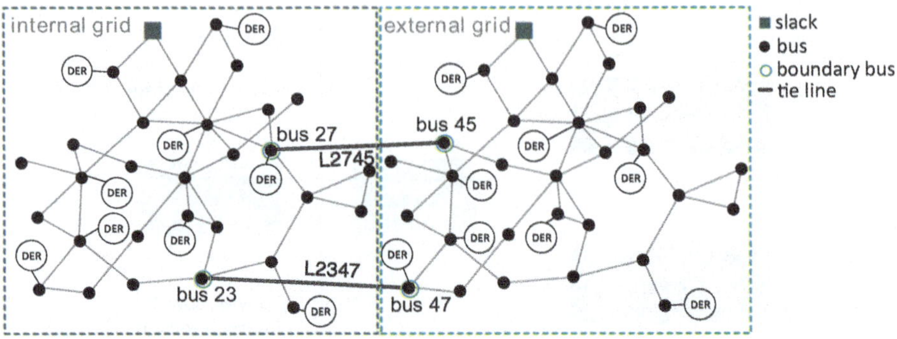

Fig. 13.13 60 bus benchmark grid used as a test case

data and power injections at buses. For the ANN training, dataset are generated systematically as presented in the following section.

13.5.3.2 Modeling

To evaluate the performance of an ANN as a grid equivalent, extended Ward and REI equivalents are implemented in pandapower. To obtain the REI equivalents, three REI buses are created to aggregate the power of generators, load, and DER in the external grid.

For an ANN to accurately approximate the power flow exchange at the interconnection for a given grid state, it is important to obtain a meaningful training dataset, which covers as many grid situations as possible. To this end, a scenario generator is defined. Regarding the test case, the two parameters that are considered are changes in load and DER power injections, which are independent of each other. Active power of the loads and DERs are scaled randomly for each class of units between 0 and 100%. Additionally, the reactive power of DERs are scaled individually and randomly in the available range ($\cos\phi = 0.95$) from -100% (inductive) to 100% (capacitive). The final training input consists of 1000 tuples with diverse scaling values. For every scaling value, a power flow calculation generates the corresponding power flow at the interconnection as the target for the training set.

Another crucial aspect of training the ANN is the determination of a feasible model architecture and the selection of a learning algorithm, which perform well for the given dataset. For this purpose, the optimization method in [10] is used to obtain the number of layers and other hyperparameters for the ANN. Table 13.2 lists the used hyperparameters for the presented task. The optimization algorithm Adam is chosen to optimize the ANN's weights, and the L1 loss function is used.

Table 13.2 Model architecture and hyperparameters for the ANN used to model a grid equivalent

Hyperparameter	Values
No./layer size/activation function	Input size/116
	Layer 1/270/PReLU layer
	2/180/PReLU layer
	3/4/linear
Epochs	243
Batch size	512
Learning rate	5e−4
Learning rate decrease	0.2

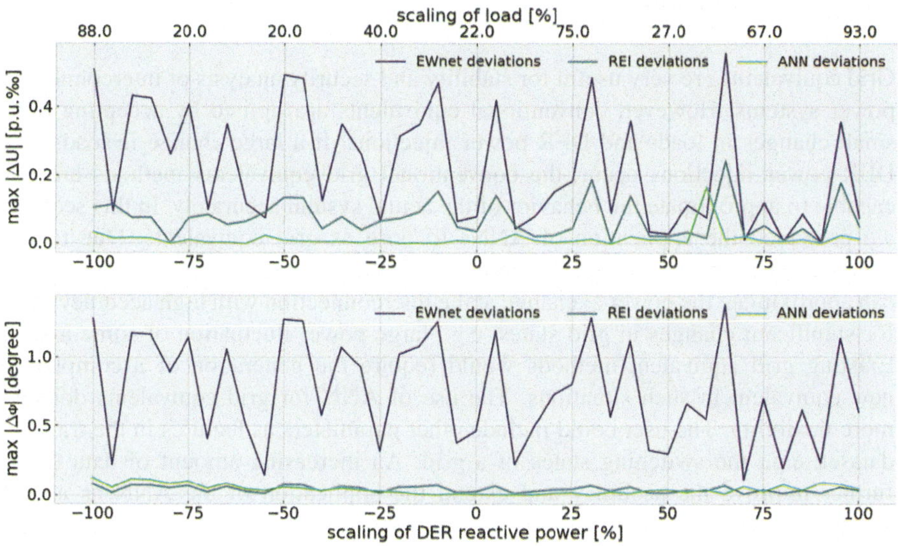

Fig. 13.14 Variation of bus voltage with change in reactive power of DER and loads

13.5.3.3 Results

The performances of the different types of grid equivalents are evaluated in terms of their ability to estimate the bus voltages of the original grid, with changes in the reactive power of DER and load. Reactive power injections of DER in the full grid (internal area and external area) are varied in steps of 5% from 100% (capacitive) to −100% (inductive). Also, load power injections in the full grid are varied randomly. All the external buses of the test grid in Fig. 13.13 are replaced by the extended Ward, REI, and ANN equivalents. The maximum deviation of voltage (magnitude and phase angle) at the existing buses (buses in the internal area) are observed. It can be seen from the plot above in Fig. 13.14 that, the per unit (p.u.) voltage magnitude errors by the three methods are differently affected by the change in reactive power of DER and loads. In most cases, extended Ward equivalent gives the worst result, while REI equivalent has generally the better accuracy because of the configurable aggregated devices at REI buses. Due to the suitable dataset and hyperparameters, the ANN equivalent is not very sensitive to the grid status change and thus provides the best accuracy overall. All of the maximum voltage errors are at the boundary bus 23 or 27. The maximum deviations of voltage phase angles from the plot below in Fig. 13.14 show that the differences between REI and ANN under various conditions are not significant. They give better result than extended Ward for the test benchmark grid.

13.5.4 Conclusion

Grid equivalents are very useful for stability and security analysis of interconnected power systems. However, conventional equivalents are limited by accepting only small changes in loads and DER power injections. If a large change in loads and DER power injections occur, the conventional grid equivalents methods are not capable to approximate the behavior of the actual system accurately. In this section, we presented the application of ANNs to pose as grid equivalents. Due to the suitable selection of data and the optimized setting of hyperparameters, the ANN can approximate the power exchange at the interconnection with high accuracy even for significant changes in grid states, e.g., large power fluctuation of some assets. Existing grid equivalent methods would require the generation of a completely new equivalent in such situations. The use of ANN for grid equivalents delivers more flexibility. The user could include other parameters as features in the training dataset, e.g., the switching states of a grid. An increasing amount of data could further improve the accuracy and extend the application of the ANN as a grid equivalent.

13.6 Estimating Power System Losses

13.6.1 Introduction

Efficient operations of power systems, which host DERs, is important to achieve a sustainable energy system. An accurate determination of energy losses in large power systems is a fundamental step. However, to evaluate the energy losses of distribution grids at large scale is a difficult task, mainly due to the huge number of grid assets and the incomplete measurement at this system level. Normally, distribution system operators rely on very limited options to determine grid losses, especially at the low-voltage (LV) level. The growing installation rate and the intermittent nature of DER pose further difficulties and uncertainties upon the loss evaluation. For this reason, an accurate evaluation method for grid losses is highly relevant. The recent developments of machine learning techniques provide a promising way to evaluate high-dimensional data in many engineering fields. In the following subsections, an application of the ANN-based regression approach to estimate the energy losses of LV grids is presented.

13.6.2 Methodology

For this application, we look at a large number of real low-voltage grids in Germany. Here, approximately 5000 LV grids in rural areas are investigated. The selected grid area and the detailed modeling procedure are described in [11]. As currently the smart meter measurement is not available in Germany, the standard load profile and the average feed-in profile for each generator technology are provided by the DSO. Using these annual time series for loads and DER, yearly power flow simulations are carried out for all grids. These calculations yield the exact annual energy losses of individual grids. To enhance the comparability among LV grids, the grid losses are analyzed in the following case study as the relative losses (w.r.t. the total injection in a grid) instead of the absolute energy losses. Secondly, we evaluate a large number of grid features. A high-dimensional dataset containing both these features and the grid losses can be formulated for all grids. To make different features comparable with each other, these grid features are normalized, e.g., in the commonly applied min-max scaling scheme. Features include, for example, the total line length of the grid, data regarding different assets like switches, distribution cabinets, transformers, or information about loads and distributed generators. Finally, the ANN method is applied to the dataset to train regression models for estimating the relative losses of these grids. The obtained estimation model can be further used for estimating energy losses of other LV grids. The ANN regression scheme is implemented with the help of a Python machine learning package scikit-learn [16].

13.6.3 Case Study

In order to validate the performance of the proposed evaluation approach, the estimations of grid losses are compared against the exact results generated by exhaustive calculations. For all the given samples, the dataset is randomly divided into three equally sized partitions. One of the partitions is considered as test dataset, while the remaining two partitions are used as training dataset, based on which the estimation model is obtained. All validation tests in the case study are following this threefolds cross-validation scheme. To assert the model's capability to generalize, 500 tests are repeated for each method and parameter set so that the influences of bad initialization and unfavorable cross-validation partitioning are minimized in the statistical analysis. In the following, we focus on the average of all 500 tests. Running a single test using the evaluation approach typically requires less than 10 s on consumer-grade computer hardware.

13.6.3.1 Parameter Selection for ANN models

The performance of machine learning methods can be significantly influenced by the setting of hyperparameters. In the first case study, the issue of selecting these parameters for ANN-based on a grid search method [16] is discussed. Due to the large space of feasible parameters, the grid search approach can only be efficiently implemented to perform large search steps for some parameters. Sample results of these grid search tests for ANNs are presented in the following.

Here, both estimation accuracy for selected parameters and linearly interpolated planes are illustrated in Fig. 13.15. Two parameters, the number of layers and the size of each hidden layer, which are both related to the construction of hidden layers, are the most important features for the ANN method. Based on the sample test results, ANNs can estimate the objective value efficiently at a large range of parameter combinations. Considering the accuracy and the model complexity of ANNs, the combination of 3 layers and a hidden layer size of 100 is selected as the optimum and recommended for further tests. Other related parameters for ANN models also include the alpha value and the activation method. In particular, the alpha value is a coefficient of an L2 regularization term that penalizes complex models. After relevant tests, alpha is selected between 1 and 10 and the rectified linear unit function is used for ANN activation.

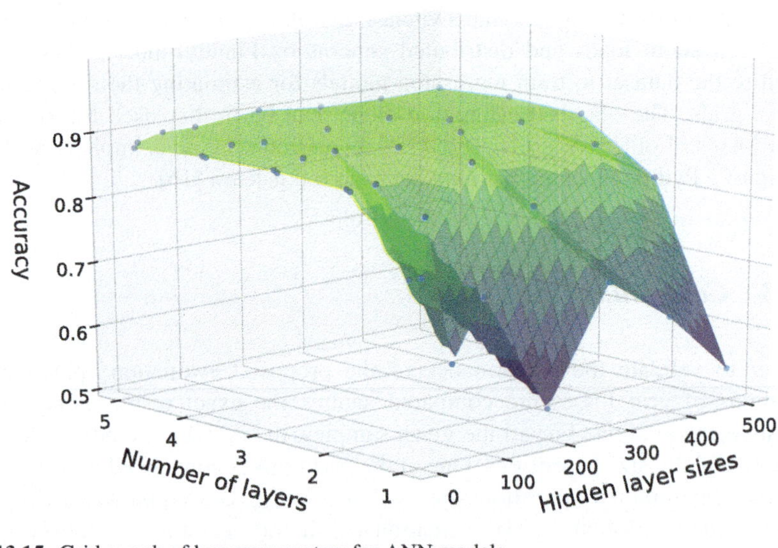

Fig. 13.15 Grid search of hyperparameters for ANN models

13.6.3.2 Feature Selection and Overall Performance

The second case study focuses on the evaluation accuracy of the proposed regression approach and the impact of feature selection on its performance. Regarding feature selection, recursive feature elimination (RFE) tests are implemented. In the RFE scheme, starting from the complete dataset containing all available features, unimportant features are eliminated step by step. One specific feature, which has the least influence on the total estimation error, is removed in each step. The lowest Mean square error (MSE) in each regression test is evaluated w.r.t. the number of selected features. These results of estimation errors MSE are also determined as the numerical average among the threefold cross-validation.

The convergence pattern of these RFE experiments is illustrated in Fig. 13.16. Firstly, very fast convergence characteristics are correlated with increasing numbers of features. The best performance using ANNs is achieved with a minimum MSE of 0.117 considering only four features among a total number of 60. Although the ANN model starts with relative high errors with the first feature (the total line length), its estimation error is significantly improved by including the second one (the median of installed power of DER). This observation indicates the large impact of DER on grid losses. By considering two additional features, the total number of nodes and the total number of distribution cabinets, the ANN-based regression method achieves its best performance. Adding further features leads to a slow increase of the evaluation error. This result shows the importance of selecting the most relevant features for representing the investigated dataset.

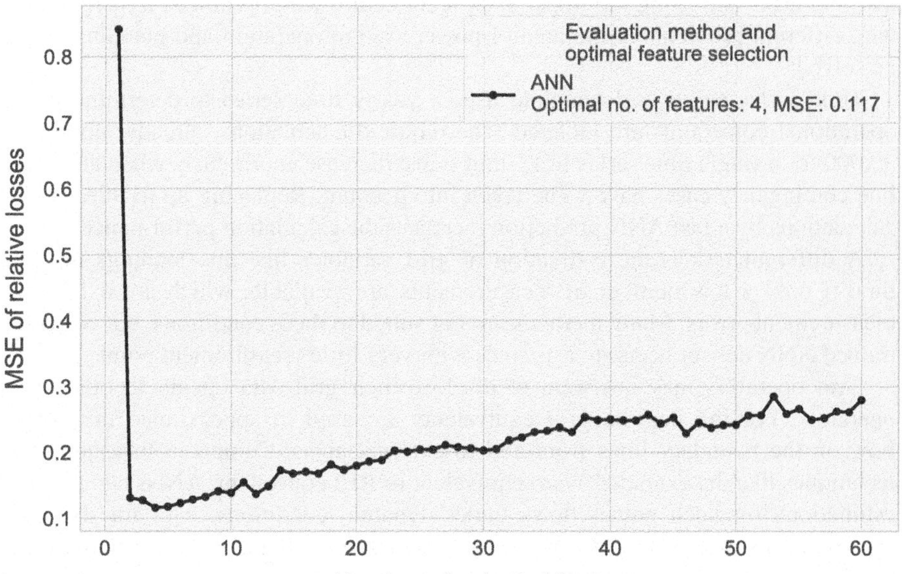

Fig. 13.16 Feature selection and MSE of relative grid losses using the proposed regression approach

13.6.4 Conclusion

In this section, we presented a typical application of machine learning techniques in power distribution system studies. In particular, we developed an ANN-based regression approach to estimate the LV grid losses from general grid data used as features. Due to the optimized setting of hyperparameters and the efficient selection of relevant features regarding loss evaluation, the proposed regression approach can estimate grid losses with high accuracy. Its performance is validated based on exhaustive simulation results of grid losses. Also, this approach provides a significant improvement regarding evaluation speed, since the high time-consuming simulation process for other grids can be saved. Moreover, the obtained regression model is capable of addressing the impact of the installation of DER on grid losses, which gives a useful tool for a currently unsolved question of grid operators. Therefore, this evaluation approach gives the grid operators a useful tool to quantify these impacts and specify the losses in individual grids. Subsequently, further technical measures for reducing loss and improving energy efficiency can be carried out more effectively.

13.7 Summary

In the last sections, we presented multiple applications of supervised learning in the field of power system operation and planning. By defining several supervised learning tasks and using the MLP ANN architecture with optimized hyperparameters, different problems in the area of power system operation and planning could be solved:

Time series-based grid planning uses a yearly time series to determine when operational constraints are violated. The required calculations, already more than 30,000 for a single time series in 15-min steps, increase enormously when all single line contingency cases have to be taken into account. Replacing up to 90% of the calculations by a fast ANN prediction increases the calculation performance.

A different task is the estimation of grid variables like line loadings in real-time if only a low number of measurements are available, which are exhibiting measurement errors. Many methods are not suited to these conditions, but correctly trained ANN deliver accurate estimates with very little measurement points.

Grid operators may not want to disclose their grid data openly to other grid operators. For this reason, grid equivalents are used to approximate the power flow on the boundary lines from and to external areas. Compared to established techniques like the extended Ward equivalent or REI equivalent, ANNs can provide estimations for such power flows under dynamic conditions, e.g., for dynamic distributed feed-in in the external grid areas. Due to its nature, the original grid topology cannot be reconstructed from the ANN's parameters.

Lastly, estimating power system losses without real-time measurements is not a trivial task. While it is possible to calculate them directly using suitable time series and power flow calculations, ANNs can take critical parameters of the grid as inputs and predict relative power losses. The effectiveness of the approach is validated by using the data of 5000 real LV grids in Germany.

The presented use cases for machine learning and especially ANNs are not exhaustive, but a collection of examples on how to effectively use machine learning for different types of tasks in power system analysis. We hope that by adopting and adjusting these examples, many more tasks can be supported by the use of machine learning.

Acronyms

AC alternating current
ANN artificial neural networks
DER distributed energy resource
HV high voltage
LODF line outage distribution factor
LV low voltage
MLP multilayer perceptron
MV medium voltage
MSE Mean square error
OPF optimal power flow
PCA principal component analysis
PF power flow
REI radial equivalent independent
ReLU rectified linear unit
RFE recursive feature elimination
SCP single contingency policy
SE state estimation
WLS weighted least squares

References

1. S.M. Ashraf, B. Rathore, S. Chakrabarti, Performance analysis of static network reduction methods commonly used in power systems, in *2014 Eighteenth National Power Systems Conference (NPSC)* (2014), pp. 1–6. https://doi.org/10.1109/NPSC.2014.7103837
2. R. Christie, IEEE 30-bus power flow test case (1961). http://www.ee.washington.edu/research/pstca/pf30/pg_tca30bus.htm
3. B.J. Claessens, P. Vrancx, F. Ruelens, Convolutional neural networks for automatic state-time feature extraction in reinforcement learning applied to residential load control. IEEE Trans. Smart Grid **9**(4), 3259–3269 (2018). https://doi.org/10.1109/TSG.2016.2629450

4. P. Dimo, Nodal analysis of power systems. Abacus Bks. Editura Academiei Republicii Socialisté România (1975). https://books.google.de/books?id=4dAiAAAAMAAJ
5. T.E. Dy Liacco, S.C. Savulescu, K.A. Ramarao, An on-line topological equivalent of a power system. IEEE Trans. Power Apparatus Syst. **PAS-97**(5), 1550–1563 (1978). https://doi.org/10.1109/TPAS.1978.354647
6. I. Goodfellow, Y. Bengio, A. Courville, *Deep Learning* (MIT Press, Cambridge, 2016). http://www.deeplearningbook.org
7. D.P. Kingma, J. Ba, Adam: a method for stochastic optimization, in *3rd International Conference on Learning Representations, ICLR 2015, San Diego, May 7–9, 2015, Conference Track Proceedings* (2015)
8. R. Leo, R.S. Milton, A. Kaviya, Multi agent reinforcement learning based distributed optimization of solar microgrid, in *2014 IEEE International Conference on Computational Intelligence and Computing Research* (2014), pp. 1–7. https://doi.org/10.1109/ICCIC.2014.7238438
9. D. Li, S.K. Jayaweera, Machine-learning aided optimal customer decisions for an interactive smart grid. IEEE Syst. J. **9**(4), 1529–1540 (2015). https://doi.org/10.1109/JSYST.2014.2334637
10. L. Li, K.G. Jamieson, G. DeSalvo, A. Rostamizadeh, A. Talwalkar, Efficient hyperparameter optimization and infinitely many armed bandits. CoRR **abs/1603.06560** (2016). http://arxiv.org/abs/1603.06560
11. C. Ma, S.R. Drauz, R. Bolgaryn, J.H. Menke, F. Schäfer, J. Dasenbrock, M. Braun, L. Hamann, M. Zink, K.H. Schmid, J. Estel, A comprehensive evaluation of the energy losses in distribution systems with high penetration of distributed generators, in *25th International Conference and Exhibition on Electricity Distribution (CIRED 2019)* (2019)
12. S. Meinecke, et al., Simbench - benchmark data set for grid analysis, grid planning and grid operation management. https://simbench.de/en. Accessed 3 July 2019
13. J.H. Menke, N. Bornhorst, M. Braun, Distribution system monitoring for smart power grids with distributed generation using artificial neural networks. Int. J. Electr. Power Energy **113**, 472–480 (2019)
14. E. Mocanu, D.C. Mocanu, P.H. Nguyen, A. Liotta, M.E. Webber, M. Gibescu, J.G. Slootweg, On-line building energy optimization using deep reinforcement learning. IEEE Trans. Smart Grid **10**, 3698–3708 (2019). https://doi.org/10.1109/TSG.2018.2834219
15. A. Paszke, S. Gross, S. Chintala, G. Chanan, E. Yang, Z. DeVito, Z. Lin, A. Desmaison, L. Antiga, A. Lerer, Automatic differentiation in pytorch, in *31st Conference on Neural Information Processing Systems (NIPS-W)* (2017)
16. F. Pedregosa, G. Varoquaux, A. Gramfort, V. Michel, B. Thirion, O. Grisel, M. Blondel, P. Prettenhofer, R. Weiss, V. Dubourg, J. Vanderplas, A. Passos, D. Cournapeau, M. Brucher, M. Perrot, E. Duchesnay, Scikit-learn: machine learning in Python. J. Mach. Learn. Res. **12**, 2825–2830 (2011)
17. F. Ruelens, B.J. Claessens, S. Vandael, S. Iacovella, P. Vingerhoets, R. Belmans, Demand response of a heterogeneous cluster of electric water heaters using batch reinforcement learning, in *2014 Power Systems Computation Conference* (2014), pp. 1–7. https://doi.org/10.1109/PSCC.2014.7038106
18. S.C. Savulescu, Equivalents for security analysis of power systems. IEEE Trans. Power Apparatus Syst. **PAS-100**(5), 2672–2682 (1981). https://doi.org/10.1109/TPAS.1981.316783
19. F. Schäfer, J.H. Menke, M. Braun, Contingency analysis of power systems with artificial neural networks, in *IEEE International Conference on Communications, Control, and Computing Technologies for Smart Grids (SmartGridComm)* (2018)
20. E. Shayesteh, B.F. Hobbs, L. Söder, M. Amelin, ATC-based system reduction for planning power systems with correlated wind and loads. IEEE Trans. Power Syst. **30**(1), 429–438 (2015). https://doi.org/10.1109/TPWRS.2014.2326615
21. D. Silver, J. Schrittwieser, K. Simonyan, I. Antonoglou, A. Huang, A. Guez, T. Hubert, L. Baker, M. Lai, A. Bolton, Y. Chen, T. Lillicrap, F. Hui, L. Sifre, G. van den Driessche, T. Graepel, D. Hassabis, Mastering the game of go without human knowledge. Nature **550**(7676), 354–359 (2017). https://doi.org/10.1038/nature24270

22. J. Stadler, H. Renner, Application of dynamic REI reduction, in *IEEE PES Innovative Smart Grid Technologies Europe 2013* (2013), pp. 1–5. https://doi.org/10.1109/ISGTEurope.2013.6695311
23. R.S. Sutton, A.G. Barto, *Reinforcement Learning: An Introduction*, 2nd edn. (The MIT Press, Cambridge, 2018). http://incompleteideas.net/book/the-book-2nd.html
24. L. Thurner, A. Scheidler, F. Schäfer, J.H. Menke, J. Dollichon, F. Meier, S. Meinecke, M. Braun, Pandapower - an open-source python tool for convenient modeling, analysis, and optimization of electric power systems. IEEE Trans. Power Syst. **33**(6), 6510–6521 (2018). https://doi.org/10.1109/TPWRS.2018.2829021
25. G.K. Venayagamoorthy, R.K. Sharma, P.K. Gautam, A. Ahmadi, Dynamic energy management system for a smart microgrid. IEEE Trans. Neural Netw. Learn. Syst. **27**(8), 1643–1656 (2016). https://doi.org/10.1109/TNNLS.2016.2514358
26. J.B. Ward, Equivalent circuits for power-flow studies. Trans. Am. Inst. Electr. Eng. **68**(1), 373–382 (1949). https://doi.org/10.1109/T-AIEE.1949.5059947
27. Z. Yan, Y. Xu, Data-driven load frequency control for stochastic power systems: a deep reinforcement learning method with continuous action search. IEEE Trans. Power Syst. **34**(2), 1653–1656 (2019). https://doi.org/10.1109/TPWRS.2018.2881359
28. D. Zhang, X. Han, C. Deng, Review on the research and practice of deep learning and reinforcement learning in smart grids. CSEE J. Power Energy Syst. **4**(3), 362–370 (2018). https://doi.org/10.17775/CSEEJPES.2018.00520

[illegible]

Index

M. Sayed-Mouchaweh (ed.), *Artificial Intelligence Techniques for a Scalable Energy Transition*, https://doi.org/10.1007/978-3-030-42726-9

W

The manufacturer's authorised representative in the EU is Springer Nature Customer Service Centre GmbH, Europaplatz 3, 69115 Heidelberg, Germany. If you have any concerns regarding our products, please contact ProductSafety@springernature.com

Printed and bound by CPI Group (UK) Ltd, Croydon, CR0 4YY
28/09/2026
02224057-0001